S. Kohy

Student Soluti

Wolfson • Pa

Physics for Scientists and Engineers
Second Edition

and

Physics with Modern Physics for Scientists and Engineers
Second Edition

MW00721254

2. Kerr

STUDENT SOLUTION MANUAL

WOLFSON • PASACHOFF

PHYSICS FOR SCIENTISTS AND ENGINEERS
SECOND EDITION

AND

PHYSICS WITH MODERN PHYSICS FOR SCIENTISTS AND ENGINEERS
SECOND EDITION

Prepared by

EDW. S. GINSBERG
UNIVERSITY OF MASSACHUSETTS - BOSTON

HarperCollinsCollegePublishers

Student Solution Manual to accompany Wolfson/Pasachoff *Physics for Scientists and Engineers*, Second Edition and *Physics with Modern Physics for Scientists and Engineers* Second Edition

Copyright © 1995 by HarperCollins College Publishers

All rights reserved. Printed in the United States of America. No part of this book may be reproduced in any manner whatsoever without written permission. For information, address HarperCollins College Publishers, 10 East 53rd Street, New York, NY 10022.

ISBN 0-06-501873-7

95 96 97 98 9 8 7 6 5 4 3 2 1

CONTENTS

PREFACE   ix

CHAPTER 1 DOING PHYSICS   1

PART 1   3

MECHANICS

CHAPTER 2 MOTION IN A STRAIGHT LINE   3
CHAPTER 3 THE VECTOR DESCRIPTION OF MOTION   8
CHAPTER 4 MOTION IN MORE THAN ONE DIRECTION   14
CHAPTER 5 DYNAMICS: WHY DO THINGS MOVE?   19
CHAPTER 6 USING NEWTON'S LAWS   23
CHAPTER 7 WORK, ENERGY, AND POWER   29
CHAPTER 8 CONSERVATION OF ENERGY   35
CHAPTER 9 GRAVITATION   39
CHAPTER 10 SYSTEMS OF PARTICLES   42
CHAPTER 11 COLLISIONS   46
CHAPTER 12 ROTATIONAL MOTION   50
CHAPTER 13 ROTATIONAL VECTORS AND ANGULAR MOMENTUM   55
CHAPTER 14 STATIC EQUILIBRIUM   60

PART 2   66

OSCILLATIONS, WAVES, AND FLUIDS

CHAPTER 15 OSCILLATORY MOTION   66
CHAPTER 16 WAVE MOTION   71
CHAPTER 17 SOUND AND OTHER WAVE PHENOMENA   74
CHAPTER 18 FLUID MOTION   78

PART 3   83

THERMODYNAMICS

CHAPTER 19 TEMPERATURE AND HEAT   83
CHAPTER 20 THE THERMAL BEHAVIOR OF MATTER   87
CHAPTER 21 HEAT, WORK, AND THE FIRST LAW
             OF THERMODYNAMICS   91
CHAPTER 22 THE SECOND LAW
             OF THERMODYNAMICS   93

PART 4   99

ELECTROMAGNETISM

CHAPTER 23 ELECTRIC CHARGE, FORCE, AND FIELD   99
CHAPTER 24 GAUSS'S LAW   103
CHAPTER 25 ELECTRIC POTENTIAL   108
CHAPTER 26 ELECTROSTATIC ENERGY
             AND CAPACITORS   112
CHAPTER 27 ELECTRIC CURRENT   116
CHAPTER 28 ELECTRIC CIRCUITS   119
CHAPTER 29 THE MAGNETIC FIELD   123
CHAPTER 30 SOURCES OF THE MAGNETIC FIELD   127
CHAPTER 31 ELECTROMAGNETIC INDUCTION   130
CHAPTER 32 INDUCTANCE AND MAGNETIC ENERGY   135
CHAPTER 33 ALTERNATING-CURRENT CIRCUITS   138
CHAPTER 34 MAXWELL'S EQUATIONS
             AND ELECTROMAGNETIC WAVES   143

PART 5   147

OPTICS

CHAPTER 35 REFLECTION AND REFRACTION   147
CHAPTER 36 IMAGE FORMATION AND OPTICAL
             INSTRUMENTS   150
CHAPTER 37 INTERFERENCE AND DIFFRACTION   153

PART 6   158

MODERN PHYSICS

CHAPTER 38 THE THEORY OF RELATIVITY   158
CHAPTER 39 INSIDE ATOMS AND NUCLEI   162

PHYSICS WITH MODERN PHYSICS   163

CHAPTER 39 LIGHT AND MATTER: WAVES OR PARTICLES?   163
CHAPTER 40 QUANTUM MECHANICS   167
CHAPTER 41 ATOMIC PHYSICS   171
CHAPTER 42 MOLECULAR AND SOLID-STATE PHYSICS   173
CHAPTER 43 NUCLEAR PHYSICS   176
CHAPTER 44 NUCLEAR ENERGY: FISSION AND FUSION   179
CHAPTER 45 FROM QUARKS TO THE COSMOS   181

The decision to compile a manual of worked-out solutions to the end-of-chapter problems in Wolfson and Pasachoff's textbook arose from a consideration of the crucial importance of homework problems in the teaching of physics. (Learning physics without doing problems is like learning to swim without going in the water!) These solutions are designed to supplement, reinforce, and sometimes replace classroom discussions.

This *Student Solution Manual* contains solutions to one-fourth of the text problems (every other odd-numbered problem). Instructors who wish their students to have access to some, but not all, of the problem solutions can ask the campus bookstore to order copies from the publisher. Please also note that this manual is intended for use with both the regular edition and the extended version (*With Modern Physics*).

Most of the solutions in this manual have been written in a form that requires the active participation of the reader. An active reader simultaneously reads and works through each solution with pencil and paper, calculator, and open textbook. Intermediate steps, such as algebraic manipulations, substitution of variables or data, diagrams, conversion of units are frequently and intentionally omitted. They must be supplied by an active reader, using the manual's solutions as guideposts and textual materials as references. Wolfson and Pasachoff intended their problems to help readers understand and use the concepts presented in their textbook. I have integrated references to the text into my solutions in fulfillment of this goal.

My solutions are not necessarily the only ones possible (or even the best, simplest, etc.). In some cases, alternate methods are indicated; in others, a different point of view from the text's is deliberately presented.

Many authors have enumerated helpful suggestions for students to follow when solving physics problems. My own version of such a list is the following:

- Carefully READ and understand the question. It may be necessary to visualize a situation or device, as described in the text, or as known from personal experience. The context of the question, i.e., the chapter or section in which it occurs, is often of help.
- THINK about how the quantities that determine the sought-for answer are physically related to the quantities that are either given in the question, or obtainable from them and other sources of data. Construction of a simple physical model, to represent the situation in the problem, may be necessary.
- Write down physical equations involving the relevant quantities, perhaps using approximations where appropriate, and SOLVE for the desired variables. For some problems, this is the most difficult step.
- CONSIDER the reasonableness of your answer. Does it make sense? Is it consistent with your initial expectations? Does it change suitably when the conditions in the problems are altered? If not, the previous steps may require repetition or verification.

In the real world of experimenhtal science, numerical results reflect the precision of actual measurements and the theoretical uncertainties in their interpretation. Such subjects are more appropriate to laboratory or advanced courses. By default, I have adopted the standard convention of regarding most numbers in problems as accurate to three significant figures, unless otherwise indicated by the context. I think common sense is a better guide than consistency at this level. On the subject of accuracy, I must admit to the responsibility for any errors which inevitably are present in a work of this size. Naturally, I will try to correct any shortcomings brought to my attention.

Although I personally have solved every problem in this manual, I acknowledge a great debt to other authors, my colleagues, former teachers, and students.

Edw. S. Ginsberg
University of Massachusetts-Boston
Boston, MA 02125-3393
ginsberg@umbsky.cc.umb.edu

● CHAPTER 1 DOING PHYSICS

Section 1-3 Measurement Systems

Problem

1. What is your mass in (a) kg; (b) g; (c) Gg; (d) fg?

Solution

(a) Most people in the United States know their weight in pounds. One lb has a mass of about .454 kg. My weight, 167 lb, is equivalent to a mass of $(167 \text{ lb})(0.454 \text{ kg/lb}) = 75.8$ kg. Using prefixes from Table 1-1 and scientific notation, we can express this mass as: (b) $(75.8 \text{ kg})(10^3 \text{ g/kg}) = 7.58 \times 10^4$ g, (c) $(7.58 \times 10^4 \text{ g})(1 \text{ Gg}/10^9 \text{ g}) = 7.58 \times 10^{-5}$ Gg, (d) 7.58×10^{19} fg.

Problem

5. How long, in nanoseconds, is the period of the cesium radiation used to define the second?

Solution

By definition, 1 s = 9,192,631,770 periods of a cesium atomic clock, so 1 period = $1 \text{ s}/9{,}192{,}631{,}770 = 1.087827757 \times 10^{-10}$ s = 0.1087827757 ns. (This is an alternative definition of the second; the other definition is really in terms of the frequency of the cesium-133 hyperfine transition, which is the reciprocal of the period.)

Problem

9. Making a turn, a jetliner flies 1.9 km on a circular path of radius 2.4 km. Through what angle does it turn?

Solution

The angle in radians is the circular arc length divided by the radius, or $\theta = s/r = 1.9 \text{ km}/2.4 \text{ km} = 0.79$ radians. This corresponds to $(0.792)(180°/\pi) = 45.4°$, or about 45°. (See the tip on intermediate results in Section 1-7.)

Section 1-4 Changing Units

Problem

13. How many cubic centimeters (cm^3) are there in a cubic meter (m^3)?

Solution

$1 \text{ m}^3 = (10^2 \text{ cm})^3 = 10^6 \text{ cm}^3$.

Problem

17. Superhighways in Canada have speed limits of 100 km/h. Does this exceed the 55 mi/h speed limit common in the United States? If so, by how much?

Solution

$(100 \text{ km/h})(1 \text{ mi}/1.609 \text{ km}) = 62.2$ mi/h. The speed limit in Canada is about 7.2 mi/h greater than in the United States.

Problem

21. An aircraft carrier goes 17 feet on a gallon of fuel. Express this result in (a) miles per gallon (mpg) and (b) kilometers per liter (km/L).

Solution

(a) $(17 \text{ ft/gal})(1 \text{ mi}/5280 \text{ ft}) = 3.2 \times 10^{-3}$ mi/gal; (b) $(17 \text{ ft/gal})(1.609 \text{ km}/5280 \text{ ft})(1 \text{ gal}/3.786 \text{ L}) = 1.4 \times 10^{-3}$ km/L. (We used conversion factors from Appendix C.)

Section 1-5 Dimensional Analysis

Problem

25. An equation you'll encounter in the next chapter is $x = \frac{1}{2}at^2$, where x is distance and t is time. Use dimensional analysis to find the dimensions of a.

Solution

We can first solve for $a = 2x/t^2$ to find its dimensions (which are usually denoted by square brackets). Thus $[a] = [x]/[t^2] = LT^{-2}$.

Section 1-6 Scientific Notation

Problem

29. Add 3.6×10^5 m and 2.1×10^3 km.

Solution

$3.6 \times 10^5 \text{ m} + 2.1 \times 10^3 \text{ km} = (0.36 + 2.10) \times 10^3 \text{ km} = 2.46 \times 10^3$ km. (Note: We displayed the manipulation of numbers in scientific notation, assuming that all quantities are to be expressed to three significant figures; to two significant figures, our result would have been 2.5×10^3 km. See Section 1-7.)

Problem

33. If there are 100,000 electronic components on a semi-conductor chip that measures 5.0 mm by 5.0 mm, (a) how much area does each component occupy? (b) If the individual components are square, how long is each on a side?

Solution

(a) The area of each (identical) component is the total area of the chip divided by the number of components, or $(5 \text{ mm} \times 5 \text{ mm})/10^5 = 2.5 \times 10^{-4} \text{ mm}^2$. (b) The side of a square is the square root of its area, so the side of one component is $\sqrt{2.5 \times 10^{-4} \text{ mm}^2} = 1.58 \times 10^{-2} \text{ mm} \approx 16 \ \mu\text{m}$.

Section 1-7 Accuracy and Significant Figures

Problem

37. A 3.6-cm long radio antenna is added to the front of an airplane 41 m long (Fig. 1-16). What is the overall length?

FIGURE 1-16 Problem 37.

Solution

The overall length of the airplane could be increased by as much as 3.6 cm, depending on how the antenna is attached. However, to two significant figures, the airplane is still 41 m long. That is because in this context, 41 m means a length greater than or equal to 40.5 m, but less than 41.5 m, and 41 m + 3.6 cm = 41.036 m satisfies this condition.

Section 1-8 Estimation

Problem

41. Paper is made from wood pulp. Estimate the number of trees that must be cut to make one day's run of a big city's daily newspaper. Assume no recycling.

Solution

A typical nightly run of a big city daily might have a circulation of about 500,000 and consist of newspapers weighing about 1 lb each, thus consuming 5×10^5 lb of paper. Newsprint is mostly made from wood pulp, a suspension of ground-up trees, so roughly 5×10^5 lb of trees are needed. (Actually, newsprint consists of about 80% wood pulp, 15% cellulose and 5% glue and fillers.) The size of trees used in the paper industry varies widely, but is smallish compared to trees used for lumber. A tree 1 ft in diameter at the base and 40 ft tall (a conical volume of $\frac{1}{3} \pi r^2 h = \frac{1}{3} \pi (\frac{1}{2} \text{ ft})^2 (40 \text{ ft}) \approx 10 \text{ ft}^3$) with

density slightly less than water (approximately 60 lb/ft³—recall that logs float) would weigh about $(60 \text{ lb/ft}^3)(10 \text{ ft}^3) = 600$ lb. Therefore, about $5 \times 10^5 \text{ lb}/600 \text{ lb} \approx 10^3$ or a thousand trees go into a weekday run of a large newspaper. (A Sunday edition might use four times this number.)

Problem

45. If you set your watch to be exactly correct now, and if it runs slow by 1 s per month, in how many years would it again read exactly the right time?

Solution

If the watch is a 12 h dial analog watch, it would have to lose 12 h before it would again read the correct time. A a rate of loss of 1 s/mo, this would take 12 h/(1 s/mo) = 12×3600 mo = 3600 y. (A 24 h dial analog or a digital watch would take twice as long.)

Problem

49. The density of interstellar space is about 1 atom per cubic cm. Stars in our galaxy are typically a few light-years apart (one light-year is the distance light travels in one year) and have typical masses of 10^{30} kg. Estimate whether there is more matter in the stars or in the interstellar gas.

Solution

From Table 1-2, the distance to the nearest star is 4×10^{16} m. Suppose there is one star of mass 10^{30} kg per cubical volume of this dimension. Then the density of star mass is $(10^{30} \text{ kg})/(4 \times 10^{16} \text{ m})^3 \sim 1.6 \times 10^{-20} \text{ kg/m}^3$. Interstellar gas is mostly hydrogen, so its density is $(1 \text{ atom/cm}^3)(1 \text{ g}/6.02 \times 10^{23} \text{ atoms})(10^3 \text{ kg/m}^3)/(1 \text{ g/cm}^3) \approx \frac{1}{6} \times 10^{-20} \text{ kg/m}^3$, or about ten times smaller than the star mass density.

Paired Problems

Problem

53. Find the dimensions of F in the formula $F = Gm_1 m_2/r^2$, where G is given in Example 1-3, both m's are masses, and r is a length.

Solution

The dimensions of F (denoted by $[F]$) are: $[F] = [G][M]^2/[r]^2 = (L^3/MT^2)(M^2/L^2) = ML/T^2$, where $[G] = L^3/MT^3$ from Example 1-3.

Supplementary Problems

Problem

57. The moon barely covers the Sun at a solar eclipse. Given that the moon is 4×10^5 km from Earth and that the Sun is 1.5×10^8 km from Earth, determine how much bigger the

Sun's diameter is than the moon's. If the moon's radius is 1800 km, how big is the Sun?

Solution

The Sun and the moon subtend the same angle (about $\frac{1}{2}°$ when viewed from Earth, therefore $\theta = s/r \approx$ diameter/distance $= d_{moon}/4{\times}10^5$ km $= d_{Sun}/1.5{\times}10^8$ km. (The small angle approximation is justified since the diameter is much smaller than the distance of each body.) Thus, the ratio of the diameters is approximately $d_{Sun}/d_{moon} = (1.5{\times}10^8/4{\times}10^5) \approx 3.8{\times}10^2$. Of course, the radii are in the same ratio as the diameters, so if $R_{moon} \approx 1800$ km, then $R_{Sun} \approx (3.8{\times}10^2){\times}(1800$ km$) \approx 680{,}000$ km.

Problem

61. A good-size nuclear weapon has an explosive yield equivalent to one million tons (1 megaton) of the chemical explosive TNT. Estimate the length of a train of boxcars needed to carry 1 megaton of TNT. (A 1-megaton nuclear weapon, in contrast, is on the order of 1 m long and may have a mass of a few hundred kg.)

Solution

A typical railroad boxcar is about 60 ft long and can carry around 100 tons. Therefore, $10^6/10^2 = 10^4$ or ten thousand boxcars would be needed to transport one megaton of TNT. A train with this many boxcars would be $60{\times}10^4$ ft ≈ 114 mi $\approx$ 180 km long, not including the engine! ●

P A R T 1 MECHANICS

● CHAPTER 2 MOTION IN A STRAIGHT LINE

Section 2-1 Distance, Time, Speed, and Velocity

Problem

1. In 1994 Leroy Burrell of the United States set a world record in the 100-m dash, with a time of 9.84 s. What was his average speed?

Solution

Burrell's average speed was (Equation 2-1) $\bar{v} = \Delta x/\Delta t = 100$ m$/9.84$ s $= 10.16$ m/s. (One can assume that the race distance was known to more than four significant figures.)

Problem

5. Starting from home, you bicycle 24 km north in 2.5 h, then turn around and pedal straight home in 1.5 h. What are your (a) displacement at the end of the first 2.5 h, (b) average velocity over the first 2.5 h, (c) average velocity for the homeward leg of the trip, (d) displacement for the entire trip, and (e) average velocity for the entire trip?

Solution

(a) $\Delta r_{out} = 24$ km (north). (b) $v_{out} = 24$ km (north)$/2.5$ h $= 9.6$ km/h (north). (c) $v_{back} = 24$ km (south)$/1.5$ h $= 16$ km/h (south). (d) $\Delta r_{out\ and\ back} = 0$. (e) $v_{round\ trip} = 0$.

Problem

9. You allow yourself 40 min to drive 25 mi to the airport, but are caught in heavy traffic and average only 20 mi/h for the first 15 min. What must your average speed be on the rest of the trip if you are to get there on time?

Solution

At an average speed of 20 mi/h for the first 15 min $= \frac{1}{4}$ h, you travel only $(20$ mi/h$)(\frac{1}{4}$ h$) = 5$ mi. Therefore, you must cover the remaining $(25 - 5)$ mi $= 20$ mi in $(40 - 15)$ min $= 25$ min $= \frac{5}{12}$ h. This implies an average speed of 20 mi$/(\frac{5}{12}$ h$) = 48$ mi/h. (Note that your overall average speed was pre-determined to be 25 mi$/(40$ h$/60) = 37.5$ mi/h, and that this equals the time-weighted average of the average speeds for the two parts of the trip: $(15$ min$/40$ min$)(20$ mi/h$) + (25$ min$/40$ min$)(48$ mi/h$)$.)

Problem

13. A fast base runner can get from first to second base in 3.4 s. If he leaves first base as the pitcher throws a 90 mi/h fastball the 61-ft distance to the catcher, and if the catcher takes 0.45 s to catch and rethrow the ball, how fast does the catcher have to throw the ball to second base to make an out? Home plate to second base is the diagonal of a square 90 ft on a side.

Solution

At 90 mi/h $= 132$ ft/s, the ball takes 61 ft$/(132$ ft/s$) = 0.462$ s to travel from the pitcher to the catcher. (We are keeping extra significant figures in the intermediate calculations as suggested in Section 1-7.) After the catcher throws the ball, it has 3.4 s $- 0.462$ s $- 0.45$ s $= 2.49$ s to reach second base at the same time as the runner. The distance is $\sqrt{2}(90$ ft$)$, so the minimum speed is $\bar{v} = \sqrt{2}(90$ ft$)/2.49$ s $= 51.2$ ft/s $= 35$ mi/h. A prudent catcher would allow extra time for the player covering second base to make the tag.

Problem

17. A jetliner leaves San Francisco for New York, 4600 km away. With a strong tailwind, its speed is 1100 km/h. At the same time, a second jet leaves New York for San Francisco. Flying into the wind, it makes only 700 km/h. When and where do the two planes pass each other?

Solution

When the planes pass, the total distance traveled by both is 4600 km. Therefore, 4600 km = (1100 km/h) Δt + (700 km/h) Δt, or Δt = 4600 km/(1800 km/h) = 2.56 h. (The planes meet 2.56 h after taking off.) The encounter occurs at a point about (700 km/h)(2.56 h) ≈ 1790 km from New York City or (1100 km/h)(2.56 h) ≈ 2810 km from San Francisco.

Section 2-2 Motion in Short Time Intervals

Problem

21. Fig. 2-20 shows the position of an object as a function of time. From the graph, determine the instantaneous velocity at (a) 1.0 s; (b) 2.0 s; (c) 3.0 s; (d) 4.5 s. (e) What is the average velocity over the interval shown?

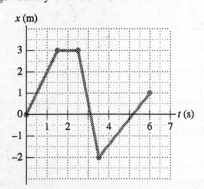

FIGURE 2-20 Problem 21.

Solution

The instantaneous velocity at a particular time is the slope of the graph of x versus t at that point, $v(t) = dx/dt$. For a straight line segment of graph, v equals the average velocity over that segment, $\bar{v} = \Delta x/\Delta t$. Each of the times specified in this problem falls on a different straight segment of the graph in Fig. 2-20, whose slopes we determine from the coordinates of the endpoints of that segment.
(a) $v(1 \text{ s}) = (3 - 0)\text{m}/(1.5 - 0)\text{s} = 2$ m/s;
(b) $v(2 \text{ s}) = (3 - 3)\text{m}/(2.5 - 1.5) \text{ s} = 0$;
(c) $v(3 \text{ s}) = (-2 - 3) \text{ m}/(3.5 - 2.5) \text{ s} = -5$ m/s;
(d) $v(4.5 \text{ s}) = [1 - (-2)]\text{m}/(6 - 3.5) \text{ s} = 1.2$ m/s.
(e) The overall average velocity is $\bar{v} =$
$[x(6 \text{ s}) - x(0)]/(6 \text{ s} - 0) = 1 \text{ m}/6 \text{ s} = 0.167$ m/s.

Problem

25. The position of an object is given by $x = bt^3 - ct^2 + dt$, with x in meters and t in seconds. The constants b, c, and d are $b = 3.0$ m/s³, $c = 8.0$ m/s², and $d = 1.0$ m/s. (a) Find all times when the object is at position $x = 0$. (b) Determine a general expression for the instantaneous velocity as a function of time, and from it find (c) the initial velocity and (d) all times when the object is instantaneously at rest. (e) Graph the object's position as a function of time, and identify on the graph the quantities you found in (a) to (d).

Solution

(a) With the aid of the quadratic formula and factorization, $x = t(bt^2 - ct + d) = 0$ implies $t = 0$, or $t = (c \pm \sqrt{c^2 - 4bd})/2b$. Substituting the given constants, $t = 0$, $t = (4 \pm \sqrt{13}) \text{ s}/3 = 0.131$ s and 2.54 s.
(b) $v(t) = dx/dt = 3bt^2 - 2ct + d$. (c) When $t = 0$, $v(0) = d = 1$ m/s. (d) $v = 3bt^2 - 2ct + d = 0$ implies $t = (c \pm \sqrt{c^2 - 3bd})/3b = (8 \pm \sqrt{55}) \text{ s}/9 = 64.9$ ms and 1.71 s. (e) The graph of this cubic has roots from part (a), slope at the origin from part (c), and relative maximum and minimum from part (d), as shown.

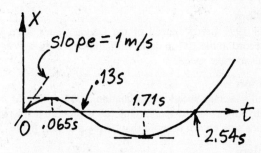

Problem 25 Solution.

Section 2-3 Acceleration

Problem

29. The 1986 explosion of the space shuttle Challenger occurred 74 s after liftoff. At that time, mission control reported a shuttle speed of 2900 ft/s (880 m/s). What was the Challenger's average acceleration during its brief flight? Compare with the acceleration of gravity.

Solution

The average acceleration of the shuttle along its trajectory, from liftoff until its tragic explosion, was (Equation 2-4)
$\bar{a} = \Delta v/\Delta t = (880 \text{ m/s} - 0)/74 \text{ s} = 11.9$ m/s² ≈ 1.2 g, where g = 9.8 m/s² is the acceleration due to gravity at the surface of the Earth.

Problem

33. Your plane reaches its takeoff runway and then holds for 4.0 min because of air-traffic congestion. The plane then heads down the runway with an average acceleration of 3.6 m/s^2. It is airborne 35 s later. What are (a) its takeoff speed and (b) its average acceleration from the time it reaches the takeoff runway until it's airborne?

Solution

(a) During the 35 s the plane is actually taking off, Equation 2-4 gives $\Delta v = v - 0 = \bar{a} \, \Delta t = (3.6 \text{ m/s}^2)(35 \text{ s}) = 126 \text{ m/s} = 454 \text{ km/h}$. (b) If we include the four minutes wait before taking off, the average acceleration for the entire interval on the runway is only $\bar{a} = \Delta v/\Delta t = (126 \text{ m/s} - 0)/(4 \text{ min} + 35 \text{ s}) = (126 \text{ m/s})/(275 \text{ s}) = 0.458 \text{ m/s}^2$.

Section 2-4 Constant Acceleration

Problem

37. A car accelerates from rest to 25 m/s in 8.0 s. Determine the distance it travels in two ways: (a) by multiplying the average velocity given in Equation 2-8 by the time and (b) by calculating the acceleration from Equation 2-7 and using the result in Equation 2-10.

Solution

(a) For constant acceleration, Equation 2-8 can be combined with Equation 2-1 to yield $\Delta x = \bar{v} \, \Delta t = \frac{1}{2}(v_0 + v) \, \Delta t = \frac{1}{2}(0 + 25 \text{ m/s})(8.0 \text{ s}) = 100 \text{ m}$. (b) Alternatively, for constant acceleration, Equation 2-7 gives $a = \bar{a} = \Delta v/\Delta t = (25 \text{ m/s} - 0)/8.0 \text{ s} = 3.13 \text{ m/s}^2$, so Equation 2-10 yields $\Delta x = x - x_0 = v_0 t + \frac{1}{2}at^2 = 0 + \frac{1}{2}(3.13 \text{ m/s}^2)(8.0 \text{ s})^2 = 100 \text{ m}$.

Section 2-5 Using the Equations of Motion

Problem

41. A rocket rises with constant acceleration to an altitude of 85 km, at which point its speed is 2.8 km/s. (a) What is its acceleration? (b) How long does the ascent take?

Solution

(a) In Equation 2-11 (with x positive upward) we are given that $x - x_0 = 85 \text{ km}$, $v_0 = 0$ (the rocket starts from rest), and $v = 2.8 \text{ km/s}$. Therefore, we can solve for the acceleration, $a = (v^2 - v_0^2)/2(x - x_0) = (2.8 \text{ km/s})^2/2(85 \text{ km}) = 46.1 \text{ m/s}^2$ (note the change of units). (b) From Equation 2-9, we can solve for the time of flight, $t = 2(x - x_0)/(v_0 + v) = 2(85 \text{ km})/(2.8 \text{ km/s}) = 60.7 \text{ s}$. (We chose to relate t directly to the given data, but once the acceleration is known, Equation 2-7 or 2-10 could have been used to find $t = v/a$ or $t = \sqrt{2(x - x_0)/a}$, respectively.)

Problem

45. A car moving initially at 50 mi/h begins decelerating at a constant rate 100 ft short of a stoplight. If the car comes to a full stop just at the light, what is the magnitude of its deceleration?

Solution

Since the car stops ($v = 0$), after traveling 100 ft $= x - x_0$, from an initial speed of $(55 \text{ mi/h})(22 \text{ ft/s}/15 \text{ mi/h}) = 73.3 \text{ ft/s}$, Equation 2-11 gives $a = -(73.3 \text{ ft/s})^2/2(100 \text{ ft}) = -26.9 \text{ ft/s}^2$. The magnitude of the deceleration is the absolute value of a.

Problem

49. A gazelle accelerates from rest at 4.1 m/s^2 over a distance of 60 m to outrun a predator. What is its final speed?

Solution

From Equation 2-11, $v^2 = 2a(x - x_0) = 2(4.1 \text{ m/s}^2)(60 \text{ m})$, or $v = 22.2 \text{ m/s}$ (almost 50 mi/h).

Problem

53. A flea extends its 0.80-mm-long legs, propelling itself 10 cm into the air. Assuming that the legs provide constant acceleration as they extend from completely folded to completely extended, find the magnitude of that acceleration. Compare with the acceleration of gravity at Earth's surface.

Solution

To reach a height $h = 10 \text{ cm}$, the flea needs an initial vertical velocity (from Equation 2-14 with $v = 0$ at the top) of $v_0^2 = 2gh$. To attain this initial speed from rest, over a leg extension of $\Delta y = .8 \text{ mm}$, requires an acceleration of $a = v_0^2/2 \, \Delta y = 2gh/2 \, \Delta y = (10 \text{ cm}/0.8 \text{ mm})g = 125 \, g = 125(9.8 \text{ m/s}^2) = 1.23 \text{ km/s}^2$.

Problem

57. The maximum deceleration of a car on a dry road is about 8 m/s^2. If two cars are moving head-on toward each other at 88 km/h (55 mi/h), and their drivers apply their brakes when they are 8.5 m apart, will they collide? If so, at what relative speed? If not, how far apart will they be when they stop? On the same graph, plot distance versus time for both cars.

Solution

The minimum distance a car needs to stop ($v = 0$) from an initial speed $v_0 = 88 \text{ km/h} = 24.4 \text{ m/s}$, with a constant acceleration $a = -8 \text{ m/s}^2$, is (Equation 2-11) $x - x_0 = -v_0^2/2a = -(24.4 \text{ m/s})^2/2(-8 \text{ m/s}^2) = 37.3 \text{ m}$ (positive in the direction of v_0). Since 85 m is greater than twice this distance, the cars can avoid a collision, and they will be $85 \text{ m} - 2(37.3 \text{ m}) = 10.3 \text{ m}$ apart when stopped.

To plot x versus t, using Equation 2-10 for each car, we need to choose an origin, say $x = 0$ at the midpoint of the separation between the cars, with positive x in the direction of the initial velocity of the first car, and $t = 0$ when the brakes are applied. Then $x_{10} = -42.5$ m $= -x_{20}$, $v_{10} = 24.4$ m/s $= -v_{20}$, and $a_1 = -8$ m/s $= -a_2$. A graph of $x_1(t)$ and $x_2(t)$ is as shown.

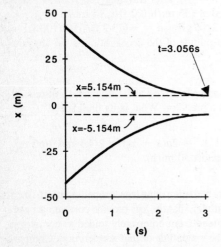

t=3.056s

x=5.154m

x=-5.154m

t (s)   Problem 57 Solution.

Problem

61. Repeat the preceding problem, now assuming your initial speed is 95 km/h.

Solution

The position as a function of time for either car, moving with constant acceleration, is given by Equation 2-10. Let us choose our origin $t = 0$ and $x = 0$ at the time and place the speeding driver in car number one notices car number two in front and applies the brakes, with the direction of initial motion positive. Then $x_{10} = 0$, $x_{20} = 10$ m, $v_{10} > v_{20} = 60$ km/h $= 16.7$ m/s, $a_1 = -4.2$ m/s^2 and $a_2 = 0$. The position of the cars is $x_1(t) = v_{10}t + \frac{1}{2}a_1t^2$ and $x_2(t) = x_{20} + v_{20}t$, valid for $0 \le t \le t^*$, where t^* is the time for which the accelerations remain constant. (Thus, t^* is either the time the cars collide, if this happens, or the time when car number one stops decelerating.) The distance between the cars is $x_{21}(t) = x_2(t) - x_1(t)$. The condition for a collision is that the quadratic equation $x_{21}(t) = 0$ have a real root (in which case the smaller root is t^*), and the condition for no collision is that this equation have no real roots.

The solution of the equation $x_{21}(t) = 0 = -\frac{1}{2}a_1t^2 - (v_{10} - v_{20})t + x_{20}$ follows from the quadratic formula, $t = [(v_{10} - v_{20}) \pm \sqrt{(v_{10} - v_{20})^2 - 2|a_1|x_{20}}]/|a_1|$. (Since a_1 is negative, we wrote it explicitly as $a_1 = -|a_1|$.) Thus, if $(v_{10} - v_{20})^2 \ge a_1|x_{20}$, there is a collision at time $t^* = [(v_{10} - v_{20}) - \sqrt{(v_{10} - v_{20})^2 - 2|a_1|x_{20}}]/|a_1|$, from which the relative velocity at collision, $v_1(t^*) - v_{20}$, can be calcu-

lated. On the other hand, if $(v_{10} - v_{20})^2 < 2|a_1|x_{20}$, there is no collision, and the minimum distance x_{21} can be found by setting the derivative of $x_{21}(t)$ equal to zero, or by physical reasoning.

When $v_{10} = 95$ km/h $= 26.4$ m/s, $(v_{10} - v_{20})^2 = (26.4$ m/s $- 16.7$ m/s$)^2 = 94.5$ m^2/s$^2 > 2(4.2$ m/s$^2)(10$ m$) = 84$ m^2/s^2, so there is a collision at $t^* = (9.72$ m/s $- \sqrt{10.5$ m^2/s$^2})/(4.2$ m/s$^2) = 1.54$ s. The relative speed at collision is $v_1(t^*) - v_{20} = v_{10} - v_{20} - |a_1|t^* = 9.72$ m/s $- (4.2$ m/s$^2)(1.54$ s$) = 3.24$ m/s $= 11.7$ km/h, where we used Equation 2-7 for the velocities.

When $v_{10} = 85$ km/h, $(v_{10} - v_{20})^2 = (25$ km/h$)^2 = 48.2$ m^2/s$^2 < 2|a_1|x_{20} = 84$ m^2/s^2, and there is no collision. The relative distance is the quadratic $x_{21}(t) = \frac{1}{2}|a_1|t^2 - (v_{10} - v_{20})t + x_{20}$. One way to obtain the distance of closest approach is to minimize this function of time. Setting the derivative equal to zero gives us $dx_{21}/dt = |a_1|t - (v_{10} - v_{20}) = 0$, or $t_{min} = (v_{10} - v_{20})/|a_1|$. Then $x_{21}(t_{min}) = \frac{1}{2}|a_1|t_{min}^2 - (v_{10} - v_{20})t_{min} + x_{20} = x_{20} - (v_{10} - v_{20})^2/2|a_1| = 10$ m $- (48.2$ m^2/s$^2)/2(4.2$ m/s$^2) = 4.26$ m. This is in fact a minimum because $d^2x_{21}/dt^2 = |a_1| > 0$.

Another way to obtain the minimum x_{21}, without using calculus, relies on purely physical reasoning. As long as the velocity of car number one, $v_1(t)$, is greater than 60 km/h (the velocity of car number two), it is gaining ground on car number two, so the relative distance x_{21} is decreasing. When $v_1(t)$ falls below 60 km/h, car number one loses ground to car number two and x_{21} starts increasing. Therefore, the closest approach occurs when $v_1(t) = v_{10} - |a_1|t = v_{20} = 60$ km/h, which gives the same t_{min} as above.

Section 2-6   The Constant Acceleration of Gravity

Problem

65. A foul ball leaves the bat going straight upward at 23 m/s. (a) How high does it rise? (b) How long is it in the air? Neglect the distance between the bat and the ground.

Solution

(a) At the maximum height, $v^2 = 0 = v_0^2 - 2g(y_{max} - y_0)$, so $y_{max} - y_0 = v_0^2/2g = (23$ m/s$)^2/2(9.8$ m/s$^2) = 27.0$ m.
(b) If we neglect the distance between the bat and the ground (and assume that the foul ball is not caught), the flight of the ball lasts until it falls back to its initial height. Then $y - y_0 = 0 = v_0t - \frac{1}{2}gt^2$, or $t = 2(23$ m/s$)/(9.8$ m/s$^2) = 4.69$ s.

Problem

69. A falling object travels one-fourth of its total distance in the last second of its fall. From what height was it dropped?

Solution

The total distance traveled by a falling object in a time t is given by Equation 2-13, with $v_0 = 0$ (the meaning of dropped). Thus $y_0 - y(t) = \frac{1}{2}gt^2$. The distance fallen during the last second (an interval from $t - 1$ s to t) is $y(t - 1\text{ s}) - y(t) = \frac{1}{2}gt^2 - \frac{1}{2}g(t - 1\text{ s})^2$. The latter is one-fourth of the former when (cancel off the common factors of $\frac{1}{2}g$) $\frac{1}{4}t^2 = t^2 - (t - 1\text{ s})^2$. Then $t - 1\text{ s} = \pm\sqrt{\frac{3}{4}}t$, or $t = 1\text{ s}/(1 - \sqrt{\frac{3}{4}}) = 7.46$ s. (We discarded the negative square root because t is obviously greater than 1 s.) Substituting this value of t into the equation for the total distance fallen, we find $y_0 - y(t) = \frac{1}{2}(9.8\text{ m/s}^2)(7.46\text{ s})^2 = 273$ m. (In a real fall from this height, air resistance should be considered.)

Problem

73. A balloon is rising at 10 m/s when its passenger throws up a ball at 12 m/s. How much later does the passenger catch the ball?

Solution

The initial (positive upward) velocity of the ball is 12 m/s relative to the passenger who throws it. Because the passenger is moving upward with constant velocity of 10 m/s, the initial velocity of the ball relative to the ground is 22 m/s. Assuming the ball is acted upon only by gravity (after being thrown at $t = 0$), we can write its vertical position as $y_B(t) = y_0 + (22\text{ m/s})t - \frac{1}{2}gt^2$. The balloon carrying the passenger is acted upon by the buoyant force of the air, in addition to gravity, so that it ascends with constant velocity (see Section 18-3). Thus, the vertical position of the passenger (in the same coordinate system used for the ball) is $y_P(t) = y_0 + (10\text{ m/s})t$. The passenger catches the ball when $y_B(t) = y_P(t)$ for $t > 0$. This implies $y_0 + (22\text{ m/s})t - \frac{1}{2}gt^2 = y_0 + (10\text{ m/s})t$, or $t = 2(12\text{ m/s})/(9.8\text{ m/s}^2) = 2.45$ s. (Because the balloon is moving with constant velocity, a coordinate system attached to the passenger, $y'_P = 0$, is an inertial frame (see Section 6-5) in which the ball's position is $y'_B = (12\text{ m/s})t - \frac{1}{2}gt^2$. Setting $y'_B = y'_P$ gives one the same time of flight.)

Problem

77. A skier starts from rest, and heads downslope with a constant acceleration of 2.3 m/s². How long does it take her to go 15 m, and what is her speed at that point?

Solution

The equations for linear motion with constant acceleration are summarized in Table 2-1. Since the initial velocity is zero, $x(t) - x_0 = \frac{1}{2}at^2$, and the time to travel 15 m is $t = \sqrt{2(15\text{ m})/(2.3\text{ m/s}^2)} = 3.61$ s. The velocity at this time is $v = at = (2.3\text{ m/s}^2)(3.61\text{ s}) = 8.31$ m/s.

Problem

81. A subway train is traveling at 80 km/h when it approaches a slower train 50 m ahead traveling in the same direction at 25 km/h. If the faster train begins decelerating at 2.1 m/s², while the slower train continues at constant speed, how soon and at what relative speed will they collide?

Solution

Take the origin $x = 0$ and $t = 0$ at the point where the first train begins decelerating, with positive x in the direction of motion. Equation 2-10 gives the instantaneous position of each train, with $x_{10} = 0$, $v_{10} = 80$ km/h, $a_1 = -2.1$ m/s², $x_{20} = 50$ m, $v_{20} = 25$ km/h, and $a_2 = 0$ given. Thus $x_1(t) = v_{10}t + \frac{1}{2}a_1t^2$ and $x_2(t) = x_{20} + v_{20}t$. The trains collide at the first time that $x_1 = x_2$, or when $x_{20} - (v_{10} - v_{20})t - \frac{1}{2}a_1t^2 = 0$. Using the quadratic formula to solve for the smaller root, we find $t = [(v_{10} - v_{20}) - \sqrt{(v_{10} - v_{20})^2 + 2a_1x_{20}}]/(-a_1) = [(55\text{ m}/3.6\text{ s}) - \sqrt{(55\text{ m}/3.6\text{ s})^2 + 2(-2.1\text{ m/s}^2)(50\text{ m})}]/(2.1\text{ m/s}^2) = 4.97$ s. The velocity of the first train at the time of the collision is $v_1 = v_{10} + a_1t = (80\text{ km/h}) - (2.1\text{ m/s}^2)(4.97\text{ s}) \times (3.6\text{ km/h/m/s}) = 42.4$ km/h. Therefore, the relative speed at impact is $v_1 - v_2 = 42.4\text{ km/h} - 25\text{ km/h} = 17.4$ km/h.

Supplementary Problems

Problem

85. Compute and rank order the average speeds, in m/s, associated with the following sporting events: (a) the 1990 Tour de France, a 2121-mi bicycle race won by Greg Lemond in 90 h 43 min 20 s; (b) the 1987 world indoor record women's 200-m dash, by Heike Dreschler in 22.27 s; (c) the 1990 world record men's indoor 400-m run, by Danny Everett in 45.04 s; (d) the 1986 world record women's 100-m freestyle, swum by Kristin Otto in 54.73 s.

Solution

It is probably most convenient to compare the average speeds for the four events in meters per second. $\bar{v}_a = (2121\text{ mi})(1609\text{ m/mi})/(90\times3600 + 43\times60 + 20)\text{ s} = 10.4$ m/s, $\bar{v}_b = 200\text{ m}/22.27\text{ s} = 8.98$ m/s, $\bar{v}_c = 400\text{ m}/45.04\text{ s} = 8.88$ m/s, and $\bar{v}_d = 100\text{ m}/54.73\text{ s} = 1.83$ m/s. Not surprisingly, cycling is faster than running, which is faster than swimming.

Problem

89. You see the traffic light ahead of you is about to turn from red to green, so you slow to a steady speed of 10 km/h and cruise to the light, reaching it just as it turns green. You then accelerate to 60 km/h in the next 12 s, then maintain constant speed. At the light, you pass a Corvette that has stopped for the red light. Just as you pass (and the light turns green) the Corvette begins accelerating, reaching 65 km/h in 6.9 s, then maintaining constant speed. (a) Plot the motions of both cars on a graph showing the 10-s period after the light turns green. (b) How long after the light turns green does the Corvette pass you? (c) How far are you from the light when the Corvette passes you? (d) How far ahead of you is the Corvette 1 min after the light turns green?

Solution

(a) Let the stoplight be at $x = 0$ and turn green at $t = 0$. Then

$$x_{\text{You}}(t) = \begin{cases} (10 \text{ km/h})t + \dfrac{1}{2}\left(\dfrac{60 \text{ km/h} - 10 \text{ km/h}}{12 \text{ s}}\right)t^2, & 0 \le t \le 12 \text{ s.} \\ 116.7 \text{ m} + (60 \text{ km/h})(t - 12 \text{ s}), & t \ge 12 \text{ s.} \end{cases}$$

$$x_{\text{Corv}}(t) = \begin{cases} \dfrac{1}{2}\left(\dfrac{65 \text{ km/h}}{6.9 \text{ s}}\right)t^2, & 0 \le t \le 6.9 \text{ s.} \\ 62.29 \text{ m} + (65 \text{ km/h})(t - 6.9 \text{ s}), & t \ge 6.9 \text{ s.} \end{cases}$$

Before plotting x versus t, we first calculate that

$x_{\text{You}}(6.9 \text{ s}) = 46.72 \text{ m}$, $x_{\text{You}}(12 \text{ s}) = 116.7 \text{ m}$,

$x_{\text{You}}(60 \text{ s}) = 916.7 \text{ m}$. $x_{\text{Corv}}(6.9 \text{ s}) = 62.29 \text{ m}$,

$x_{\text{Corv}}(12 \text{ s}) = 154.4 \text{ m}$, $x_{\text{Corv}}(60 \text{ s}) = 1021 \text{ m}$.

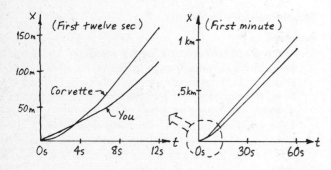

(b) Evidentially, the Corvette passes you before 6.9 s, while both cars are accelerating, so $x_{\text{You}} = x_{\text{Corv}}$ implies:

$$(10 \text{ km/h})t + \frac{1}{2}\left(\frac{50 \text{ km/h}}{12 \text{ s}}\right)t^2 = \frac{1}{2}\left(\frac{65 \text{ km/h}}{6.9 \text{ s}}\right)t^2, \text{ or}$$

$t = 3.81 \text{ s}$.

(c) When the cars pass, both are at the same position:

$$x_{\text{You}}(3.81 \text{ s}) = x_{\text{Corv}}(3.81 \text{ s}) = \frac{1}{2}\left(\frac{65 \text{ km/h}}{6.9 \text{ s}}\right)(3.81 \text{ s})^2 = 19.0 \text{ m}$$

from the green light. (d) After 1 min, the Corvette is $x_{\text{Corv}}(60 \text{ s}) - x_{\text{You}}(60 \text{ s}) = 1021 \text{ m} - 916.7 \text{ m} = 104 \text{ m}$ ahead.

Problem

93. You drop a rock into a well. 2.7 s later, you hear the splash. (a) How far down is the water? The speed of sound is 340 m/s. (b) What percentage error would be introduced by neglecting the travel time for the sound?

Solution

(a) The time for a rock to fall a height h is $t_1 = \sqrt{2h/g}$; the time for the sound to arrive is $t_2 = h/v_s$, where v_s is the speed of sound. The total time is 2.7 s $= t_1 + t_2 = \sqrt{2h/g} + h/v_s$. Solving for $\sqrt{h}$ and squaring, we find

$$h = \left[\frac{v_s}{2}\left(\sqrt{\frac{2}{g} + \frac{4(2.7 \text{ s})}{v_s}} - \sqrt{\frac{2}{g}}\right)\right]^2 = 33.2 \text{ m},$$

when $v_s = 340 \text{ m/s}$ and $g = 9.8 \text{ m/s}^2$ are substituted. (b) If we had neglected t_2, as in Problem 46, $h' = 35.7 \text{ m}$, with a percent error of $100(35.7 - 33.2)/33.2 = 7.5\%$. ●

● **CHAPTER 3** THE VECTOR DESCRIPTION OF MOTION

Sections 3-2 and 3-3 Adding and Subtracting Vectors

Problem

1. You walk west 220 m, then turn 45° toward the north and walk another 50 m. How far and in what direction from your starting point do you end up?

Solution

We can find the magnitude and direction of the vector sum of the two displacements either using geometry and a diagram, or by adding vector components.

From the law of cosines:

$$C = \sqrt{A^2 + B^2 - 2AB\cos\gamma}$$
$$= \sqrt{(220 \text{ m})^2 + (5 \text{ m})^2 - 2(220 \text{ m})(50 \text{ m})\cos 135°}$$
$$\approx 258 \text{ m}.$$

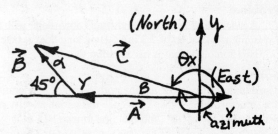

Problem 1 Solution.

From the law of sines: $C/\sin\gamma = B/\sin\beta$, or

$$\beta = \sin^{-1}\left(\frac{B\sin\gamma}{C}\right) = \sin^{-1}\left[\left(\frac{50 \text{ m}}{258 \text{ m}}\right)\sin 135°\right] = 7.88°.$$

The direction of $\mathbf{C}$ can be specified as 7.88° N of W, or 82.12° W of N, or by the azimuth 277.88° (CW from N).

The x and y components of the vectors above are:
$\mathbf{A} = (-220\text{ m})\hat{\imath}$, $\mathbf{B} = (50\text{ m})(\hat{\imath}\cos 135° + \hat{\jmath}\cos 45°) = (50\text{ m})(-\hat{\imath} + \hat{\jmath})/\sqrt{2}$. Then $\mathbf{C} = \mathbf{A} + \mathbf{B} = -(220\text{ m} + 50\text{ m}/\sqrt{2})\hat{\imath} + (50\text{ m}/\sqrt{2})\hat{\jmath} = -225\text{ m}\hat{\imath} + 35.4\text{ m}\hat{\jmath}$, and $C = \sqrt{C_x^2 + C_y^2} = 258$ m, $\theta_x = \tan^{-1}(C_y/C_x) = 180° - 7.88° = 172.12°$ (where, since $C_y > 0$ and $C_x < 0$, θ_x is in the second quadrant).

Problem

5. The vector $\mathbf{A}$ is 12 units long and points 30° north of east. The vector $\mathbf{B}$ is 18 units long and points 45° west of north. Find a vector $\mathbf{C}$ such that $\mathbf{A} + \mathbf{B} + \mathbf{C} = 0$.

Solution

The desired vector is $\mathbf{C} = -(\mathbf{A} + \mathbf{B})$. In a coordinate system with x axis east and y axis north, $\mathbf{A} = 12(\hat{\imath}\cos 30° + \hat{\jmath}\sin 30°) = 6(\sqrt{3}\hat{\imath} + \hat{\jmath})$, $\mathbf{B} = 18(\hat{\imath}\cos 135° + \hat{\jmath}\sin 135°) = 9\sqrt{2}(-\hat{\imath} + \hat{\jmath})$. Thus $\mathbf{C} = (9\sqrt{2} - 6\sqrt{3})\hat{\imath} - (9\sqrt{2} + 6)\hat{\jmath} = 2.34\,\hat{\imath} - 18.7\,\hat{\jmath}$, with magnitude $\sqrt{(2.34)^2 + (-18.7)^2} = 18.9$ units and direction $\theta = \tan^{-1}(-18.7/2.34) = 277°$ or $-82.9°$. (A direction lying in the fourth quadrant, $C_x > 0$ and $C_y < 0$, can be represented by a negative angle, measured CW from the x axis.) The same result follows from the laws of cosines and sines, but components are easier to use, especially with more than two vectors: $|\mathbf{C}| = \sqrt{(12)^2 + (18)^2 - 2(12)(18)\cos 75°}$, $\alpha = \sin^{-1}(12 \sin 75°/18.9) = 37.9°$, and $\theta = -(45° + \alpha)$.

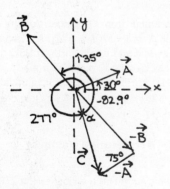

Problem 5 Solution.

Problem

9. A direct flight from Orlando, Florida, to Atlanta, Georgia, covers 660 km and heads at 29° west of north. Your flight, however, stops at Charleston, South Carolina, on the way to Atlanta. Charleston is 510 km from Orlando, in a direction 9.3° east of north. What are the magnitude and direction of the Charleston-to-Atlanta leg of your flight?

Solution

The displacements between the three cities are as shown in the diagram. Evidently, the displacement from Charleston to Atlanta is $\mathbf{C} = \mathbf{A} - \mathbf{B} = (660\text{ km})(\hat{\imath}\cos(90° + 29°) + \hat{\jmath}\sin 119°) - (510\text{ km})(\hat{\imath}\cos(90° - 9.3°) + \hat{\jmath}\sin 80.7°) = (-320\hat{\imath} + 577\hat{\jmath} - 82.4\hat{\imath} - 503\hat{\jmath})$ km, $= (-402\hat{\imath} + 74.0\hat{\jmath})$ km. Its magnitude is $\sqrt{(-402)^2 + (74.0)^2}$ km $= 409$ km and its direction is $\theta = \tan^{-1}(74.0/(-402)) = 170°$ CCW from the x axis, or $\theta - 90° = 79.6°$ CCW from the y axis (Note: For those solving this problem from the laws of cosines and sines $\sqrt{(660)^2 + (510)^2 - 2(660)(510)\cos 38.3°} = 409$, $\sin^{-1}(660 \sin 38.3°/409) = 91.1°$, and $90° - (91.1° - 80.7°) = 79.6°$.)

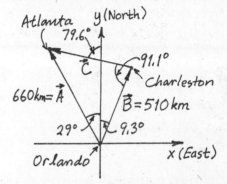

Problem 9 Solution.

Section 3-4 Multiplying Vectors by Scalars

Problem

13. Two vectors $\mathbf{A}$ and $\mathbf{B}$ have the same length A. $\mathbf{A}$ points horizontally to the right, while $\mathbf{B}$ makes a 30° angle with the horizontal. What are the length and direction of the vector $\mathbf{A} - 2\mathbf{B}$?

Solution

Let the x axis be parallel to $\mathbf{A}$ and the y axis be on the same side of the horizontal as $\mathbf{B}$, as shown. Then $\mathbf{A} = A\hat{\imath}$, $\mathbf{B} = |\mathbf{B}|(\hat{\imath}\cos 30° + \hat{\jmath}\sin 30°) = \frac{1}{2}A(\sqrt{3}\hat{\imath} + \hat{\jmath})$, and $\mathbf{A} - 2\mathbf{B} = A[(1 - \sqrt{3})\hat{\imath} - \hat{\jmath}]$. The magnitude of $\mathbf{A} - 2\mathbf{B}$ is $A\sqrt{(1 - \sqrt{3})^2 + (-1)^2} = 1.24A$, and its direction (CCW from the x axis and in the third quadrant) is $\theta = \tan^{-1}(1/(\sqrt{3} - 1)) = 234°$

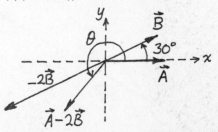

Problem 13 Solution.

Section 3-5 Coordinate Systems, Vector Components, and Unit Vectors

Problem

17. Repeat Problem 2, using unit vector notation.

Solution

Take the x axis to the right and the y axis $90°$ counterclockwise from it. Then:

$$\mathbf{A} = 10(\hat{\imath}\cos 35° + \hat{\jmath}\sin 35°) = 8.19\hat{\imath} + 5.74\hat{\jmath}$$
$$\mathbf{B} = 6(\hat{\imath}\cos 235° + \hat{\jmath}\sin 235°) = -3.44\hat{\imath} - 4.91\hat{\jmath}$$
$$\mathbf{C} = 8(\hat{\imath}\cos 115° + \hat{\jmath}\sin 115°) = -338\hat{\imath} + 7.25\hat{\jmath}$$

A component of the sum (or difference) is the sum (or difference) of the components, therefore:

$$\mathbf{A} + \mathbf{B} = 4.75\hat{\imath} + 0.821\hat{\jmath} = 4.82(\hat{\imath}\cos 9.80° + \hat{\jmath}\sin 9.80°)$$
$$\mathbf{A} - \mathbf{B} = 11.6\hat{\imath} + 10.7\hat{\jmath} = 15.8(\hat{\imath}\cos 42.5° + \hat{\jmath}\sin 42.5°)$$
$$\mathbf{A} + \mathbf{C} = 4.81\hat{\imath} + 13.0\hat{\jmath} = 13.8(\hat{\imath}\cos 69.7° + \hat{\jmath}\sin 69.7°)$$
$$\mathbf{A} + \mathbf{B} + \mathbf{C} = 1.37\hat{\imath} + 8.07\hat{\jmath}$$
$$= 8.19(\hat{\imath}\cos 80.4° + \hat{\jmath}\sin 80.4°)$$

The first form is the vector in components, the second gives the magnitude ($\sqrt{x^2 + y^2}$) and direction $\theta_x = \tan^{-1}(y/x)$.

Problem

21. A vector $\mathbf{A}$ is 10 units long and points $30°$ counterclockwise (CCW) from horizontal. What are the x and y components on a coordinate system (a) with the x axis horizontal and the y axis vertical; (b) with the x axis at $45°$ CCW from horizontal and the y axis $45°$ CCW from vertical; and (c) with the x axis at $30°$ CCW from horizontal and the y axis $90°$ CCW from the x axis?

Solution

The component of a vector along any direction equals the magnitude of the vector times the cosine of the angle ($\leq 180°$). Thus:
(a) $A_x = 10\cos 30° = 8.66$, $A_y = 10\cos 60° = 5.00$
(b) $A'_x = 10\cos 15° = 9.66$, $A'_y = 10\cos 105° = -2.59$
(c) $A''_x = 10\cos 0° = 10$, $A''_y = 10\cos 90° = 0$.

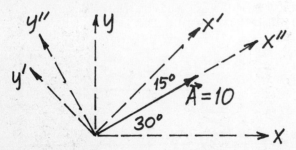

Problem 21 Solution.

Problem

25. In Fig. 3-13, suppose that vectors $\mathbf{A}$ and $\mathbf{C}$ both make $30°$ angles with the horizontal while $\mathbf{B}$ makes a $60°$ angle, and that $A = 2.3$ km, $B = 1.0$ km, and $C = 2.9$ km. (a) Express the displacement vector $\Delta\mathbf{r}$ from start to summit in each of the coordinate systems shown, and (b) determine its length.

Solution

Using the x-y system in Fig. 3-13, we have $\theta_A = \theta_C = 30°$ and $\theta_B = 60°$. Then $\mathbf{A} = A(\hat{\imath}\cos\theta_A + \hat{\jmath}\sin\theta_A) = (2.3 \text{ km})(\hat{\imath}\cos 30° + \hat{\jmath}\sin 30°) = (1.99\hat{\imath} + 1.15\hat{\jmath})$ km, $\mathbf{B} = (1 \text{ km})(\hat{\imath}\cos 60° + \hat{\jmath}\sin 60°) = (0.50\hat{\imath} + 0.87\hat{\jmath})$ km, and $\mathbf{C} = (2.9 \text{ km})(\hat{\imath}\cos 30° + \hat{\jmath}\sin 30°) = (2.51\hat{\imath} + 1.45\hat{\jmath})$ km. Thus, $\Delta r = \mathbf{A} + \mathbf{B} + \mathbf{C} = (5.00\hat{\imath} + 3.47\hat{\jmath})$ km and $\Delta r = \sqrt{(5.00)^2 + (3.47)^2}$ km $= 6.09$ km. A similar calculation in the x'-y' system, with $\theta'_A = \theta'_C = 0$ and $\theta'_B = 30°$, yields $\mathbf{A} = 2.3\hat{\imath}'$ km, $\mathbf{B} = (1 \text{ km})(\hat{\imath}'\cos 30° + \hat{\jmath}'\sin 30°) = (0.87\hat{\imath}' + 0.50\hat{\jmath}')$ km, $\mathbf{C} = 2.9\hat{\imath}'$ km, $\Delta r = (6.07\hat{\imath}' + 0.50\hat{\jmath}')$ km, and $\sqrt{(6.07)^2 + (0.50)^2} = 6.09$ of course.

Section 3-6 Velocity Vectors

Problem

29. The Orlando-to-Atlanta flight described in Problem 9 takes 2.5 h. What is the average velocity vector? Express (a) as a magnitude and direction, and (b) in unit vector notation with the x axis east and the y axis north.

Solution

(b) The displacement from Orlando to Atlanta calculated in Problem 9 was $\mathbf{A} = (-320\hat{\imath} + 577\hat{\jmath})$ km in a coordinate system with x axis east and y axis north. If this trip took 2.5 h, the average velocity was $\overline{\mathbf{V}} = \mathbf{A}/2.5$ h $= (-128\hat{\imath} + 231\hat{\jmath})$ km/h. (a) This has magnitude $\sqrt{(-128)^2 + (231)^2}$ km/h $= 264$ km/h and direction $\theta = \tan^{-1}(231/-128) = 119°$ (which was given).

Section 3-7 Acceleration Vectors

Problem

33. A supersonic aircraft is traveling east at 2100 km/h. It then begins to turn southward, emerging from the turn 2.5 min later heading due south at 1800 km/h. What are the magnitude and direction of its average acceleration during the turn?

Solution

In a coordinate system with x axis east and y axis north, the initial velocity of the airplane at the beginning of its turn is $v_1 = 2100\hat{\imath}$ km/h, and the final velocity at the end is $v_2 = -1800\hat{\jmath}$ km/h. The average acceleration (Equation 3-8) is

$\mathbf{a} = (\mathbf{v}_2 - \mathbf{v}_1)/(t_2 - t_1) = (-1800\hat{\jmath} - 2100\hat{\imath})(km/h)/2.5 \text{ min}$
$= -(3.89\hat{\imath} + 3.33\hat{\jmath}) \text{ m/s}^2$, with magnitude
$\sqrt{(-3.89)^2 + (-3.33)^2} \text{ m} = 5.12 \text{ m/s}^2$ and direction $\theta = \tan^{-1}(-3.33/3.89) = 221°$ (in the third quadrant, nearly southwest).

Problem

37. Attempting to stop on a slippery road, a car moving at 80 km/h skids across the road at a 30° angle to its initial motion, coming to a stop in 3.9 s. Determine the average acceleration in m/s², using a coordinate system with the x axis in the direction of the car's original motion and y axis toward the side of the road to which the car skids.

Solution

The car's acceleration is opposite to the direction of the skid, since it comes to a stop (if $\mathbf{v} = 0$, $\mathbf{a} = -\mathbf{v}_0/\Delta t$). Its magnitude is given by Equation 3-8, $|\mathbf{a}| = |\mathbf{v} - \mathbf{v}_0|/\Delta t = |-\mathbf{v}_0|/\Delta t = (80 \text{ m}/3.6 \text{ s})/3.9 \text{ s} = 5.70 \text{ m/s}^2$. In relation to the coordinate system specified,
$\mathbf{a} = (5.70 \text{ m/s}^2)(\hat{\imath} \cos 210° + \hat{\jmath} \sin 210°) = -(5.70 \text{ m/s}^2)(\sqrt{3}\hat{\imath} + \hat{\jmath})/2 = -(4.93\hat{\imath} + 2.85\hat{\jmath}) \text{ m/s}^2$. (Note that $\mathbf{v}_0$, the velocity at the start of the skid, is not in the direction of the initial motion before the skid.)

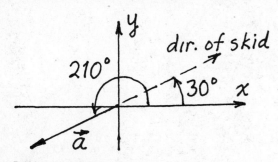

Problem 37 Solution.

Section 3-8 Relative Motion

Problem

41. A dog paces around the perimeter of a rectangular barge that is headed up a river at 14 km/h relative to the riverbank. The current in the water is at 3.0 km/h. If the dog walks at 4.0 km/h, what are its speeds relative to (a) the shore and (b) the water as it walks around the barge?

Problem 41 Solution.

Solution

(a) Let S be a frame of reference fixed on the shore, with x axis upstream, and let S' be a frame attached to the barge. The velocity of S' relative to S is $\mathbf{V} = (14 \text{ km/h})\hat{\imath}$. The velocity of the dog relative to the shore is (from Equation 3-25) $\mathbf{v} = \mathbf{v}' + \mathbf{V}$, and its speed is $v = |\mathbf{v}' + \mathbf{V}|$, where $v' = 4$ km/h. When the dog is walking upstream, $\mathbf{v}' \parallel \mathbf{V}$, $v = (4 + 14) \text{ km/h} = 18 \text{ km/h}$, and when walking downstream, $-\mathbf{v}' \parallel \mathbf{V}$, and $v = (14 - 4) \text{ km/h} = 10 \text{ km/h}$. When $\mathbf{v}' \perp \mathbf{V}$, $v = \sqrt{14^2 + 4^2} = 14.7 \text{ km/h}$. (In general, $v^2 = v'^2 + V^2 + 2v'V\cos\theta'$, where θ' is the angle between $\mathbf{v}'$ and $\mathbf{V}$ in S'.) (b) Since the current flows downstream, according to Equation 3-25:

$$\begin{pmatrix} \text{vel. of barge} \\ \text{rel. to water} \end{pmatrix} = \begin{pmatrix} \text{vel. of barge} \\ \text{rel. to shore} \end{pmatrix} - \begin{pmatrix} \text{vel. of water} \\ \text{rel. to shore} \end{pmatrix}$$
$$= 14\hat{\imath} - (-3\hat{\imath}) \text{ km/h}.$$

Going through the same steps as in part (a), for a new frame S moving with the water, with a new relative velocity $\mathbf{V} = (17 \text{ km/h})\hat{\imath}$, we find the speed of the dog relative to the water to be $(4 + 17) = 21 \text{ km/h}$, $(17 - 4) = 13 \text{ km/h}$, and $\sqrt{4^2 + 17^2} = 17.4 \text{ km/h}$, for the corresponding segments of the barge's perimeter.

Problem

45. You're on an airport "people mover," a conveyor belt going at 2.2 m/s through a level section of the terminal. A button falls off your coat and drops freely 1.6 m, hitting the belt 0.57 s later. What are the magnitude and direction of the button's displacement and average velocity during its fall in (a) the frame of reference of the "people mover" and (b) the frame of reference of the airport terminal? (c) As it falls, what is its acceleration in each frame of reference?

Solution

(c) Let S be the frame of reference of the airport and S' the frame of reference of the conveyor belt. If the velocity of S' relative to S is a constant, 2.2 m/s in the x-x' direction, then the acceleration of gravity is the same in both systems. (a) In S', the initial velocity of the button is zero and it falls vertically downward. Its displacement is simply $\Delta y' = -1.6$ m and its average velocity is $\Delta y'/\Delta t = -1.6 \text{ m}/0.57 \text{ s} = 2.81 \text{ m/s}$. (b) In S, the initial velocity of the button is not zero (it is 2.2 m/s in the x direction) and so the button follows a projectile trajectory to be described in Chapter 4. Here, we observe that while the button falls vertically through a displacement $\Delta y = \Delta y' = -1.6$ m, it also moves horizontally (in the direction of the conveyor belt) through a displacement $\Delta x = (2.2 \text{ m/s})(0.57 \text{ s}) = 1.25$ m. Its net displacement in S is $\Delta \mathbf{r} = \Delta x\hat{\imath} + \Delta y\hat{\jmath} = (1.25\hat{\imath} - 1.60\hat{\jmath})$ m, so its average velocity is $\bar{\mathbf{v}} = \Delta\bar{\mathbf{r}}/\Delta t = (1.25\hat{\imath} - 1.60\hat{\jmath}) \text{ m}/0.57 \text{ s} = (2.20\hat{\imath} - 2.8\hat{\jmath}) \text{ m/s}$. These have magnitudes $|\Delta \mathbf{r}| =$

$\sqrt{(1.25)^2 + (-1.60)^2}$ m = 2.03 m and $|\mathbf{v}|$ = $\sqrt{(2.20)^2 + (-2.81)^2}$ m/s = 3.57 m/s, and the same direction $\theta = \tan^{-1}(-1.60/1.25) = -51.9°$ from the horizontal, or 38.1° from the downward vertical.

Paired Problems

Problem

49. The three displacement vectors shown in Fig. 3-22 are each 10.0 m long. (a) Write each in unit vector notation. (b) Find a vector **D** such that **A** + **B** + **C** + **D** = **0**. (c) Find the length of **D**.

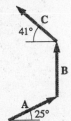

FIGURE 3-22 Problem 49.

Solution

(a) The angle each vector makes CCW from the x axis is given, $\theta_A = 25°$, $\theta_B = 90°$, $\theta_C = 180° - 41° = 139°$, as is the magnitude of each vector, $|\mathbf{A}| = |\mathbf{B}| = |\mathbf{C}| = 10.0$ m. Thus, each can be expressed in terms of the unit vectors and its components: $\mathbf{A} = A_x\hat{\mathbf{i}} + A_y\hat{\mathbf{j}} = |\mathbf{A}|(\hat{\mathbf{i}}\cos\theta_A + \hat{\mathbf{j}}\sin\theta_A) =$ (10.0 m)$(\hat{\mathbf{i}}\cos 25° + \hat{\mathbf{j}}\sin 25°) = (9.06\hat{\mathbf{i}} + 4.23\hat{\mathbf{j}})$ m, $\mathbf{B} = (10.0$ m$)(\hat{\mathbf{i}}\cos 90° + \hat{\mathbf{j}}\sin 90°) = 10.0\,\hat{\mathbf{j}}$ m, and $\mathbf{C} =$ (10.0 m)$(\hat{\mathbf{i}}\cos 139° + \hat{\mathbf{j}}\sin 139°) = (-7.55\hat{\mathbf{i}} + 6.56\hat{\mathbf{j}})$ m. (b) The vector equation **A** + **B** + **C** + **D** = **0** has solution **D** = $-($**A** + **B** + **C**$) = -[(9.06 - 7.55)\hat{\mathbf{i}} +$ $(4.23 + 10.0 + 6.56)\hat{\mathbf{j}}]$ m = $-(1.52\hat{\mathbf{i}} + 20.8\hat{\mathbf{j}})$ m. (c) $|\mathbf{D}| =$ $\sqrt{(-1.52)^2 + (-20.8)^2}$ m = 20.8 m. (Note: We did not round off any intermediate calculations.)

Problem

53. The sweep-second hand of a clock is 3.1 cm long. What are the magnitude of (a) the average velocity and (b) the average acceleration of the hand's tip over a 5.0-s interval? (c) What is the angle between the average velocity and acceleration vectors?

Solution

There will be numerous occasions to use vector components to analyse circular motion in later chapters (or see the solutions to Problems 20, 30, 35 and 36), so let's use geometry to solve this problem. (a) The angular displacement of the hand during a 5 s interval is $\theta = (5/60)(360°) = 30°$. The position vectors (from the center hub) of the tip at the beginning and end of the interval, $\mathbf{r}_1$ and $\mathbf{r}_2$, form the sides of an isosceles triangle whose base is the magnitude of the displacement,

$|\Delta\mathbf{r}| = 2|\mathbf{r}|\sin\frac{1}{2}\theta = 2(3.1$ cm$)\sin(30°/2) = 1.60$ cm, and whose base angle is $\frac{1}{2}(180° - 30°) = 75°$. Thus, the average velocity has magnitude $|\Delta\mathbf{r}|/\Delta t = 1.60$ cm/5 s = 0.321 cm/s and direction $180° - 75° = 105°$ CW from $\mathbf{r}_1$. (b) The instantaneous speed of the tip of the second-hand is a constant and equal to the circumference divided by 60 s, or $v = 2\pi(3.1$ cm$)/60$ s = 0.325 cm/s. The direction of the velocity of the tip is tangent to the circumference, or perpendicular to the radius, in the direction of motion (CW). The angle between two tangents is the same as the angle between the two corresponding radii, so $\mathbf{v}_1$, $\mathbf{v}_2$ and $\Delta\mathbf{v}$ form an isosceles triangle similar to the one in part (a). Thus $|\Delta\mathbf{v}| =$ $2|\mathbf{v}|\sin\frac{1}{2}\theta = 2(0.325$ cm/s$)\sin(\frac{1}{2}\times 30°) = 0.168$ cm/s. The magnitude of the average acceleration is $|\Delta\mathbf{v}|/\Delta t =$ $(0.168$ cm/s$)/5$ s = 3.36×10^{-2} cm/s², and its direction is 105° CW from the direction of $\mathbf{v}_1$, or 195° CW from the direction of $\mathbf{r}_1$. (c) The angle between $\overline{\mathbf{a}}$ and $\overline{\mathbf{v}}$, from parts (a) and (b), is $195° - 105° = 90°$. (Note: This is the geometry used in Section 4.4 to discuss centripetal acceleration.)

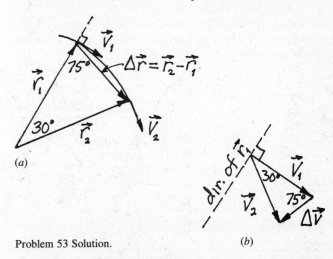

Problem 53 Solution.

Supplementary Problems

Problem

57. A satellite is in a circular orbit 240 km above Earth's surface, moving at a constant speed of 7.80 km/s. A tracking station picks up the satellite when it is 5.0° above the horizon, as shown in Fig. 3-24. The satellite is tracked until it is directly overhead. What are the magnitudes of (a) its displacement, (b) its average velocity, and (c) its average acceleration during the tracking interval? Is the value of the average acceleration approximately familiar?

Solution

Once we know the angular displacement $\Delta\theta$, of the satellite in its orbit, we can calculate the magnitudes of the displacement

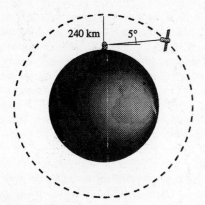

FIGURE 3-24 Problem 57 (figure is not to scale).

from its initial position P_1 to its final position P_2, its average velocity and acceleration, during the same tracking interval, by using the geometrical analysis in the solution to Problem 53, to which the reader is referred. In the diagram based on Fig. 3-24, O is the center of the Earth, $R_E = 6370$ km is the average radius of the Earth, A is the tracking station, and $r = R_E + h = 6370$ km $+ 240$ km $= 6610$ km is the radius of the orbit.

We apply the law of cosines to triangle OAP_1 to find AP_1, and then the law of sines to find $\Delta\theta$. Thus,

$$(6610 \text{ km})^2 = (6370 \text{ km})^2 + (AP_1)^2$$
$$- 2(6370 \text{ km})(AP_1)\cos 95°.$$

This is a quadratic equation with (positive) solution

$$AP_1 = (6370 \text{ km})\cos 95°$$
$$+\sqrt{(6370 \text{ km})^2 \cos^2 95° + (6610 \text{ km})^2 - (6370 \text{ km})^2}$$
$$= 1295 \text{ km}.$$

(We are keeping four significant figures in the intermediate results, but will round off to three at the end.) Then $\sin \Delta\theta/AP_1 = \sin 95°/r$ gives $\Delta\theta = \sin^{-1}(1295 \sin 95°/6610) = 11.26°$.

(a) Now that we know $\Delta\theta$, the magnitude of the displacement can be found from the isosceles triangle OP_1P_2 as in Problem 53. $P_1P_2 = 2r \sin\frac{1}{2}\Delta\theta = 2(6610 \text{ km})\sin\frac{1}{2}(11.26°) = 1296$ km ≈ 1300 km.

(b) To find the magnitude of the average velocity, we first need to find the tracking interval Δt. The time for a complete orbit (called the period) is the orbital circumference divided by the speed, or $T = 2\pi r/v = 2\pi(6610 \text{ km})/(7.8 \text{ km/s}) = 5.325\times10^3$ s. During the tracking interval, the satellite completes only a fraction $\Delta\theta/360°$ of a complete orbit, so $\Delta T = (11.26°/360°)(5.325\times10^3 \text{ s}) = 166.5$ s. Thus, $|\overline{\mathbf{v}}| = P_1P_2/\Delta t = 1296$ km$/166.5$ s $= 7.78$ km/s.

(c) As shown in the solution to Problem 53, the velocity vectors at P_1 and P_2 form an isosceles triangle similar to OP_1P_2. Thus $|\Delta\mathbf{v}| = 2|\mathbf{v}|\sin\frac{1}{2}\Delta\theta =$

$2(7.8 \text{ km/s})\sin\frac{1}{2}(11.26°) = 1.530$ km/s, and the magnitude of the average acceleration is $|\overline{\mathbf{a}}| = |\Delta\mathbf{v}|/\Delta t = (1.530 \text{ km/s})/166.5$ s $= 9.19$ m/s^2. When we recall that the Earth's gravity holds the satellite in its orbit, it is not surprising that this magnitude is close to g. Of course, at an altitude of 240 km, the Earth's gravitational field is only 9.11 m/s^2 (compared to 9.81 m/s^2 at the surface). We must also remember that there is a discrepancy between the average and instantaneous accelerations due to the finite size of Δt.

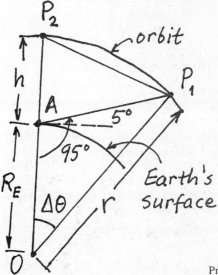

Problem 57 Solution.

Problem

61. Write an expression for a unit vector that lies at 45° between the positive x and y axes.

Solution

A vector of unit magnitude, making a 45° angle CCW with the x axis, can be expressed as $1\cdot\cos 45° \, \hat{\mathbf{i}} + 1\cdot\sin 45° \, \hat{\mathbf{j}} = (\hat{\mathbf{i}} + \hat{\mathbf{j}})/\sqrt{2}$. (A unit vector in any direction in the x-y plane is therefore $\hat{\mathbf{n}} = \hat{\mathbf{i}} \cos\theta + \hat{\mathbf{j}} \sin\theta$.)

Problem

65. Town B is located across the river from town A and at a 40.0° angle upstream from A, as shown in Fig. 3-26. A ferryboat travels from A to B; it sails at 18.0 km/h relative to the water. If the current in the river flows at 5.60 km/h, at what angle should the boat head? What will be its speed relative to the ground? *Hint:* Set up Equation 3-10 for this situation. Each component of Equation 3-10 yields two equations in the unknowns v, the magnitude of the boat's velocity relative to the ground, and ϕ, the unknown angle. Solve the x equation for $\cos \phi$ and substitute into the second equation, using the relation $\sin \phi = \sqrt{1 - \cos^2 \phi}$. You can then solve for v, then go back and get ϕ from your first equation.

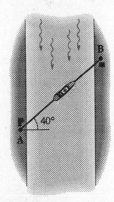

FIGURE 3-26 Problem 65.

Solution

The diagram shows the velocities and coordinate axes added to Fig. 3-26. $\mathbf{v}'$ (the velocity of the boat relative to the water), $\mathbf{v}$ (the velocity of the boat relative to the ground), and $\mathbf{V}$ (the velocity of the water relative to the ground) are related by

Equation 3-10, $\mathbf{v}' = \mathbf{v} - \mathbf{V}$. This vector equation is equivalent to two scalar equations, one for each component, $v'_x = v_x - V_x$ and $v'_y = v_y - V_y$. The components of each vector, in terms of its magnitude and angle, are $v'_x = v' \cos\theta'$, $v'_y = v' \sin\theta'$, $v_x = v \cos 40°$, $v_y = v \sin 40°$, and $V_x = 0$, $V_y = -V$. Therefore, the x and y component equations are $v' \cos\theta' = v \cos 40°$ and $v' \sin\theta' = v \sin 40° + V$. We can eliminate θ' by squaring and adding (since $\sin^2 + \cos^2 = 1$): $v'^2(\cos^2\theta' + \sin^2\theta') = v^2(\cos^2 40° + \sin^2 40°) + 2vV \sin 40° + V^2$, or $v^2 + 2vV \sin 40° - v'^2 + V^2 = 0$. The positive root of this quadratic for v (appropriate for a magnitude) is $v = -V \sin 40° + \sqrt{V^2 \sin^2 40° + v'^2 - V^2}$. If the given values $v' = 18.0$ km/h and $V = 5.60$ km/h are substituted, we find the speed relative to the ground is $v = 13.9$ km/h. Going back to the x component equation, we find the heading $\theta' = \cos^{-1}(v \cos 40°/v') = \cos^{-1}(13.9 \cos 40°/18.0) = 53.8°$. (Equations similar to these will be solved when collisions in 2-dimensions and the conservation of momentum are discussed in Chapter 11.)

● CHAPTER 4 MOTION IN MORE THAN ONE DIMENSION

Section 4-1 Velocity and Acceleration

Problem

1. A skater is gliding along the ice at 2.8 m/s, when she undergoes an acceleration of magnitude 1.1 m/s² for 2.0 s. At the end of that time she is moving at 5.0 m/s. What must be the angle between the acceleration vector and the initial velocity vector?

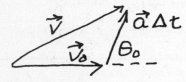

Problem 1 Solution.

Solution

For constant acceleration, Equation 4-3 shows that the vectors $\mathbf{v_0}$, $\mathbf{a}\Delta t$ and $\mathbf{v}$ form a triangle as shown. The law of cosines gives $v^2 = v_0^2 + (a\Delta t)^2 - 2v_0 a \Delta t \cos(180° - \theta_0)$. When the given magnitudes are substituted, one can solve for θ_0: $(5.0 \text{ m/s})^2 = (2.8 \text{ m/s})^2 + (1.1 \text{ m/s}^2)(2.0 \text{ s})^2 + 2(2.8 \text{ m/s}) \times (1.1 \text{ m/s}^2)(2.0 \text{ s})\cos\theta_0$, or $\cos\theta_0 = 1.00$ (exactly) and $\theta_0 = 0°$. Since $\mathbf{v_0}$ and $\mathbf{a}$ are colinear, the change in speed is maximal.

Section 4-2 Constant Acceleration

Problem

5. The position of an object as a function of time is given by $\mathbf{r} = (2.4t + 1.2t^2)\hat{\mathbf{i}} + (0.89t - 1.9t^2)\hat{\mathbf{j}}$ m, where t is the time in seconds. What are the magnitude and direction of the acceleration?

Solution

One can always find the acceleration by taking the second derivative of the position, $\mathbf{a}(t) = d^2\mathbf{r}(t)/dt^2$. However, collecting terms with the same power of t, one can write the position in meters as $\mathbf{r}(t) = (2.4\hat{\mathbf{i}} + 0.89\hat{\mathbf{j}})t + (1.2\hat{\mathbf{i}} - 1.9\hat{\mathbf{j}})t^2$. Comparison with Equation 4-4 shows that this represents motion with constant acceleration equal to twice the coefficient of the t^2 term, or $\mathbf{a} = (2.4\hat{\mathbf{i}} - 3.8\hat{\mathbf{j}})$ m/s². The magnitude and direction of $\mathbf{a}$ are $\sqrt{(2.4)^2 + (-3.8)^2}$ m/s² = 4.49 m/s² and $\tan^{-1}(-3.8/2.4) = 302°$ (CCW from the x axis, in the fourth quadrant) or $-57.7°$.

Problem

9. A hockey puck is moving at 7.15 m/s when a stick imparts a constant acceleration of 63.5 m/s² at a 90.0° angle to the original direction of motion. If the acceleration lasts 0.132 s, how far does the puck move during this time?

Solution

Take the x axis in the direction of the initial velocity, $\mathbf{v_0} = 7.15\hat{\mathbf{i}}$ m/s, and the y axis in the direction of the acceleration, $\mathbf{a} = 63.5\hat{\mathbf{j}}$ m/s². The displacement during the 0.132 s

interval of constant acceleration is (Equation 4-4)
$\Delta r = r - r_0 = v_0 t + \frac{1}{2}at^2 = (7.15$ m/s$)(0.132$ s$)\hat{i} + \frac{1}{2}(63.5$ m/s$^2)(0.132$ s$)^2\hat{j} = (0.944\hat{i} + 0.553\hat{j})$ m. This has magnitude $\sqrt{(0.944)^2 + (0.553)^2}$ m $= 1.09$ m (and makes an angle of $\tan^{-1}(0.553/0.944) = 30.4°$ with the direction of the initial velocity).

Problem

13. Figure 4-24 shows a cathode-ray tube, used to display electrical signals in oscilloscopes and other scientific instruments. Electrons are accelerated by the electron gun, then move down the center of the tube at 2.0×10^9 cm/s. In the 4.2-cm-long deflecting region they undergo an acceleration directed perpendicular to the long axis of the tube. The acceleration "steers" them to a particular spot on the screen, where they produce a visible glow. (a) What acceleration is needed to deflect the electrons through 15°, as shown in the figure? (b) What is the shape of an electron's path in the deflecting region?

Solution

(a) With x-y axes as drawn on Fig. 4-24, the electrons emerge from the deflecting region with velocity $\mathbf{v} = v_0\hat{i} + at\,\hat{j}$, after a time $t = x/v_0$, where $x = 4.2$ cm and $v_0 = 2\times10^9$ cm/s. The angle of deflection (direction of $\mathbf{v}$) is 15°, so $\tan 15° = v_y/v_x = at/v_0 = ax/v_0^2$. Thus, $a = v_0^2 \tan 15°/x = 2.55\times10^{17}$ cm/s^2 (when values are substituted). (b) Since the acceleration is assumed constant, the electron trajectory is parabolic in the deflecting region.

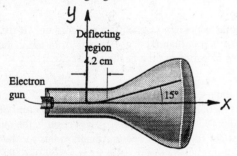

FIGURE 4-24 Problem 13 Solution.

Section 4-3 Projectile Motion

Problem

17. A kid fires a blob of water horizontally from a squirt gun held 1.6 m above the ground. It hits another kid 2.1 m away square in the back, at a point 0.93 m above the ground (see Fig. 4-25). What was the initial speed of the blob?

Solution

Since the blob was fired horizontally ($v_{0y} = 0$), the time it takes to fall from $y_0 = 1.6$ m to $y = 0.93$ m is given by Equation 4-8, $t = \sqrt{2(y_0 - y)/g} =$

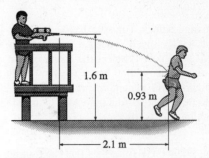

FIGURE 4-25 Problem 17.

$\sqrt{2(1.6 - 0.93)$ m$/(9.8$ m/s$^2)} = 0.370$ s. Its initial speed, $v_0 = v_{0x}$, can be found from Equation 4-7, $v_0 = (x - x_0)/t = 2.1$m/0.370 s $= 5.68$ m/s.

Problem

21. Ink droplets in an ink-jet printer are ejected horizontally at 12 m/s, and travel a horizontal distance of 1.0 mm to the paper. How far do they fall in this interval?

Solution

From $x - x_0 = v_{0x}t$, the time of flight can be found. Substitution into $y_0 - y = \frac{1}{2}gt^2$ (recall that $v_{0y} = 0$) yields $y_0 - y = \frac{1}{2}g(x - x_0)^2/v_0^2 = \frac{1}{2}(9.8$ m/s$^2)(10^{-3}$ m$)^2/(12$ m/s$)^2 = 3.40\times10^{-8}$ m $= 34$ nm for the distance fallen, practically negligible. Note that this analysis is equivalent to using Equation 4-9 with $\theta = 0$.

Problem

25. A car moving at 40 km/h strikes a pedestrian a glancing blow, breaking both the car's front signal light lens and the pedestrian's hip. Pieces of the lens are found 4.0 m down the road from the center of a 1.2-m wide crosswalk, and a lawsuit hinges on whether or not the pedestrian was in the crosswalk at the time of the accident. Assuming that the lens was initially 63 cm off the ground, and that the lens pieces continued moving horizontally with the car's speed at the time of the impact, was the pedestrian in the crosswalk?

Solution

What is an issue here is the horizontal range of a piece of signal light lens in projectile motion, starting from a height of $y_0 - y = 0.63$ m off the ground, with an initial horizontal velocity of $v_0 = v_{0x} = (40$ m/3.6 s$)$, and $v_{0y} = 0$. Eliminating the time of flight from Equations 4-7 and 8 (see the solution to Problem 21), one obtains $x - x_0 = v_0\sqrt{2(y_0 - y)/g} = (11.1$ m/s$) \sqrt{2(0.63$ m$)/(9.8$ m/s$^2)} = 3.98$ m. If the pieces of lens did not bounce very far from the point where they hit the ground, this places the point of impact of the accident just 4.00 m $- 3.98$ m $= 2$ cm from the center of the crosswalk. (Forensic physics is crucial to the prosecution's case.)

Problem

29. A submarine-launched missile has a range of 4500 km.
(a) What launch speed is needed for this range when the launch angle is 45°? (Neglect the distance over which the missile accelerates.) (b) What is the total flight time? (c) What would be the minimum launch speed at a 20° launch angle, used to "depress" the trajectory so as to foil a space-based antimissile defense?

Solution

(a) Assuming Equation 4-10 applies (i.e. the trajectory begins and ends at the same height, or $y(t) = y_0$), one finds $v_0 = \sqrt{xg/\sin 2\theta} = \sqrt{(4500 \text{ km})(.0098 \text{ km/s}^2)/\sin 90°} = 6.64$ km/s. (b) The time of flight is the positive solution of Equation 4-8 when $y(t) - y_0 = 0 = v_{0y}t - \frac{1}{2}gt^2$. Thus $t = 2v_{0y}/g = 2(6.64 \text{ km/s})\sin 45°/(9.8 \text{ m/s}^2) = 958$ s $= 16.0$ min. (c) At a 20° launch angle, $v_0 = \sqrt{(4500 \text{ km})(0.0098 \text{ km/s}^2)/\sin 40°} = 8.28$ km/s.

Problem

33. If you can hit a golf ball 180 m on Earth, how far can you hit it on the moon? (Your answer is an underestimate, because the distance on Earth is restricted by air resistance as well as by a larger g.)

Solution

For given v_0, the horizontal range is inversely proportional to g. With surface gravities from Appendix E, we find $x_{moon} = (g_{Earth}/g_{moon})x_{Earth} = (9.81/1.62)(180 \text{ m}) = 1090$ m.

Problem

37. A circular fountain has jets of water directed from the circumference inward at an angle of 45°. Each jet reaches a maximum height of 2.2 m. (a) If all the jets converge in the center of the circle and at their initial height, what is the radius of the fountain? (b) If one of the jets is aimed at 10° too low, how far short of the center does it fall?

Solution

(a) The radius is the horizontal range, $r = v_0^2/g$ (Equation 4-10 with $\theta = 45°$). The maximum height is $h = v_{0y}^2/2g = v_0^2/4g$ (Equation 2-14 with $v_y = 0$ and $v_{0y} = v_0\cos 45° = v_0/\sqrt{2}$). Therefore, $r = (4gh)/g = 4h = 4(2.2 \text{ m}) = 8.8$ m.
(b) If one jet is directed at 35° with the same initial speed ($v_0^2 = rg$), it would fall short by $r - x$, where x is given by

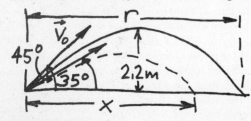

Problem 37 Solution.

Equation 4-10. Therefore, $r - x = r - (v_0^2/g)\sin(2\times35°) = (8.8 \text{ m})(1 - \sin 70°) = 0.531$ m.

Problem

41. Show that, for a given initial speed, the horizontal range of a projectile is the same for launch angles $45° + \alpha$ and $45° - \alpha$, where α is between 0° and 45°.

Solution

The trigonometric identity in Appendix A for the sine of the sum of two angles shows that $\sin 2(45° \pm \alpha) = \sin(90° \pm 2\alpha) = \sin 90° \cos 2\alpha \pm \cos 90° \sin 2\alpha = \cos 2\alpha$, so the horizontal range formula (Equation 4-10) gives the same range for either launch angle, at the same initial speed.

Section 4-4 Circular Motion

Problem

45. Estimate the acceleration of the moon, which completes a nearly circular orbit of 385,000 km radius in 27 days.

Solution

The centripetal acceleration is given in terms of the period for uniform circular motion by Equation 4-12 in Example 4-7. In the case of the moon, $a = 4\pi^2 r/T^2 = 4\pi^2(3.85\times10^8 \text{ m})/(27.3\times86,400 \text{ s})^2 = 2.73\times10^{-3} \text{ m/s}^2$, where we used more accurate data from Appendix E. (Note: "centripetal" is a purely kinematic adjective descriptive of circular motion. In this case, the origin of the moon's centripetal acceleration is the gravitational attraction of the Earth.)

Problem

49. A beetle can cling to a phonograph record only if the acceleration is less than 0.25 g. How far from the center of a $33\frac{1}{3}$-rpm record can the beetle stand?

Solution

With the imposed acceleration limit, Equation 4-12 implies $(4\pi^2 r/T^2) < 0.25g$, or $r < 0.25g(T/2\pi)^2$. The period of an LP record is $(33\frac{1}{3})^{-1}$ min $= 1.8$ s. (The period is the number of minutes per revolution.) The maximum radius for the beetle is $(.25)(9.8 \text{ m/s}^2)(1.8 \text{ s}/2\pi)^2 = 20.1$ cm $= 7.92$ in. Since this is greater than the radius of a 12-in record, the beetle can stand anywhere on a normal LP record.

Problem

53. A runner rounds the semicircular end of a track whose curvature radius is 35 m. The runner moves at constant speed, with an acceleration of 1.8 m/s². How long does it take to complete the turn?

Solution

Since the runner has a constant speed along a circular arc, the acceleration must be purely a centripetal acceleration, namely $a_c = 1.8 \text{ m/s}^2 = v^2/r$, and the speed is $v = \sqrt{(1.8 \text{ m/s}^2)(35 \text{ m})} = 7.94$ m/s. At this speed, it takes time

$t = \pi r/v = \pi(35 \text{ m})/(7.94 \text{ m/s}) = 13.9$ s to complete a semicircle. (Of course, Equation 4-12 would give the same result, $\frac{1}{2}T = \pi \sqrt{r/a_c}$.)

Section 4-5 Nonuniform Circular Motion

Problem

57. An object is set into motion on a circular path of radius r by giving it a constant tangential acceleration a_t. Derive an expression for the time t when the acceleration vector points at 45° to the direction of motion.

Solution

The tangential acceleration (in the direction of motion) is perpendicular to the radial acceleration, so the resultant total acceleration (their vector sum) is at 45° between them at the instant when $a_t = a_r = v^2/r$. The linear speed along the circle depends on a_t only (since a_c is perpendicular to the velocity) so $v = a_t t$ for constant a_t (provided the object is "set into motion" with $v_0 = 0$ at $t = 0$). Thus, $a_t = (a_t t)^2/r$ or $t = \sqrt{r/a_t}$.

Problem

61. If you can throw a stone straight up to a height of 16 m, how far could you throw it horizontally over level ground? Assume the same throwing speed and optimum launch angle.

Solution

To throw an object vertically to a maximum height of $h = 16$ m $= y_{\max} - y_0$ requires an initial speed of $v_0 = \sqrt{2g(y_{\max} - y_0)} = \sqrt{2gh}$. With this value of v_0 and the optimum launch angle $\theta_0 = 45°$, Equation 4-10 gives a maximum horizontal range on level ground of $x = v_0^2/g = 2h = 32$ m. (The maximum horizontal range on level ground is twice the maximum height for vertical motion with the same initial speed. This result holds in the approximation of constant g and no air resistance.)

Problem

65. A fireworks rocket is 73 m above the ground when it explodes. Immediately after the explosion, one piece is moving at 51 m/s at 23° to the upward vertical direction. A second piece is moving at 38 m/s at 11° below the horizontal direction. At what horizontal distance from the explosion site does each piece land?

Solution

In the trajectory equation (Equation 4-9) with origin at the position of the explosion, the coefficients are known for each piece. One can solve this quadratic equation for x, when $y = -73$ m (ground level), and select the positive root (since the trajectories start at $x = 0$ and end on the ground in the direction of v_{0x}, which is chosen positive).

For the first piece, $\tan 23° = 0.424$, and $g/2v_{0x}^2 = (9.8 \text{ m/s}^2)/2(51 \cos 23° \text{ m/s})^2 = 2.22 \times 10^{-3} \text{ m}^{-1}$ and so the

quadratic is -73 m $= 0.424x - (2.22 \times 10^{-3} \text{ m}^{-1})x^2$. This has positive root $x = [0.424 + \sqrt{(0.424)^2 + 4(73 \text{ m})(2.22 \times 10^{-3} \text{ m}^{-1})}]/(4.44 \times 10^{-3} \text{ m}^{-1}) = 300$ m.

Similarly, for the second piece, $\tan(-11°) = -0.194$ and $g/2v_{0x}^2 = (9.8 \text{ m/s}^2)/2(38 \cos(-11°) \text{ m/s})^2 = 3.52 \times 10^{-3} \text{ m}^{-1}$, so $x = [-0.194 + \sqrt{(-0.194)^2 + 4(73 \text{ m})(3.52 \times 10^{-3} \text{ m}^{-1})}]/(7.04 \times 10^{-3} \text{ m}^{-1}) = 119$ m.

Problem

69. After takeoff, a plane makes a three-quarter circle turn of radius 7.1 km, maintaining constant altitude but steadily increasing its speed from 390 to 740 km/h. Midway through the turn, what is the angle between the plane's velocity and acceleration vectors?

Solution

The plane's velocity is tangent to its circular path in the direction of motion and so is the tangential acceleration a_t. The radial (centripetal) acceleration a_c is perpendicular to this, so the angle between the total acceleration and the velocity is $\theta = \tan^{-1}(a_c/a_t)$, inclined toward the center of the turn. For the linear motion over a circular distance of $s = \frac{3}{4}(2\pi r) = 1.5\pi(7.1 \text{ km}) = 33.5$ km, Equation 2-11 can be used to find the constant a_t, $a_t = (v_f^2 - v_i^2)/2s = [(740/3.6)^2 - (390/3.6)^2](\text{m}^2/\text{s}^2)/2(33.5 \text{ km}) = 0.456 \text{ m/s}^2$. We can find $a_c = v^2/r$ midway through the turn by using Equation 2-11 again, since $v^2 = v_i^2 + 2a_t(s/2) = v_i^2 + \frac{1}{2}(v_f^2 - v_i^2) = \frac{1}{2}(v_i^2 + v_f^2)$. Then $a_c = \frac{1}{2}[(390/3.6)^2 + (740/3.6)^2](\text{m}^2/\text{s}^2)/(7.1 \text{ km}) = 3.80 \text{ m/s}^2$. Finally, $\theta = \tan^{-1}(3.80/0.456) = 83.2°$. (Instead of working out each component of acceleration numerically, we could have written the final result symbolically as follows:

$$\frac{a_c}{a_t} = \left(\frac{v_i^2 + v_f^2}{2r}\right) \cdot \frac{2(3/4)(2\pi r)}{(v_f^2 - v_i^2)} = \frac{3\pi}{2}\left(\frac{v_f^2 + v_i^2}{v_f^2 - v_i^2}\right),$$

and $\theta = \tan^{-1}[3\pi(740^2 + 390^2)/2(740^2 - 390^2)]$ as before.)

Supplementary Problems

Problem

73. A monkey is hanging from a branch a height h above the ground. A naturalist stands a horizontal distance d from a point directly below the monkey. The naturalist aims a tranquilizer dart directly at the monkey, but just as he fires the monkey lets go. Show that the dart will nevertheless hit the monkey, provided its initial speed exceeds $\sqrt{(d^2 + h^2)g/2h}$.

Solution

Gravity accelerates the dart and the monkey equally, so both fall the same vertical distance from the point of aim (the

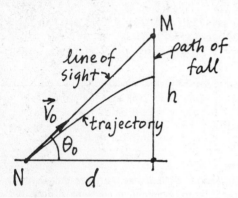

Problem 73 Solution.

monkey's original position) resulting in a hit, provided the initial speed of the dart is sufficient to reach the monkey before the monkey reaches the ground. To prove this assertion, let the dart be fired from ground level ($y = 0$) with speed v_0 and direction $\theta_0 = \tan^{-1}(h/d)$ (line of sight from naturalist N to monkey M) while the monkey drops from height h at $t = 0$. The vertical height of each is $y_{\text{monkey}} = h - \frac{1}{2}gt^2$ and $y_{\text{dart}} = v_{0y}t - \frac{1}{2}gt^2$, where $v_{0y} = v_0 \sin\theta_0 = v_0 h/\sqrt{d^2 + h^2}$. (The term $-\frac{1}{2}gt^2$ represents the effect of gravity, which appears the same way in both y-coordinate equations.) The dart strikes the monkey when $y_{\text{monkey}} = y_{\text{dart}}$, which implies $h = v_{0y}t$, or $t = \sqrt{d^2 + h^2}/v_0$. This must be less than the time required for the monkey to fall to the ground, which is $\sqrt{2h/g}$ (from $y_{\text{monkey}} = 0$). Thus $\sqrt{d^2 + h^2}/v_0 < \sqrt{2h/g}$ or $v_0 > \sqrt{g(d^2 + h^2)/2h}$. (This condition can also be understood from the horizontal range formula, Equation 4-10. The range of the dart has to be greater than the horizontal distance to the monkey, $d < v_0^2 \sin 2\theta_0/g = v_0^2 \, hd/(d^2 + h^2)g$.)

Problem

77. (a) Use the result of the preceding problem to find an expression for the launch angles on level ground (i.e., $y = 0$). (b) Taking an approximate value for the range for the 30° and 60° launch angles from the graph in Fig. 4-13, and the inital speed from the caption of that figure, show that your result yields approximately those angles.

Solution

(a) Setting $y = 0$ in the expression for $\tan\theta_0$ in the previous problem, one obtains $\tan\theta_0 = b \pm \sqrt{b^2 - 1}$, where $b = v_0^2/gx$ (this is equivalent to the horizontal range formula, Equation 4-10). (b) In Fig. 4-13, $x \approx 220$ m for $\theta_0 = 30°$ and 60°, so $b = (50 \text{ m/s})^2/(9.8 \text{ m/s}^2)(220 \text{ m}) = 1.16$. Then $\theta_0 = \tan^{-1}[1.16 \pm \sqrt{(1.16)^2 - 1}] = \tan^{-1}(0.573)$ or $\tan^{-1}(1.75) = 29.8°$ or 60.2°, entirely consistent with the estimate of x from the graph.

Problem

81. At the peak of its trajectory, a projectile is moving horizontally with speed v_x, and its acceleration (g, downward) is obviously perpendicular to its velocity. Use these facts to find an expression for the radius of curvature at the peak of the parabolic trajectory.

Solution

At the apex of a parabolic trajectory, the acceleration is perpendicular to the velocity ($v = v_x$, $v_y = 0$) and is therefore entirely radial ($a = a_r$, $a_t = 0$). In projectile motion, the acceleration has constant magnitude g, so at the apex, the radius of curvature has magnitude $v^2/a_r = v_x^2/g$.

(Actually, $a_r = -g$ and the radius of curvature is negative, $R = -v_x^2/g$ at the apex. It is only the magnitude of R that is discussed in the text and specified in this problem. A negative R means that the slope of the tangent decreases as the trajectory is traced in the direction of increasing arc length. For a plane curve $y = f(x)$, the radius of curvature is defined as $R = \pm[1 + (dy/dx)^2]^{3/2}/(d^2y/dx^2)$, where the positive (negative) sign is used if the arc length increases in the positive (negative) x direction. (See any comprehensive calculus text, e.g. Seeley, Harcourt Brace.) For the trajectory of Equation 4-9, $dy/dx = \tan\theta_0 - x(g/v_0^2\cos^2\theta_0)$, $d^2y/dx^2 = -g/v_0^2\cos^2\theta_0$, and the positive sign in the expression for R is used when $\cos\theta_0 \geq 0$. Of course, $dy/dx = v_y/v_x$ gives the direction of the instantaneous velocity, which is zero at the apex of the trajectory. Then $R = (d^2y/dx^2)^{-1} = -v_0^2\cos^2\theta_0/g = -v_x^2/g$ at the apex, as above.)

Problem

85. While increasing its speed, a train enters a 90° circular turn of radius r with speed v_0 and with tangential acceleration equal to half its radial acceleration. If its tangential acceleration remains constant, show that when the train leaves the turn its tangential and radial accelerations are related by $a_t = a_r/(2 + \pi)$.

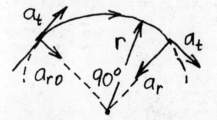

Problem 85 Solution.

Solution

At the beginning of the turn, $2a_t = a_{r0} = v_0^2/r$. At the end of the turn, $a_r = v^2/r$, where $v^2 = v_0^2 + 2a_t(\pi r/2)$ (since a_t is constant and the distance traveled is $\frac{1}{4}$ of the circumference). Putting this together, we find $a_r = (v_0^2 + \pi a_t r)/r = (2a_t r + \pi a_t r)/r = a_t(2 + \pi)$.

● **CHAPTER 5**   DYNAMICS: WHY DO THINGS MOVE?

Section 5-5   Newton's Second Law

Problem

1. A subway train has a mass of 1.5×10^6 kg. What force is required to accelerate the train at 2.5 m/s^2?

Solution

$F = ma = (1.5 \times 10^6 \text{ kg})(2.5 \text{ m/s}^2) = 3.75$ MN. (This is the magnitude of the net force acting; see Table 1-1 for SI prefixes.)

Problem

5. In an x-ray tube, electrons are accelerated to speeds on the order of 10^8 m/s, then slammed into a target where they come to a stop in about 10^{-18} s. Estimate the average stopping force on each electron.

Solution

The magnitude of the average force is $\overline{F} = m\overline{a} = m|\Delta v/\Delta t| = (9.11 \times 10^{-31} \text{ kg})(10^8 \text{ m/s})/(10^{-18} \text{ s}) \simeq 9 \times 10^{-5}$ N. Compared to the TV tube in Example 5-2, the electron in an x-ray tube experiences a force billions of times greater. It is a result of the violence of this interaction that x rays, called bremsstrahlung, are emitted (see Problem 2-46).

Problem

9. Object A accelerates at 8.1 m/s^2 when a 3.3-N force is applied. Object B accelerates at 2.7 m/s^2 when the same force is applied. (a) How do the masses of the two objects compare? (b) If A and B were stuck together and accelerated by the 3.3-N force, what would be the acceleration of the composite object?

Solution

In this idealized one-dimensional situation, the applied force of $F = 3.3$ N is the only force acting. (a) When applied to either object, Newton's second law gives $F = m_A a_A$ and $F = m_B a_B$, so $m_B/m_A = a_A/a_B = (8.1 \text{ m/s}^2)/(2.7 \text{ m/s}^2) = 3$. (For constant net force, mass is inversely proportional to acceleration.) (b) When F is applied to the combined object, $F = (m_A + m_B)a$. Since $F = m_A a_A$ and $m_B = 3m_A$, one finds $a = F/(m_A + m_B) = m_A a_A/4m_A = \frac{1}{4}(8.1 \text{ m/s}^2) = 2.03$ m/s^2.

Problem

13. The maximum braking force of a 1400-kg car is about 8000 N. Estimate the stopping distance when the car is traveling (a) 40 km/h; (b) 60 km/h; (c) 80 km/h; (d) 55 mi/h.

Solution

The maximum braking acceleration is $a = F/m = -8000$ N/1400 kg $= -5.71$ m/s^2. (We expressed the braking force as negative because it is opposite to the direction of motion.) (a) On a straight horizontal road, a car traveling at a velocity of $v_{0x} = 40$ km/h can stop in a minimum distance found from Equation 2-11 and the maximum deceleration just calculated: $x - x_0 = -v_{0x}^2/2a = -(40 \text{ m/3.6 s})^2/2(-5.71 \text{ m/s}^2) = 10.8$ m. For the other initial velocities, the stopping distance is (b) 24.3 m, (c) 43.2 m, and (d) 52.9 m.

Problem

17. A 1.25-kg object is moving in the x direction at 17.4 m/s. 3.41 s later, it is moving at 26.8 m/s at 34.0° to the x axis. What are the magnitude and direction of the force applied during this time?

Solution

Newton's second law says that the average force acting is equal to the rate of change of momentum, $\mathbf{F}_{av} = m \, \Delta\mathbf{v}/\Delta t$ (as explained following Equation 5-2). The initial velocity is $17.4\hat{\mathbf{i}}$ m/s and the final velocity is (26.8 m/s) $(\hat{\mathbf{i}}\cos 34° + \hat{\mathbf{j}}\sin 34°) = (22.2\hat{\mathbf{i}} + 15.0\,\hat{\mathbf{j}})$ m/s, so $\mathbf{F}_{av} = (1.25 \text{ kg})[(22.2 - 17.4)\hat{\mathbf{i}} + 15.0\,\hat{\mathbf{j}}](\text{m/s})/(3.41 \text{ s}) = (1.77\hat{\mathbf{i}} + 5.49\,\hat{\mathbf{j}})$ N. This has magnitude 5.77 N and direction 72.2° CCW to the x axis.

Section 5-6   Mass and Weight: The Force of Gravity

Problem

21. A cereal box says "net weight 340 grams." What is the actual weight (a) in SI units? (b) In ounces?

Solution

(a) The actual weight (equal to the force of gravity at rest on the surface of the Earth) is $mg = (0.340 \text{ kg})(9.81 \text{ m/s}^2) = 3.33$ N. (b) With reference to Appendix E, $(3.33 \text{ N})(0.2248 \times 16 \text{ oz/N}) = 12.0$ oz. (The word "net" in net weight means just the weight of the contents; gross weight includes the weight of the container, etc. This may be compared with the use of the word in net force, which means the sum of all the forces or the resultant force. A net weight, profit, or amount is the resultant after all corrections have been taken into account.)

Problem

25. A neutron star is a fantastically dense object with the mass of a star crushed into a region about 10 km in diameter. If my mass is 65 kg, and if I would weigh 5.4×10^{14} N on a certain neutron star, what is the acceleration of gravity on the neutron star?

Solution

If we define weight on a neutron star analogously to its definition on Earth, the surface gravity of the neutron star is an enormous $g = W/m = (5.4\times10^{14}$ N$)/(65$ kg$) = 8.3\times10^{12}$ m/s^2, nearly 10^{12} times g.

Section 5-7 Adding Forces

Problem

29. An elevator accelerates downward at 2.4 m/s^2. What force does the floor of the elevator exert on a 52-kg passenger?

Solution

The passenger also accelerates downward with $a_y = -2.4$ m/s^2 (y axis positive upward), so the vertical component of the net force on the passenger is $F_{net,y} = ma_y$. The only vertical forces acting on the passenger are the force of gravity, $F_{g,y} = -W = -mg$, and the normal force of the floor, $F_{norm,y} = N$. Therefore, $F_{net,y} = -mg + N = ma_y$, or $N = m(g + a_y) = (52$ kg$)(9.8 - 2.4)$ m/s$^2 = 385$ N.

Problem

33. At liftoff, a space shuttle with 2.0×10^6 kg total mass undergoes an upward acceleration of $0.60g$. (a) What is the total thrust force developed by its engines? (b) What force does the seat back exert on a 60-kg astronaut during liftoff?

Solution

(a) At liftoff, the only significant vertical forces on the space shuttle are gravity ($F_{g,y} = -mg$ downward) and the thrust ($F_y > 0$ upward). (Air resistance can be neglected because the initial velocity is zero.) Therefore, $F_{net,y} = F_y - mg = ma_y$, or $F_y = m(g + a_y) = mg(1 + 0.6) = (2\times10^6$ kg$)(9.8$ m/s$^2)(1.6) = 31.4MN$. (b) The vertical forces on an astronaut are her/his weight and the normal force of the seat back ($N > 0$ upward; the astronauts are seated facing upward at liftoff). The acceleration of the astronaut is the same as that of the shuttle, so with reasoning analogous to part (a) above, $N = m(g + a_y) = (60$ kg$)(9.8$ m/s$^2)(1.6) = 941$ N.

Problem

37. An elevator moves upward at 5.2 m/s. What is the minimum stopping time it can have if the passengers are to remain on the floor?

Solution

Newton's second law and the analysis in Problem 35 show that passengers remain in contact with the floor as long as the elevator's acceleration satisfies $a_y \geq -g$ (a downward acceleration must have a magnitude less than g). To come to a stop ($v_y = 0$) from an initial upward velocity ($v_{0y} = 5.2$ m/s) requires a time $t = (0 - v_{0y})/a_y = v_{0y}/(-a_y)$, and therefore $t \geq v_{0y}/g = (5.2$ m/s$)/(9.8$ m/s$^2) = 0.531$ s is the condition for the passengers to stay on the floor.

Section 5-8 Newton's Third Law

Problem

41. Repeat the preceding problem, now with the left-right order of the blocks reversed. (That is, find the force on the rightmost mass, now 1.0 kg.)

Problem 41 Solution.

Solution

Assume that the table surface is horizontal and frictionless so that the only horizontal forces are the applied force and the contact forces between the blocks (positive to the right). The latter we denote by F_{12}, etc., which means the force block 1 exerts on block 2, etc. Since the blocks are in contact, they all have the same acceleration a, to the right. Newton's second law applied to each block separately is $F_{app} + F_{23} = m_3a$, $F_{32} + F_{12} = m_2a$ and $F_{21} = m_1a$ (where we just consider horizontal components). Similarly, Newton's third law applied to each pair of contact forces is $F_{32} + F_{23} = 0$ and $F_{21} + F_{12} = 0$. Adding the second law equations and using the third law equations, we find $F_{app} + (F_{23} + F_{32}) + (F_{12} + F_{21}) = F_{app} = (m_1 + m_2 + m_3)$ a, or $a = F_{app}/(m_1 + m_2 + m_3)$. Substituting this into the second law equation for m_1, we find $F_{21} = m_1a = m_1 F_{app}/(m_1 + m_2 + m_3) = (1$ kg$)(12$ N$)/(1 + 2 + 3)$ kg $= 2$ N. This is the force that the middle block exerts on the block to its right; the other contact force, $F_{23} = -6$ N, can be found from a similar procedure.

Problem

45. A 2200-kg airplane is pulling two gliders, the first of mass 310 kg and the second of mass 260 kg, down the runway

with an acceleration of 1.9 m/s² (Fig. 5-34). Neglecting the mass of the two ropes and any frictional forces, determine (a) the horizontal thrust of the plane's propeller; (b) the tension force in the first rope; (c) the tension force in the second rope; and (d) the net force on the first glider.

Solution

Assuming a level runway (as shown in Fig. 5-34), we may write the horizontal component (positive in direction of **a**) of the equations of motion (Newton's second law) for the three planes (all assumed to have the same **a**) as follows: $F_{th} - T_1 = m_1 a$ (airplane), $T_1 - T_2 = m_2 a$ (first glider), $T_2 = m_3 a$ (second glider). (Note: the tension has the same magnitude at every point in a rope of negligible mass.) (a) Add all the equations of motion (the tensions cancel in pairs due to Newton's third law): $F_{th} = (m_1 + m_2 + m_3)a = (2200 + 310 + 260)$ kg (1.9 m/s²) = 5.26 kN. (b) $T_1 = F_{th} - m_1 a = (m_2 + m_3)a = (570$ kg)(1.9 m/s²) = 1.08 kN. (c) $T_2 = m_3 a = (260$ kg)(1.9 m/s²) = 494 N. (d) $m_2 a = (310$ kg)(1.9 m/s²) = 589 N.

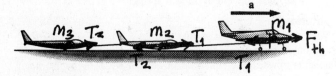

FIGURE 5-34 Problem 45 Solution.

Problem

49. A spring sketches 22 cm when a 40-N force is applied. If a 6.1-kg mass is suspended from the spring, how much will it stretch?

Solution

Hooke's law relates the magnitude of the applied force (the "reaction" to the spring force in Equation 5-8) to the magnitude of the stretch, hence $k = 40$ N/0.22 m = 182 N/m is the spring constant for this spring. If a 6.1 kg mass is suspended from the spring, it will exert a force equal to its weight, so the stretch produced in the spring will be $x = (6.1$ kg)(9.8 m/s²)/(182 N/m) = 32.9 cm.

Problem

53. A biologist is studying the growth of rats in an orbiting space station. To determine a rat's mass, she puts it in a 320-g cage, attaches a spring scale, and pulls so the scale reads 0.46 N. If the resulting acceleration of the rat and cage is 0.40 m/s², what is the rat's mass?

Solution

According to the scale, a force of 0.46 N applied to the cage and rat produces an acceleration of 0.40 m/s², so their combined mass is $F/a = (0.46$ N)/(0.40 m/s²) = 1.15 kg. The rat's mass is this minus the cage's, or 1150 − 320 = 830 g.

Paired Problems

Problem

57. Starting from rest, a 940-kg racing car covers 400 m in 4.95 s. What is the average force acting on the car?

Solution

The car's average acceleration (for straight line motion) is $a_{av} = 2(x - x_0)/t^2$ (see Equation 2-10 with $v_0 = 0$). Newton's second law gives the average net force on the car as $F_{av} = ma_{av} = 2(940$ kg)(400 m)/(4.95 s)² = 30.7 kN, in the direction of the motion.

Problem

61. You step into an elevator, and it accelerates to a downward speed of 9.2 m/s in 2.1 s. How does your apparent weight during this acceleration time compare with your actual weight?

Solution

Your apparent weight is the force you exert on the floor of the elevator, equal and opposite to the upward force the floor exerts on you. The latter and gravity, which equals your actual weight, are the two forces which determine your vertical acceleration, a_y. If we take the positive direction vertically upward, then $W_{app} - W = ma_y = (W/g)a_y$, or $W_{app}/W = 1 + a_y/g$. (We expressed your mass in terms of your actual weight in order to facilitate comparison with your apparent weight.) As expected, $W_{app}/W > 1$ for upward acceleration $a_y > 0$, and vice versa for downward acceleration as in this problem. If the elevator starts from rest, its acceleration is $(v - 0)/t = (-9.2$ m/s)/(2.1 s) = −4.38 m/s² and $W_{app}/W = 1 - 4.38/9.8 = 55.3\%$.

Problem

65. A 2.0-kg mass and a 3.0-kg mass are on a horizontal frictionless surface, connected by a massless spring with spring constant $k = 140$ N/m. A 15-N force is applied to the larger mass, as shown in Fig. 5-36. How much does the spring stretch from its equilibrium length?

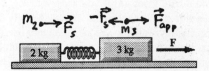

FIGURE 5-36 Problem 65 Solution.

Solution

The spring stretches until the acceleration of both masses is the same (positive in the direction of the applied force in Fig. 5-36). Since the spring is assumed massless, the tension in it is the same at both ends. (If this were not so, there would be a non-zero net horizontal force on the spring and hence its acceleration would be infinite, $a = (F \neq 0)/(m = 0)$.) The magnitude of the spring tension is given by Hooke's law, $F_s = k|x|$, where $|x|$ is the stretch. The horizontal component of Newton's second law applied to each mass is $F_{app} - F_s = m_3 a$ and $F_s = m_2 a$, as indicated in the sketch. Eliminating the acceleration between these equations gives us $F_{app} - F_s = m_3(F_s/m_2)$ or $F_s = k|x| = F_{app}/(1 + m_3/m_2)$, from which the stretch is easily found: $|x| = F_{app}/k(1 + m_3/m_2) = (15 \text{ N}/140 \text{ N/m})/(1 + 3/2) = 4.29$ cm.

Supplementary Problems

Problem

69. In throwing a 200-g ball, your hand exerts a constant upward force of 9.4 N for 0.32 s. How high does the ball rise after leaving your hand?

Solution

The maximum height to which the ball rises after leaving your hand can be calculated from its initial velocity v_{0y} (since $v_y = 0$ when $y = y_{max}$ in Equation 2-14, and we neglect air resistance), $y_{max} - y_c = v_{0y}^2/2g$. To find v_{0y}, the vertical forces acting on the ball during the 0.32 s it is thrown must be considered. These are the applied force of your hand, $F_{app} = 9.4$ N (positive upward), and the weight of the ball, $-mg$. They impart an upward acceleration of $(F_{app} - mg)/m$ to the ball (Newton's second law), which results in an initial upward speed for the throw of $v_{0y} = (F_{app} - mg)t/m$ (definition of acceleration). Putting this together and substituting the numerical values given, we get

$$y_{max} - y_0 = \left(\frac{F_{app}}{m} - g\right)^2 \frac{t^2}{2g}$$

$$= \left(\frac{9.4 \text{ N}}{0.2 \text{ kg}} - 9.8\frac{\text{m}}{\text{s}^2}\right)^2 \frac{(0.32 \text{ s})^2}{2(9.8 \text{ m/s}^2)} = 7.23 \text{ m}.$$

Problem

73. You have a mass of 60 kg, and you jump from a 78-cm-high table onto a hard floor. (a) If you keep your legs rigid, you come to a stop in a distance of 2.9 cm, as your body tissues compress slightly. What force does the floor exert on you? (b) If you bend your knees when you land, the bulk of your body comes to a stop over a distance of 0.54 m. Now estimate the force exerted on you by the floor. Neglect the fact that your legs stop in a shorter distance than the rest of you.

Solution

Suppose the motion is purely vertical. You would hit the floor with velocity $v = -\sqrt{2gh}$ (positive upward), where $h = 78$ cm. Your average acceleration while stopping is $a = v^2/2d$, where d is the distance over which your body comes to rest. The net force exerted on you is $ma = m(v^2/2d) = m(2gh/2d) = mg(h/d)$, while the force exerted on you by the floor is $F - mg = ma$, or $F = (1 + h/d)mg$, expressed as a factor times your weight, $mg = (60 \text{ kg})(9.8 \text{ m/s}^2) = 588$ N. (a) With your legs kept rigid, $F = (588 \text{ N})(1 + 78/2.9) = 16.4$ kN (within one order of magnitude of producing certain injury). (b) Under the conditions stated, a much safer value of $F = (588 \text{ N})(1 + 78/54) = 1.44$ kN is sustained.

Problem

77. Three identical massless springs of unstretched length ℓ and spring constant k are connected to three equal masses m as shown in Fig. 5-38. A force is applied at the top of the upper spring to give the whole system the same acceleration a. Determine the length of each spring.

Solution

Neglect the mass of the springs and let T_n, $n = 1, 2, 3$, be the tensions, starting from the bottom spring. With positive components upward, the equations of motion are $T_1 - mg = ma$, $T_2 - T_1 - mg = ma$, and $T_3 - T_2 - mg = ma$. Therefore $T_1 = m(g + a)$, $T_2 = T_1 + m(g + a) = 2m(g + a)$, and $T_3 = T_2 + m(g + a) = 3m(g + a)$. The stretch in each spring is $x_n = T_n/k$, and its length is $\ell + x_n = \ell + nm(g + a)/k$, where $n = 1, 2, 3$ is the spring's number.

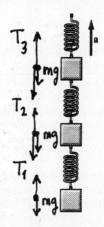

FIGURE 5-38 Problem 77 Solution.

Problem

81. Although we usually write Newton's law for one-dimensional motion in the form $F = ma$, the most basic version of the law reads $F = d(mv)/dt$. The simpler form holds only when the mass is constant. (a) Consider an object whose mass may be changing, and show that the rule for the derivative of a product (see Appendix A) can be used to write Newton's law in the form $F = ma + v\,(dm/dt)$. (b) A railroad car is being pulled beneath a grain elevator that dumps in grain at the rate of 450 kg/s. What force must be applied to keep the car moving at a constant 2.0 m/s?

Solution

(a) $F_{net} = d(mv)/dt = m(dv/dt) + (dm/dt)v = ma + v(dm/dt)$. (b) We can apply Newton's second law in the above form to the horizontal motion of the rail car. At a steady speed, $a = dv/dt = 0$, so the net horizontal force that must act on the car, whose mass is increasing at a rate of $dm/dt = 450$ kg/s, is $F_{net} = v(dm/dt) = (2$ m/s$)(450$ kg/s$) = 900$ N. If the grain does not exert any horizontal force on the car as it falls in, then all of the net force above must be applied to the car by a locomotive.

●

● CHAPTER 6   USING NEWTON'S LAWS

Section 6-1   Using Newton's Second Law

Problem

1. Two forces, both in the x-y plane, act on a 1.5-kg mass, which accelerates at 7.3 m/s² in a direction 30° counterclockwise from the x axis. One force has magnitude 6.8 N and points in the $+x$ direction. Find the other force.

Solution

Newton's second law for this mass says $\mathbf{F}_{net} = \mathbf{F}_1 + \mathbf{F}_2 = m\mathbf{a}$, where we assume no other significant forces are acting. Since the acceleration and the first force are given, one can solve for the second, $\mathbf{F}_2 = m\mathbf{a} - \mathbf{F}_1 = (1.5$ kg$)(7.3$ m/s²$)(\hat{\mathbf{i}}\cos 30° + \hat{\mathbf{j}}\sin 30°) - (6.8$ N$)\hat{\mathbf{i}} = (2.68\hat{\mathbf{i}} + 5.48\hat{\mathbf{j}})$ N. This has magnitude 6.10 N and direction 63.9° CCW from the x axis.

Problem

5. A block of mass m slides with acceleration a down a frictionless slope that makes an angle θ to the horizontal; the only forces acting on it are the force of gravity $\mathbf{F}_g$ and the normal force $\mathbf{N}$ of the slope. Show that the magnitude of the normal force is given by $N = m\sqrt{g^2 - a^2}$.

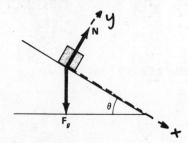

Problem 5 Solution.

Solution

Choose the x axis down the slope (parallel to the acceleration) and the y axis parallel to the normal. Then $a_x = a$, $a_y = 0$, $N_x = 0$, $N_y = N$, $F_{gx} = F_g \cos(90° - \theta) = mg\sin\theta$, and $F_{gy} = -mg\cos\theta$. Newton's second law, $\mathbf{N} + \mathbf{F}_g = m\mathbf{a}$, in components gives:

$$mg\sin\theta = ma, \quad \text{and} \quad N - mg\cos\theta = 0.$$

Eliminate θ (using $\sin^2\theta + \cos^2\theta = 1$) to find:

$$(a/g)^2 + (N/mg)^2 = 1, \quad \text{or} \quad N = m\sqrt{g^2 - a^2}.$$

Problem

9. A 15-kg monkey hangs from the middle of a massless rope as shown in Fig. 6-50. What is the tension in the rope? Compare with the monkey's weight.

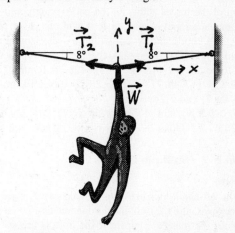

FIGURE 6-50   Problem 9 Solution.

Solution

The sum of the forces at the center of the rope (shown on Fig. 6-50) is zero (if the monkey is at rest), $\mathbf{T}_1 + \mathbf{T}_2 + \mathbf{W} = 0$. The x comparison of this equation requires that the tension is the same on both sides: $T_1 \cos 8° + T_2 \cos 172° = 0$, or $T_1 = T_2$. The y comparison gives $2T \sin 8° = W$, or $T = W/2 \sin 8° = 3.59W = 3.59(15 \text{ kg})(9.8 \text{ m/s}^2) = 528$ N.

Problem

13. A camper hangs a 26-kg pack between two trees, using two separate pieces of rope of different lengths, as shown in Fig. 6-54. What is the tension in each rope?

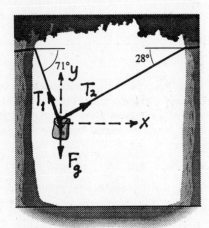

FIGURE 6-54 Problem 13. Solution.

Solution

The sum of the forces acting on the pack (gravity and the tension along each rope) is zero, since it is at rest, $\mathbf{F}_g + \mathbf{T}_1 + \mathbf{T}_2 = 0$. In a coordinate system with x axis horizontal to the right and y axis vertical upward, the x and y components of the net force are $-T_1 \cos 71° + T_2 \cos 28° = 0$, and $-26 \times 9.8 \text{ N} + T_1 \sin 71° + T_2 \sin 28° = 0$ (see Example 6-3). Solving for T_1 and T_2, one finds $T_1 = (26 \times 9.8 \text{ N})/(\sin 71° + \tan 28° \cos 71°) = 228$ N and $T_2 = (26 \times 9.8 \text{ N})/(\sin 28° + \tan 71° \cos 28°) = 84.0$ N.

Section 6-2 Multiple Objects

Problem

17. If the left-hand slope in Fig. 6-46 makes a 60° angle with the horizontal, and the right-hand slope makes a 20° angle, how should the masses compare if the objects are not to slide along the frictionless slopes?

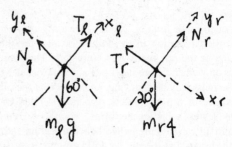

Problem 17 Solution.

Solution

The free-body force diagrams for the left- and right-hand masses are shown in the sketch, where there is only a normal contact force since each slope is frictionless, and we indicate separate parallel and perpendicular x-y axes. If the masses don't slide, the net force on each must be zero, or $T_\ell - m_\ell g \sin 60° = 0$, and $m_r g \sin 20° - T_r = 0$ (we only need the parallel components in this problem). If the masses of the string and pulley are negligible and there is no friction, then $T_\ell = T_r$. Adding the force equations, we find $m_r g \sin 20° - m_\ell g \sin 60° = 0$, or the mass ratio must be $m_r/m_\ell = \sin 60°/\sin 20° = 2.53$ for no motion.

Problem

21. In a florist's display, hanging plants of mass 3.85 kg and 9.28 kg are suspended from an essentially massless wire, as shown in Fig. 6-57. Find the tension in each section of the wire.

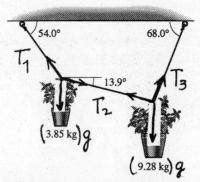

FIGURE 6-57 Problem 21 Solution.

Solution

Let the tensions in each section of wire be denoted by T_1, T_2, and T_3 as shown on the figure (Fig. 6-57 needed). The horizontal and vertical components of the net force on the junction of the wire with each plant are equal to zero, since the system is stationary. Thus:

$$T_1 \sin 54.0° - T_2 \sin 13.9° - (3.85 \times 9.8) \text{ N} = 0$$

$$-T_1 \cos 54.0° + T_2 \cos 13.9° = 0$$

$$T_2 \sin 13.9° + T_3 \sin 68.0° - (9.28 \times 9.8) \text{ N} = 0$$

$$-T_2 \cos 13.9° + T_3 \cos 68.0° = 0$$

One can solve any three of these equations for the unknown tensions, perhaps using the fourth equation as a check (if you do, remember not to round off). For example, $T_1 = (3.85 \times 9.8)$ N/$(\sin 54.0° - \cos 54.0° \tan 13.9°) = 56.9$ N, $T_2 = T_1 \cos 54.0°/\cos 13.9° = 34.4$ N, and $T_3 = T_2 \cos 13.9°/\cos 68.0° = 89.2$ N. (Note that the given angles and weights are not independent of one another.)

Section 6-3   Circular Motion

Problem

25. Show that the force needed to keep a mass m in a circular path of radius r with period T is $4\pi^2 mr/T^2$.

Solution

For an object of mass m in uniform circular motion, the net force has magnitude mv^2/r (Equation 6-4). The period of the motion (time for one revolution) is $T = 2\pi r/v$, so the centripetal force can also be written as $m(2\pi r/T)^2/r = mr(2\pi/T)^2 = 4\pi^2 mr/T^2$ (see Equation 4-18).

Problem

29. A subway train rounds an unbanked curve at 67 km/h. A passenger hanging onto a strap notices that an adjacent unused strap makes an angle of 15° to the vertical. What is the radius of the turn?

Solution

The net force on the unused strap is the vector sum of the tension in the strap (acting along its length at 15° to the vertical) and its weight. This must equal the mass times the horizontal centripetal acceleration. The situation is the same as Example 6-15. Thus, $T \cos\theta = mg$ and $T \sin\theta = mv^2/r$. Dividing these equations to eliminate T, and solving for the radius of the turn, one finds $r = v^2/g \tan\theta = (67 \text{ m}/3.6 \text{ s})^2/(9.8 \tan 15° \text{ m/s}^2) = 132$ m.

Problem

33. An indoor running track is square-shaped with rounded corners; each corner has a radius of 6.5 m on its inside edge. The track includes six 1.0-m-wide lanes. What should be the banking angles on (a) the innermost and (b) the outermost lanes if the design speed of the track is 24 km/h?

Solution

The banking angle is $\theta = \tan^{-1}(v^2/gr)$ (see Example 6-7). A competitive runner rounds a turn on the inside edge of his or her lane. (a) $\theta = \tan^{-1}[(24 \text{ m}/3.6 \text{ s})^2/(9.8 \text{ m/s}^2)(6.5 \text{ m})] = 34.9°$. (b) The radius of the inside edge of the outermost lane is 6.5 m + 5(1 m) = 11.5 m, so $\theta = 21.5°$.

Problem

37. A 1200-kg car drives on the country road shown in Fig. 6-60. The radius of curvature of the crests and dips is 31 m. What is the maximum speed at which the car can maintain road contact at the crests?

Solution

If air resistance is ignored, the forces on the car are gravity and the contact force of the road, which is represented by the sum of the normal force (perpendicularly away from the road) and friction between the tires and the road (parallel to the road in the direction of motion). Newton's second law for the car is $\mathbf{F}_g + \mathbf{N} + \mathbf{f}_s = m\mathbf{a}$. At a crest, $\mathbf{N}$ is vertically upward, $\mathbf{f}_s$ is horizontal, and the vertical component of $\mathbf{a}$ is the radical acceleration $-v^2/r$ (downward in this case). The vertical component of Newton's second law is then $-mg + N = -mv^2/r$. As long as the car is in contact with the road, $N \geq 0$, thus, $v \leq \sqrt{gr} = \sqrt{(9.8 \text{ m/s}^2)(31 \text{ m})} = 17.4$ m/s = 62.7 km/h.

Section 6-4   Friction

Problem

41. A 380-N force is required to push a table at constant velocity across a floor where the coefficient of sliding friction is 0.86. What is the mass of the table?

Solution

For constant velocity, the acceleration is zero. If the floor is level and the applied force horizontal, then $N = mg$ and $F_{app} - f_k = ma = 0 = F_{app} - \mu_k mg$. Thus, $m = F_{app}/\mu_k g = (380 \text{ N})/(0.86 \times 9.8 \text{ m/s}^2) = 45.1$ kg.

Problem

45. The handle of a 22-kg lawnmower makes a 35° angle with the horizontal. If the coefficient of friction between lawnmower and ground is 0.68, what magnitude of force is required to push the mower at constant velocity? Assume the force is applied in the direction of the handle. Compare with the mower's weight.

Solution

Assuming the ground is also horizontal, we may depict the forces on the lawnmower as shown. At constant velocity (constant speed in a straight line) $\mathbf{a} = 0$, and $\mathbf{F} + \mathbf{f}_k + \mathbf{N} + m\mathbf{g} = 0$. The x and y components of this equation are: $F\cos 35° - f_k = 0$, and $N - F\sin 35° - mg = 0$. Using $f_k = \mu_k N = \mu_k(F\sin 35° + mg)$, with $\mu_k = 0.68$, we find $F = \mu_k mg/(\cos 35° - \mu_k \sin 35°) = 1.58mg = 342$ N.

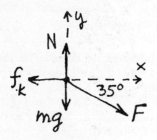

Problem 45 Solution.

Problem

49. In a factory, boxes drop vertically onto a conveyor belt moving horizontally at 1.7 m/s. If the coefficient of kinetic friction is 0.46, how long does it take each box to come to rest with respect to the belt?

Solution

Kinetic friction accelerates each box up to the speed of the belt: $f_k = \mu_k N = \mu_k mg = ma$ (if we suppose the belt to be horizontal). This takes time $t = v/a = v/\mu_k g = (1.7 \text{ m/s})/(0.46)(9.8 \text{ m/s}^2) = .377$ s.

Problem

53. A bug walks outward from the center of a turntable rotating at $33\frac{1}{3}$ revolutions per minute. If the coefficient of friction is 0.15, how far does the bug get before slipping?

Solution

Assume that the turntable is level. Then the frictional forces produces the (centripetal) acceleration of the bug, and the normal force equals its weight. Thus, $f_s = m(v^2/r) =$

$mr(2\pi/T)^2 \le \mu_s N = \mu_s mg$, or $r \le \mu_s g(T/2\pi)^2 = (0.15)(9.8 \text{ m/s}^2)(60 \text{ s}/2\pi(33.3))^2 = 12.1$ cm. Note that the period of revolution is 60 s divided by the number of revolutions per minute. (The maximum radius attainable without the bug's slipping is less than the radius of a 12-inch long-playing record, a popular item prior to the introduction of compact disks.)

Problem

57. In a typical front-wheel-drive car, 70% of the car's weight rides on the front wheels. If the coefficient of friction between tires and road is 0.61, what is the maximum acceleration of the car?

Solution

On a level road, the maximum acceleration from static friction between the tires and the road is $a_{max} = \mu_s N/m$ (see Example 6-13). In this case, the normal force on the front tires (the ones producing the frictional force which accelerates the car) is 70% of mg, whereas the whole mass must be accelerated. Thus, $a_{max} = (0.61)(0.70)(9.8 \text{ m/s}^2) = 4.18 \text{ m/s}^2$.

Problem

61. Starting from rest, the skier of the previous problem traverses a 1.8-km-long trail in 65 s. If the trail is inclined at 12°, what is the coefficient of kinetic friction?

Solution

If she just glides down slope all the way, the skier's acceleration is $a = g(\sin\theta - \mu_k \cos\theta) = 2\ell/t^2$, where ℓ is the length of the slope. Solving for μ_k, we obtain

$$\mu_k = \tan\theta - \frac{2\ell}{gt^2 \cos\theta}$$

$$= \tan 12° - \frac{2(1800 \text{ m})}{(9.8 \text{ m/s}^2)(65 \text{ s})^2 \cos 12°} = 0.12.$$

Problem

65. A block is shoved down a 22° slope with an initial speed of 1.4 m/s. If it slides 34 cm before stopping, what is the coefficient of friction?

Solution

The acceleration down the slope can be found from kinematics (Equation 2-11) and from Newton's second law (as in Example 6-10): $a = g\sin\theta - \mu_k g\cos\theta = -v_0^2/2\ell$. In this equation, the values of θ, v_0, and ℓ are given, therefore we can solve for μ_k:

$$\mu_k = \tan 22° + \frac{(1.4 \text{ m/s})^2}{2(0.34 \text{ m})(9.8 \text{ m/s}^2)\cos 22°} = 0.72$$

Section 6-5   Accelerated Reference Frames

Problem

69. A space station is in the shape of a hollow ring with an outer diameter of 150 m. How fast should it rotate to simulate Earth's surface gravity—that is, so that the force exerted on an object by the outside wall is equal to the object's weight on Earth's surface?

Solution

The force exerted by the outer wall of the space station must supply a centripetal acceleration equal to the surface gravity on Earth, $a_c = (2\pi/T)^2 r = g$ (see Problem 25). Thus, the period of rotation of the station should be $T = 2\pi\sqrt{r/g} = 2\pi\sqrt{(75\text{ m})/(9.8\text{ m/s}^2)} = 17.4$ s. (Equivalently, the angular velocity should be 1 rev/17.4 s = 3.45 revolutions per minute.)

Problem

73. A spring scale hangs from the ceiling of a railroad car, and a 10.0-N weight is attached to the scale. What is the scale reading when the train is (a) moving in a straight line at a constant 90 km/h; (b) rounding a 215-m-radius turn on level tracks, at a constant 90 km/h; (c) rounding a banked 215-m-radius turn at a constant 90 km/h, assuming the banking angle is optimum for this speed? (d) What angle does the scale make with the ceiling of the car in each case?

Solution

Under any of the circumstances in this problem, only two forces act on the weight, gravity and the spring force. Thus, Newton's second law for the weight says $\mathbf{F}_g + \mathbf{F}_{spr} = m\mathbf{a}$. If the train has only horizontal motion (a banked curve simply rotates the train but does not move it up or down), and the weight is hanging (but not oscillating up and down), and then its vertical acceleration is zero and we can choose the x axis in the direction of $\mathbf{a}$. Thus, the x and y components of Newton's second law become $F_{spr,x} = ma_x$ and $F_{spr,y} - mg = 0$. We conclude that the scale reading (magnitude of the scale force) is $\sqrt{F_{spr,x}^2 + F_{spr,y}^2} = \sqrt{(ma_x)^2 + (mg)^2} = mg\sqrt{1 + (a_x/g)^2} = (10.0\text{ N})\sqrt{1 + (a_x/g)^2}$. Also, the angle the scale makes with the horizontal is $\theta = \tan^{-1}(F_{spr,y}/F_{spr,x}) = \tan^{-1}(g/a_x)$. Now let us consider the different cases. (a) The train's velocity is constant, so $a_x = 0$, $F_{spr} = 10.0$ N, and $\theta = \tan^{-1}(\infty) = 90°$. Of course, the train's ceiling is horizontal, so θ is also the angle the scale makes with the ceiling. (b) In this case, a_x is the centripetal acceleration $v^2/r = (90\text{ m}/3.6\text{ s})^2/(215\text{ m}) = 2.91\text{ m/s}^2$. Thus, $F_{spr} = (10.0\text{ N})\sqrt{1 + (2.91/9.8)^2} = 10.4$ N, $\theta = \tan^{-1}(9.8/2.91) = 73.5°$ and the ceiling is still horizontal. (c) The acceleration is the same for a banked or level curve (it's the source of the centripetal force on the train which is different) so $F_{spr} = 10.4$ N and $\theta = 73.5°$. However,

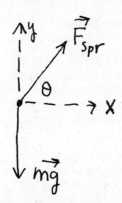

Problem 73 Solution.

since the curve is banked at the optimum angle, the ceiling makes an angle of $\tan^{-1}(v^2/gr) = \tan^{-1}(a_x/g)$ with the horizontal. This is just the complement of the angle θ above that the scale makes with the horizontal, in such a direction that the angle between the scale and the ceiling is 90° (the curve is banked in the direction of the centripetal acceleration).

Paired Problems

Problem

77. A tetherball on a 1.7-m rope is struck so it goes into circular motion in a horizontal plane, with the rope making a 15° angle to the horizontal. What is the ball's speed?

Solution

The tetherball whirling in a horizontal circle is analogous to the mass on a string in Example 6-6. From step 6, $v = \sqrt{g\ell\cos^2\theta/\sin\theta} = \sqrt{(9.8\text{ m/s}^2)(1.7\text{ m})\cos^2 15°/\sin 15°} = 7.75$ m/s.

Problem

81. A car moving at 40 km/h negotiates a 130-m radius banked turn designed for 60 km/h. (a) What coefficient of friction is needed to keep the car on the road? (b) To which side of the curve would it move if it hit an essentially frictionless icy patch?

Solution

The forces on a car (in a plane perpendicular to the velocity) rounding a banked curve at arbitrary speed are analyzed in detail in the solution to Problem 89 below. (a) It is shown there, that to prevent skidding, $\mu_s \geq |v^2 - v_d^2|/gR(1 + v^2v_d^2/g^2R^2)$, where R is the radius of the curve and v_d is the design speed for the proper banking angle, $\tan\theta_d = v_d^2/gR$. In this problem, $v_d = (60/3.6)$ m/s, $v = (40/3.6)$ m/s, and $R = 130$ m, so $\mu_s \geq 0.12$. (b) Since $v < v_d$, the car would slide down the bank of the curve in the absence of friction.

Supplementary Problems

Problem

85. In the loop-the-loop track of Fig. 6-21, show that the car leaves the track at an angle ϕ given by $\cos\phi = v^2/rg$, where ϕ is the angle made by a vertical line through the center of the circular track and a line from the center to the point where the car leaves the track.

Solution

The angle ϕ and the forces acting on the car are shown in the sketch. The radial component of the net force (towards the center of the track) equals the mass times the centripetal acceleration, $N + mg\cos\phi = mv^2/r$. (The tangential component is not of interest in this problem.) The car leaves the track when $N = (mv^2/r) - mg\cos\phi = 0$ (no more contact) or $\cos\phi = v^2/gr$. This implies that the car leaves the track at real angles for $v^2 < gr$; otherwise, the car never leaves the track, as in Example 6-8.

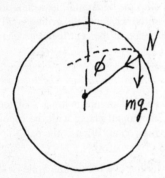

Problem 85 Solution.

Problem

89. A highway turn of radius R is banked for a design speed v_d. If a car enters the turn at speed $v = v_d + \Delta v$, where Δv can be positive or negative, show that the minimum coefficient of static friction needed to prevent slipping is

$$\mu_s = \frac{|\Delta v|}{gR}\frac{(2v_d + \Delta v)}{[1 + (v_d v/gR)^2]}.$$

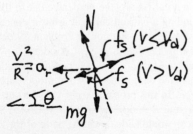

Problem 89 Solution.

Solution

The equation of motion for a car rounding a banked turn is $\mathbf{N} + \mathbf{mg} + \mathbf{f}_s = m\mathbf{a}_r$, where $a_r = v^2/R$ is the radical acceleration (assumed horizontal and constant in magnitude) and the forces are as shown. Note that the frictional force changes direction for v greater or less than the design speed. Taking components parallel and perpendicular to the road, we find $N - mg\cos\theta = m(v^2/R)\sin\theta$, $mg\sin\theta \pm f_s = (v^2/R)\cos\theta$, where the upper sign is for $v > v_d$, and the lower for $v < v_d$. (We chose these components because the solution for N and f_s is direct.) This argument applies if the car does not skid (otherwise $\mathbf{a} \neq \mathbf{a}_r$) so $f_s \leq \mu_s N$. Therefore:

$$\mu_s \geq \frac{f_s}{N} = \frac{\mp g\sin\theta \pm v^2\cos\theta/R}{g\cos\theta + v^2\sin\theta/R}$$

$$= \frac{\mp g\tan\theta \pm v^2/R}{g + v^2\tan\theta/R}$$

$$= \frac{\mp v_d^2/v^2}{gR(1 + v^2 v_d^2/g^2R^2)}$$

since $v_d^2/gR = \tan\theta$. If we set $\Delta v = v - v_d$, the condition on μ_s becomes:

$$\mu_s \geq \pm\frac{\Delta v(2v_d + \Delta v)}{gR(1 + v_d^2 v^2/g^2R^2)} = \frac{|\Delta v|(2v_d + \Delta v)}{gR(1 + v_d^2 v^2/g^2R^2)}.$$

(Note that $|\Delta v|$ is $+\Delta v$ for $v > v_d$ and $-\Delta v$ for $v < v_d$.) The expression for the minimum coefficient of friction is not particularly simple, but for $v = 0$ (car at rest) it reduces to $|-v_d|(2v_d - v_d)/gR = v_d^2/gR = \tan\theta$, as in Example 6-12.

Problem

93. A 2.1-kg mass is connected to a spring of spring constant $k = 150$ N/m and unstretched length 18 cm. The pair are mounted on a frictionless air table, with the free end of the spring attached to a frictionless pivot. The mass is set into circular motion at 1.4 m/s. Find the radius of its path.

Solution

Since the airtable is frictionless, the only horizontal force acting on the mass is the spring force, of magnitude $k(\ell - \ell_0)$ and in the direction of the centripetal acceleration v^2/ℓ. Here, the radius of the circle is ℓ, the length of the spring, while ℓ_0 is the unstretched length. Therefore, $k(\ell - \ell_0) = mv^2/\ell$. This is a quadratic equation for ℓ, $\ell^2 - \ell_0\ell - mv^2/k = 0$, with positive solution $\ell = \frac{1}{2}[\ell_0 + \sqrt{\ell_0^2 + 4mv^2/k}] = \frac{1}{2}[0.18\text{ m} + \sqrt{(0.18\text{ m})^2 + 4(2.1\text{ kg})(1.4\text{ m/s})^2/(150\text{ N/m})}] = 27.9$ cm.

● CHAPTER 7 WORK, ENERGY, AND POWER

Section 7-1 Work

Problem

1. How much work do you do as you exert a 95-N force to push a 30-kg shopping cart through a 14-m-long supermarket aisle?

Solution

If the force is constant, and parallel to the displacement, $W = \mathbf{F} \cdot \Delta\mathbf{r} = F\Delta r = (95 \text{ N})(14 \text{ m}) = 1.33 \text{ kJ}$.

Problem

5. The world's highest waterfall, the Cherun-Meru in Venezuela, has a total drop of 980 m. How much work does gravity do on a cubic meter of water dropping down the Cherun-Meru?

Solution

The force of gravity at the Earth's surface on a cubic meter of water is $F_g = mg = 9.8 \text{ kN}$ vertically downward (see the inside book cover of the text for the density of water). The displacement of the water is parallel to this, so the work done by gravity on the water is $W_g = F_g \, \Delta y = (9.8 \text{ kN})(980 \text{ m}) = 9.6 \text{ MJ}$.

Problem

9. A locomotive does 8.8×10^{11} J of work in pulling a 2×10^5-kg train 150 km. What is the average force in the coupling between the locomotive and the rest of the train?

Solution

If we define the average force by $W = F_{av} \, \Delta r$, then $F_{av} = 8.8 \times 10^{11}$ J/150 km = 5.87 MN.

Section 7-2 Work and the Scalar Product

Problem

13. Show that the scalar product obeys the distributive law: $\mathbf{A} \cdot (\mathbf{B} + \mathbf{C}) = \mathbf{A} \cdot \mathbf{B} + \mathbf{A} \cdot \mathbf{C}$.

Solution

This follows easily from the definition of the scalar product in terms of components: $\mathbf{A} \cdot (\mathbf{B} + \mathbf{C}) = A_x(B_x + C_x) + A_y(B_y + C_y) + A_z(B_z + C_z) = A_x B_x + A_y B_y + A_z B_z + A_x C_x + A_y C_y + A_z C_z = \mathbf{A} \cdot \mathbf{B} + \mathbf{A} \cdot \mathbf{C}$. (With more effort, it also follows from trigonometry. First:

$$D\sin(\theta_D - \theta_B) = C\sin(\theta_C - \theta_B) \quad \text{(law of sines)},$$

$$D = \sqrt{B^2 + C^2 + 2BC\cos(\theta_C - \theta_B)} \quad \text{(law of cosines)},$$

and

$$\cos\alpha = \sqrt{1 - \sin^2\alpha},$$

together give

$$D\cos(\theta_D - \theta_B) = B + C\cos(\theta_C - \theta_B).$$

Second:

$$
\begin{aligned}
\mathbf{A} \cdot \mathbf{D} &= AD\cos\theta_D \\
&= AD[\cos(\theta_D - \theta_B)\cos\theta_B - \sin(\theta_D - \theta_B)\sin\theta_B] \\
&= A[B + C\cos(\theta_C - \theta_B)]\cos\theta_B - \\
&\qquad A[C\sin(\theta_C - \theta_B)]\sin\theta_B \\
&= AB\cos\theta_B + \\
&\qquad AC[\cos(\theta_C - \theta_B)\cos\theta_B - \sin(\theta_C - \theta_B)\sin\theta_B] \\
&= \mathbf{A} \cdot \mathbf{B} + \mathbf{A} \cdot \mathbf{C}.)
\end{aligned}
$$

(We used the identity for $\cos(\alpha + \beta)$ from Appendix A.)

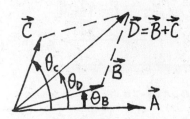

Problem 13 Solution.

Problem

17. Given the following vectors:

   A has length 10 and points 30° above the x axis
   B has length 4.0 and points 10° to the left of the y axis
   C = $5.6\hat{\imath} - 3.1\hat{\jmath}$
   D = $1.9\hat{\imath} + 7.2\hat{\jmath}$,
   compute the scalar products (a) **A** · **B**; (b) **C** · **D**; (c) **B** · **C**.

Solution

(a) The angle between **A** and **B** is 60° + 10° = 70°, so $\mathbf{A} \cdot \mathbf{B} = (10)(4)\cos 70° = 13.7$ (Note: units are ignored in this problem.) (b) In terms of components, $\mathbf{C} \cdot \mathbf{D} = (5.6)(1.9) + (-3.1)(7.2) = -11.7$. (c) Express **B** in components $(4\cos 100°\hat{\imath} + 4\sin 100°\hat{\jmath})$, or **C** by magnitude and direction (6.40 at 29.0° below the x axis). In either case, $\mathbf{B} \cdot \mathbf{C} = (4\cos 100°)(5.6) + (4\sin 100°)(-3.1) = 4(6.4)\cos 129° = -16.1$.

Problem

21. Given that $\mathbf{A} = 2\hat{\imath} + 2\hat{\jmath}$, $\mathbf{B} = 5\hat{\imath}$, and $\mathbf{C} = \sqrt{2}\hat{\imath} - \pi\hat{\jmath}$, find the angles between (a) $\mathbf{A}$ and $\mathbf{B}$; (b) $\mathbf{A}$ and $\mathbf{C}$; (c) $\mathbf{B}$ and $\mathbf{C}$. *Hint:* See previous problem.

Solution

(a) and (c) Since $\mathbf{B}$ is along the x axis, the angle between $\mathbf{A}$ and $\mathbf{B}$ is $\theta_{AB} = \tan^{-1}(2/2) = 45°$, and between $\mathbf{B}$ and $\mathbf{C}$, $\theta_{BC} = \tan^{-1}|-\pi/\sqrt{2}| = 65.8°$. (b) In general, the angle between two vectors is:

$$\theta_{AC} = \cos^{-1}\left(\frac{\mathbf{A} \cdot \mathbf{C}}{AC}\right) = \cos^{-1}\left(\frac{2\sqrt{2} + 2(-\pi)}{2\sqrt{2}\sqrt{2 + \pi^2}}\right) = 111°$$

Problem

25. A rope pulls a box a horizontal distance of 23 m, as shown in Fig. 7-26. If the rope tension is 120 N, and if the rope does 2500 J of work on the box, what angle does it make with the horizontal?

Solution

$W = 2500 \text{ J} = (120 \text{ N})(23 \text{ m})\cos\theta$, so $\theta = 25.1°$.

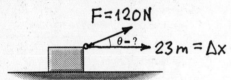

FIGURE 7-26 Problem 25 Solution.

Section 7-3 A Varying Force

Problem

29. A force F acts in the x direction, its magnitude given by $F = ax^2$, where x is in meters and a is exactly 5 N/m². (a) Find an exact value for the work done by this force as it acts on a particle moving from $x = 0$ to $x = 6$ m. Now find approximate values for the work by dividing the area under the force curve into rectangles of width (b) $\Delta x = 2$ m; (c) $\Delta x = 1$ m; (d) $\Delta x = \frac{1}{2}$ m with height equal to the magnitude of the force in the center of the interval. Calculate the percent error in each case.

Solution

(a) $W = \int_{x_1}^{x_2} F\,dx = \int_0^{6\,\text{m}} ax^2\,dx = \frac{1}{3}a|x^3|_0^{6\,\text{m}} = \frac{1}{3}(5 \text{ N/m}^2)(6 \text{ m})^3 = 360$ J. (b) $W \approx \sum_{i=1}^{3} F(x_i)\,\Delta x_i$, where $x_i = 1$ m, 3 m, 5 m are the midpoints and $\Delta x_i = 2$ m. Then: $W \approx (5 \text{ N/m}^2)(1^2 + 3^2 + 5^2) \text{ m}^2(2 \text{ m}) = 350$ J. The percent error is only $\delta = 100(360 - 350)/360 = 2.78\%$. (c) Now, $x_i = .5, 1.5, 2.5, 3.5, 4.5$, and 5.5 (in meters), and $\Delta x_i = 1$ m. $W \approx \sum_{i=1}^{6} F(x_i)\,\Delta x_i =$ (5 N/m²)(0.5 m)²$(1^2 + 3^2 + 5^2 + 7^2 + 9^2 + 11^2)(1$ m$) =$ 357.5 J and $\delta = 100(2.5)/360 = 0.694\%$. (d) $W \approx \sum_{i=1}^{12} F(x_i)\,\Delta x_i =$ (5 N/m²)(0.25 m)²$(1^2 + 3^2 + \cdots + 23^2)$

(0.5 m) = 359.375 J with $\delta = 0.174\%$. (The direct calculation of the sum is tedious, but we can use the formula for the sum of the squares of the first n numbers, namely $\sum_1^n k^2 = \frac{1}{6}n(n + 1)(n + 2)$. The sum in question is $\sum_{k=1}^{12} (2k - 1)^2 = \sum_{k=1}^{23} k^2 - \sum_{k=1}^{11} (2k)^2 = 4324 - 2024 = 2300$.)

Problem

33. A force $\mathbf{F}$ acts in the x direction, its x component given by $F = F_0\cos(x/x_0)$, where $F_0 = 51$ N and $x_0 = 13$ m. Calculate the work done by this force acting on an object as it moves from $x = 0$ to $x = 37$ m. *Hint:* Consult Appendix A for the integral of the cosine function and treat the argument of the cosine as a quantity in radians.

Solution

From Equation 7-8:

$$W = \int_0^{37\,\text{m}} F_0\cos\frac{x}{x_0}\,dx = F_0\,x_0\sin\frac{x}{x_0}\Big|_0^{37\,\text{m}}$$
$$= (51 \text{ N})(13 \text{ m})\sin(37/13)$$
$$= 193 \text{ J}.$$

Problem

37. The force exerted by a rubber band is given approximately by

$$F = F_0\left[\frac{\ell_0 + x}{\ell_0} - \frac{\ell_0^2}{(\ell_0 + x)^2}\right],$$

where ℓ_0 is the unstretched length, x the stretch, and F_0 is a constant (although F_0 varies with temperature). Find the work needed to stretch the rubber band a distance x.

Solution

$$W = \int_0^x F_0\left[\frac{\ell_0 + x'}{\ell_0} - \frac{\ell_0^2}{(\ell_0 + x')^2}\right]dx'$$
$$= F_0\left|\frac{1}{\ell_0}\left(\ell_0 x' + \frac{x'^2}{2}\right) + \frac{\ell_0^2}{\ell_0 + x'}\right|_0^x$$
$$= F_0\left(x + \frac{x^2}{2\ell_0} + \frac{\ell_0^2}{\ell_0 + x} - \ell_0\right).$$

(x' is a dummy variable)

Section 7-4 Force and Work in Three Dimensions

Problem

41. A cylindrical log of radius R lies half buried in the ground, as shown in Fig. 7-31. An ant of mass m climbs to the top of the log. Show that the work done by gravity on the ant is $-mgR$.

FIGURE 7-31 Problem 41.

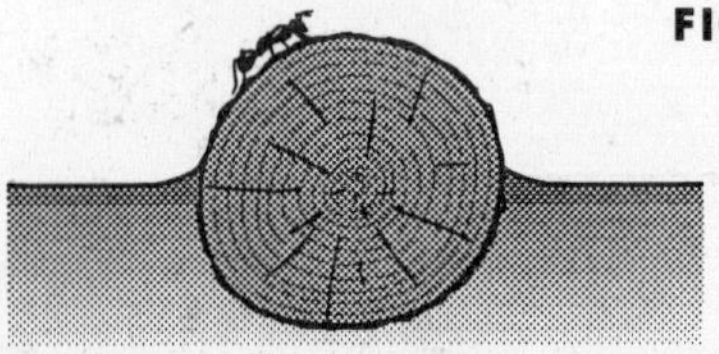

Solution

If we let $d\mathbf{r} = dx\hat{\imath} + dy\hat{\jmath}$, as in Example 7-9 (where $\hat{\imath}$ is horizontal to the right and $\hat{\jmath}$ is vertical upward) then $W_g = \int_1^2 \mathbf{F}_g \cdot d\mathbf{r} = \int_1^2 (-mg\hat{\jmath}) \cdot (dx\hat{\imath} + dy\hat{\jmath}) = -mg\int_{y_1}^{y_2} dy = -mg(y_2 - y_1) = -mgR$. (The difference in height going from the ground to the top of a half-buried log is just the radius.) Another way of obtaining this result is to use reasoning similar to that in the solution to Problem 39. The force of gravity is constant and can be taken outside the integral, $W_g = \int_1^2 \mathbf{F}_g \cdot d\mathbf{r} = \mathbf{F}_g \cdot \int_1^2 d\mathbf{r}$. The integral left is the total displacement, $\int_1^2 d\mathbf{r} = R\hat{\jmath} - R\hat{\imath}$ (if we take origin at the center of the log's cross-section), so $W_g = (-mg\hat{\jmath}) \cdot (R\hat{\jmath} - R\hat{\imath}) = -mgR$. Finally, one could use Equation 7-3 for the dot product, as shown:

$$W_g = \int_1^2 \mathbf{F}_g \cdot d\mathbf{r} = -\int_1^2 mg\cos\theta \,|d\mathbf{r}|$$

$$= -\int_0^{\pi/2} mg\cos\theta \cdot R\,d\theta = -mgR \left|\sin\theta\right|_0^{\pi/2}$$

$$= -mgR.$$

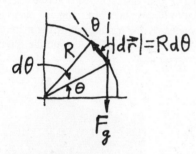

Problem 41 Solution.

Section 7-5   Kinetic Energy

Problem

45. A 3.5×10^5-kg jumbo jet is cruising at 1000 km/h. (a) What is the kinetic energy relative to the ground? A 85-kg passenger strolls down the aisle at 2.9 km/h. What is the passenger's kinetic energy (b) relative to the ground and (c) relative to the plane?

Solution

(a) $K_{jet} = \frac{1}{2}(3.5\times10^5 \text{ kg})(1000 \text{ m}/3.6 \text{ s})^2 = 13.5$ GJ (see Table 1-1). (b) The passenger's speed relative to the ground is (1000 ± 2.9) km/h, depending on his or her direction, therefore: $K_{pass} = \frac{1}{2}(85 \text{ kg})(1000 \pm 2.9)^2(1 \text{ m}/3.6 \text{ s})^2 = 3.26$ or 3.30 MJ. (c) Relative to the plane, $K'_{pass} = \frac{1}{2}(85 \text{ kg})(2.9 \text{ m}/3.6 \text{ s})^2 = 27.6$ J.

Problem

49. A 60-kg skateboarder comes over the top of a hill at 5.0 m/s, and reaches 10 m/s at the bottom of the hill. Find the total work done on the skateboarder between the top and bottom of the hill.

Solution

The work-energy theorem, Equation 7-16, gives $W_{net} = \Delta K = \frac{1}{2}m(v_2^2 - v_1^2) = \frac{1}{2}(60 \text{ kg})(10^2 - 5^2)(\text{m/s})^2 = 2.25$ kJ.

Problem

53. You drop a 150-g baseball from a sixth-story window 16 m above the ground. What are (a) its kinetic energy and (b) its speed when it hits the ground? Neglect air resistance.

Solution

(b) The speed of the baseball (magnitude of the velocity) follows from Equation 2-14, the initial conditions (with $y = 0$ at ground level) and the neglect of air resistance: $v = \sqrt{-2g(y - y_0)} = \sqrt{-2(9.8 \text{ m/s}^2)(-16 \text{ m})} = 17.7$ m/s. (a) The kinetic energy is $K = \frac{1}{2}mv^2 = \frac{1}{2}(0.150 \text{ kg})(17.7 \text{ m/s})^2 = 23.5$ J. Alternatively, $K = \frac{1}{2}m(2gy_0) = mgy_0$, and $v = \sqrt{2K/m}$.

Problem

57. A block of mass m slides from rest without friction down the slope shown in Fig. 7-32. (a) How much work is done on the block by the normal force of the slope? (b) Show that the final speed is $\sqrt{2gh}$ regardless of the details of the slope.

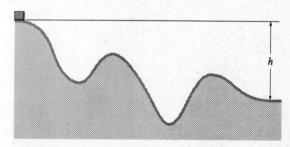

FIGURE 7-32   Problem 57.

Solution

(a) The normal force is perpendicular to the displacement and does no work, $W_N = \int \mathbf{N} \cdot d\mathbf{r} = 0$. (b) In the absence of friction, only gravity and the normal force act, so $W_{net} = W_g + W_N = W_g = \Delta K$. The work done by gravity is $-mg\,\Delta y = mgh$, and $\Delta K = \frac{1}{2}mv^2 - 0$, so $v = \sqrt{2gh}$, independent of the shape of the slope.

Section 7-6   Power

Problem

61. How much work can a 3.5-hp lawnmower engine do in 1 h?

Solution

Working at constant power output, Equation 7-17 gives the total work (energy output) as $\Delta W = \bar{P} \, \Delta t =$ (3.5 hp)(746 W/hp)(3600 s) = 9.40 MJ. (Note the change to appropriate SI units.)

Problem

65. Estimate your power output as you do deep knee bends at the rate of one per second.

Solution

The work done against gravity in raising or lowering a weight through a height, h, has magnitude mgh. The body begins and ends each deep knee bend at rest ($\Delta K = 0$), so the muscles do a total work (down and up) of $2mgh$ for each complete repetition. If we assume that the lower extremities comprise 35% of the body mass, and are not included in the moving mass, then mg, for a 75 kg person, is about 0.65(75 kg)(9.8 m/s²) = 480 N. We guess that h is somewhat greater than 25% of the body height, or about 45 cm, so the muscle power output for one repetition per second is about 2(480 N)(0.45 m)/s = 430 W.

Problem

69. The rate at which the United States imports oil, expressed in terms of the energy content of the imported oil, is very nearly 500 GW. Using the "Energy Content of Fuels" table in Appendix C, convert this figure to gallons per day.

Solution

Appendix C lists the energy content of oil as 39 kWh/gal. Therefore, the import rate is (500 GW)(1 gal/39 kWh) × (24 h/d), or roughly 300 million gallons per day.

Problem

73. A 1400-kg car ascends a mountain road at a steady 60 km/h. The force of air resistance on the car is 450 N. If the car's engine supplies energy to the drive wheels at the rate of 38 kW, what is the slope angle of the road?

Solution

At constant velocity, there is no change in kinetic energy, so the net work done on the car is zero. Therefore, the power supplied by the engine equals the power expended against gravity and air resistance. The latter can be found from Equation 7-21, since gravity makes an angle of $\theta + 90°$ with the velocity (where θ is the slope angle to the horizontal), while air resistance makes an angle of 180° to the velocity.

Then 38 kW $= -\mathbf{F}_g \cdot \mathbf{v} - \mathbf{F}_{air} \cdot \mathbf{v} = -mgv \cos(\theta + 90°) - F_{air} \, v \cos(180°) = mgv \sin\theta + F_{air} \, v$, or $\theta =$ $\sin^{-1}[((38 \text{ kW}/60 \text{ km/h}) - 450 \text{ N})/(1400 \text{ kg} \times 9.8 \text{ m/s}^2)] = 7.67°$. (See Example 7-14 and use care with SI units and prefixes.)

Paired Problems

Problem

77. A force pointing in the x direction is given by $F = F_0(x/x_0)$, where F_0 and x_0 are constants and x is the position. Find an expression for the work done by this force as it acts on an object moving from $x = 0$ to $x = x_0$.

Solution

The work done in this one-dimensional situation is given by Equation 7-8:

$$W = \int_0^{x_0} \left(\frac{F_0}{x_0}\right) x \, dx = \left(\frac{F_0}{x_0}\right)\frac{x_0^2}{2} = \frac{1}{2} F_0 x_0.$$

Problem

81. A 460-kg piano is pushed at constant speed up a ramp, raising it a vertical distance of 1.9 m (see Fig. 7-33). If the coefficient of friction between ramp and piano is 0.62, find the work done by the agent pushing the piano if the ramp angle is (a) 15° and (b) 30°. Assume the force is applied parallel to the ramp.

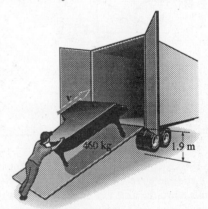

FIGURE 7-33   Problem 81.

Solution

The usual relevant forces on an object pushed up an incline of length $\ell = h/\sin\theta$, by an applied force parallel to the slope, are shown in the sketch. At constant velocity, the acceleration is zero, so the parallel and perpendicular components of Newton's second law, together with the empirical relation for kinetic friction, give $N = mg\cos\theta$, $f_k = \mu_k N$, and $F_{app} =$

$mg \sin\theta + f_k = mg(\sin\theta + \mu_k \cos\theta)$. Thus, the work done by the applied force is $W_{app} = F_{app}\ell = F_{app}\, h/\sin\theta = mgh(1 + \mu_k \cot\theta)$, where h is the vertical rise. (a) Evaluating the above expression using the data supplied, we find $W_{app} = (460 \text{ kg})(9.8 \text{ m/s}^2)(1.9 \text{ m})(1 + 0.62 \cot 15°) = 28.4 \text{ kJ}$. (b) When $15°$ is replaced by $30°$ in the above calculation, we find $W_{app} = 17.8 \text{ kJ}$. (The work done against gravity is the same in parts (a) and (b) since h is the same, but the work done against friction is greater in (a) because the incline is longer and the normal force is greater; however F_{app} is less.)

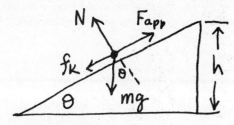

Problem 81 Solution.

Supplementary Problems

Problem

85. The power output of a machine of mass m increases linearly with time, according to the formula $P = bt$, where b is a constant. (a) Find an expression for the work done between $t = 0$ and some arbitrary time t. (b) Suppose the machine is initially at rest and all the work it supplies goes into increasing its own speed. Use the work-energy theorem to show that the speed increases linearly with time, and find an expression for the acceleration.

Solution

(a) From Equation 7-20, $W = \int_0^t P\, dt' = \int_0^t bt'\, dt' = \frac{1}{2} bt^2$. (We used t' for the dummy variable of integration.) (b) If we assume that $W = W_{net} = \Delta K$, then $\frac{1}{2} bt^2 = \frac{1}{2} mv^2$, since the machine starts from rest. Thus $v = \sqrt{b/m}\, t$ and $a = dv/dt = \sqrt{b/m}$. (v is the speed and a is the tangential acceleration along the path of the machine.)

Problem

89. If the ant of Problem 41 climbs with constant speed v along the log surface, (a) what is its power output as a function of time? Integrate your expression for power to show that the total work to climb the log is the same as in Problem 41.

Solution

(a) With reference to the diagram in the solution of Problem 41, $P_{ant} = \mathbf{F} \cdot \mathbf{v} = mgv \cos\theta$. At constant speed, the distance traveled on the arc of a circle in time t is $R\theta = vt$ (angle in radians), so $P_{ant} = mgv \cos(vt/R)$. (b) $W = \int P\, dt$, so:

$$W_{ant} = \int_0^{\pi R/2v} mgv \cos(vt/R)\, dt = mgv \left. \frac{R}{v} \sin\left(\frac{vt}{R}\right) \right|_0^{\pi R/2v}$$
$$= mgR.$$

Problem

93. A 2.3-kg particle's position as a function of time is given by $x = bt^3 - ct$, where $b = 0.41 \text{ m/s}^3$ and $c = 1.9 \text{ m/s}$. Find the work done on the particle between the times $t = 0$ and $t = 2.0 \text{ s}$.

Solution

The net work is $W = \Delta K = \frac{1}{2} m[v(t = 2s)^2 - v(t = 0)^2]$. Since $v(t) = dx/dt = 3bt^2 - c$, we find $W = \frac{1}{2}(2.3 \text{ kg})[(3.02 \text{ m/s})^2 - (-1.9 \text{ m/s})^2] = 6.34 \text{ J}$.

● **CHAPTER 8** CONSERVATION OF ENERGY

Section 8-1 Conservative and Nonconservative Forces

Problem

1. Determine the work done by the frictional force in moving a block of mass m from point 1 to point 2 over the two paths shown in Fig. 8-23. The coefficient of friction has the constant value μ over the surface. (The diagram lies in a horizontal plane.)

Solution

Fig. 8-23 is a plane view of the horizontal surface over which the block is moved, showing the paths (a) and (b). The force of friction is μmg opposite to the displacement ($\mathbf{f} \cdot d\mathbf{r} = -f dr$)

so $W_{(a)} = -\mu mg(\ell + \ell) = -2\mu mg\ell$, and $W_{(b)} = -\mu mg\sqrt{\ell^2 + \ell^2} = -\sqrt{2}\mu mg\ell$. Since the work done depends on the path, friction is not a conservative force.

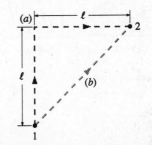

FIGURE 8-23 Problem 1.

Section 8-2 Potential Energy

Problem

5. Find the potential energy of a 9000-kg camper when it's (a) atop New Hampshire's Mount Washington, 1900 m above sea level, and (b) in Death Valley, California, 86 m below sea level. Take the zero of potential energy at sea level.

Solution

If we define the zero of potential energy to be at zero altitude $(y = 0)$, then $U(0) = 0$ and Equation 8-3 (for the gravitational potential energy near the surface of the Earth, $|y| \ll 6370$ km) gives $U(y) - U(0) = U(y) = mg(y - 0) = mgy$. Therefore, (a) $U(1900 \text{ m}) = (9000 \times 9.8 \text{ N})(1900 \text{ m}) = 168$ MJ, and (b) $U(-86 \text{ m}) = (9000 \times 9.8 \text{ N})(-86 \text{ m}) = -7.59$ MJ.

Problem

9. A 1.50-kg brick measures 20.0 cm $\times$ 8.00 cm $\times$ 5.50 cm. Taking the zero of potential energy when the brick lies on its broadest face, what is the potential energy (a) when the brick is standing on end and (b) when it is balanced on its 8-cm edge, with its center directly above that edge? Note: You can treat the brick as though all its mass is concentrated at its center.

Solution

The center of the brick is a distance $\Delta y = 10$ cm $-$ 2.75 cm $= 7.25$ cm above the zero of potential energy in position (a), and $\Delta y = \frac{1}{2}\sqrt{(20 \text{ cm})^2 + (5.5 \text{ cm})^2} - 2.75$ cm $= 10.4$ cm $- 2.75$ cm $= 7.62$ cm in position (b) (see sketch; the center is midway along the diagonal of the face of the brick). From Equation 8-3, the gravitational potential energy is $U_a = mg \, \Delta y = (1.5 \times 9.8 \text{ N})(7.25 \text{ cm}) = 1.07$ J and $U_b = (1.5 \times 9.8 \text{ N})(7.62 \text{ cm}) = 1.12$ J above the zero energy.

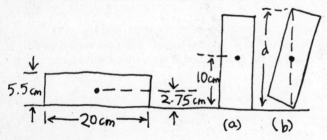

Problem 9 Solution.

Problem

13. How far would you have to stretch a spring of spring constant $k = 1700$ N/m until it stored 250 J of energy?

Solution

Assuming one starts stretching from the unstretched position $(x = 0)$, Equation 8-4 gives $x = \sqrt{2U/k} = \sqrt{2(250 \text{ J})/(1700 \text{ N/m})} = 54.2$ cm.

Problem

17. A particle moves along the x axis under the influence of a force $F = ax^2 + b$, where a and b are constants. Find its potential energy as a function of position, taking $U = 0$ at $x = 0$.

Solution

Equation 8-2a, with $U(0) = 0$, gives

$$U(x) = -\int_0^x F_x \, dx' = -\int_0^x (ax'^2 + b) \, dx' = -\frac{1}{3}ax^3 - bx.$$

Section 8-3 Conservation of Mechanical Energy

Problem

21. A Navy jet of mass 10,000 kg lands on an aircraft carrier and snags a cable to slow it down. The cable is attached to a spring with spring constant 40,000 N/m. If the spring stretches 25 m to stop the plane, what was the landing speed of the plane?

Solution

If we assume no change in gravitational potential energy, $K_{\text{initial}} = \frac{1}{2}mv^2 = U_{\text{final}} = \frac{1}{2}kx^2$, or $v = \sqrt{k/m} \, x$. Thus, $v = \sqrt{(40,000 \text{ N/m})/(10,000 \text{ kg})} \, (25 \text{ m}) = 50$ m/s.

Problem

25. Derive Equation 2-14 using the conservation of energy principle.

Solution

Gravity is a conservative force, so for free fall near the Earth's surface, $U_0 + K_0 = U + K$, or $\frac{1}{2}mv_0^2 + mgy_0 = \frac{1}{2}mv^2 + mgy$. (We neglect the effects of air resistance.) Canceling m, and rearranging terms, one recaptures Equation 2-14.

Problem

29. An initial speed of 2.4 km/s (the "escape speed") is required for an object launched from the moon to get arbitrarily far from the moon. At a mining operation on the moon, 1000-kg packets of ore are to be launched to a smelting plant in orbit around the Earth. If they are launched with a large spring whose maximum compression is 15 m, what should be the spring constant of the spring?

Solution

For an ideal spring (without losses): $\frac{1}{2}ky^2 = \frac{1}{2}mv_B^2 + mgy$. (This is Equation 8-8 for points A and B in the sketch.) Since the gravitational potential energy change is negligible compared to the other terms, $k \approx m(v_B/y)^2 = 10^3$ kg $\times$ (2.4 km/s/15 m)^2 = 25.6 MN/m. (Note: The surface gravity on the moon is 1.62 m/s^2, so the maximum change in potential energy of a packet is only mgy = 24.3 kJ, while its kinetic energy, $\frac{1}{2}mv_B^2$ = 2.88 GJ, is more than 10^5 times larger. Likewise, the gravitational potential energy of the spring is negligible.)

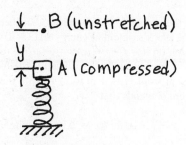

Problem 29 Solution.

Problem

33. Show that the pendulum string in Example 8-6 can remain taut all the way to the top of its smaller loop only if $a \leq \frac{2}{5}\ell$. (Note that the maximum release angle is 90° for the string to be taut on the way down.)

Solution

For the string to be taut at the top of the small circle, the tension must be greater than (or equal to) zero; $T = mv_{top}^2/a - mg \geq 0$, or $v_{top}^2 \geq ga$. Therefore, the mechanical energy at the top is greater than (or equal to) a corresponding value; $E = K + U = \frac{1}{2}mv_{top}^2 + mg(2a) \geq \frac{1}{2}m(ga) + 2mga = \frac{5}{2}mga$, where the zero of potential energy is the lowest point (as in Example 8-6). Since energy is conserved, $E = U_0 = mg\ell(1 - \cos\theta_0)$, where θ_0 is the release angle (see Example 8-6 again), so $a \leq 2E/5mg = \frac{2}{5}\ell(1 - \cos\theta_0)$. The greatest upper limit for the radius a corresponds to the maximum release angle, as stated in the question above, since $\cos 90° = 0$.

Problem

37. A mass m is dropped from a height h above the top of a spring of constant k that is mounted vertically on the floor (Fig. 8-29). Show that the maximum compression of the spring is given by $(mg/k)(1 + \sqrt{1 + 2kh/mg})$. What is the significance of the other root of the quadratic equation?

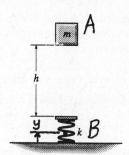

FIGURE 8-29 Problem 37 Solution.

Solution

If the maximum compression is y, as shown, and we measure gravitational potential energy from the lowest point, B, then the conservation of energy between points A and B requires that $mg(h + y) = \frac{1}{2}ky^2$. The quadratic formula can be used to find $y = (mg/k)(1 \pm \sqrt{1 + 2kh/mg})$. Only positive values of y are physically meaningful in this problem, because the spring is not compressed unless $y > 0$.

Section 8-4 Potential Energy Curves

Problem

41. The potential energy associated with a conservative force is shown in Fig. 8-31. Consider particles with total energies $E_1 = -1.5$ J, $E_2 = -0.5$ J, $E_3 = 0.5$ J, $E_4 = 1.5$ J, and $E_5 = 3.0$ J. Discuss the subsequent motion, including the approximate location of any turning points, if the particles are initially at point $x = 1$ m and moving in the $-x$ direction.

Solution

All the particles start in the left-hand potential well, and reverse direction when they hit the left-most (infinite) potential barrier. The first four particles have insufficient energy to escape from this well (the height of the next barrier is about

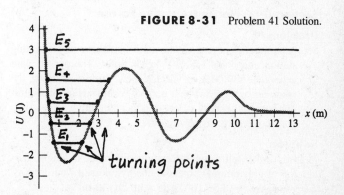

FIGURE 8-31 Problem 41 Solution.

2.0 J $> E_4$) and experience a second turning point (between $x = 2.0$ m and 4.0 m, depending on energy). The fifth particle's motion is unbounded.

Section 8-5 Force and Potential Energy

Note: In the following problems, motion is restricted to one dimension.

Problem

45. A particle is trapped in a potential well described by $U(x) = 2.6x^2 - 4$, where U is in joules and x is in meters. Find the force on the particle when it's at (a) $x = 2.1$ m; (b) $x = 0$ m; and (c) $x = -1.4$ m.

Solution

For one-dimensional motion, Equation 8-9 gives the force $F_x = -dU/dx = -d/dx(1.6x^2 - 4) = -2(1.6)x = -3.2x$, where F_x is in newtons for x in meters. Therefore: (a) $F_x(2.1$ m$) = -(3.2)(2.1)$ N $= -6.72$ N (force in the negative x direction), (b) $F_x(0) = 0$; (c) $F_x(-1.4$ m$) = -(3.2)(-1.4)$ N $= 4.48$ N.

Problem

49. The potential energy of a spring is given by $U = ax^2 - bx + c$, where $a = 5.20$ N/m, $b = 3.12$ N, and $c = 0.468$ J, and where x is the *overall* length of the spring (not the stretch). (a) What is the equilibrium length of the spring? (b) What is the spring constant?

Solution

(a) The natural, or unstretched, length of the spring is the value of x for which the spring force is zero. Thus, $F_x = -dU/dx = -2ax + b = 0$, when $x = b/2a = (3.12$ N$)/2(5.2$ N/m$) = 30$ cm. (b) Hooke's Law defines the spring constant in terms of the stretch (in this case $x - b/2a$). Since $F_x = -2a(x - b/2a)$, the spring constant is $k = 2a = 10.4$ N/M. (Alternatively, $k = -dF_x/dx = d^2U/dx^2$.)

Section 8-6 Nonconservative Forces

Problem

53. A pumped-storage reservoir sits 140 m above its generating station and holds 8.5×10^9 kg of water. The power plant generates 330 MW of electric power while draining the reservoir over an 8.0-hour period. What fraction of the initial potential energy is lost to nonconservative forces (i.e., does not emerge as electricity)?

Solution

If all the water fell through the same difference in height, the amount of gravitational potential energy released would be $\Delta U = mg \, \Delta y = (8.5 \times 10^9$ kg$)(9.8$ m/s$^2)(140$ m$) =$

1.17×10^{13} J. The energy generated by the power plant at an average power output of 330 MW over an 8 h period is $(33$ MW$)(8 \times 3600$ s$) = 9.50 \times 10^{12}$ J, so the fraction lost is $(11.7 - 9.50)/11.7 = 18.5\%$.

Problem

57. A surface is frictionless except for a region between $x = 1$ m and $x = 2$ m, where the coefficient of friction is given by $\mu = ax^2 + bx + c$, with $a = -2$ m^{-2}, $b = 6$ m^{-1}, and $c = -4$. A block is sliding in the $-x$ direction when it encounters this region. What is the minimum speed it must have to get all the way across the region?

Solution

Assume that the surface is horizontal, so that there are no changes in the block's potential energy ($\Delta U = 0$) and the force of friction on it is $f_k = -\mu_k N = -\mu_k mg$ (opposite to the direction of motion along the x axis). If the block crosses the entire region from $x_1 = 1$ m to $x_2 = 2$ m, the work-energy theorem demands that

$$W_{nc} = -\int_{x_1}^{x_2} \mu_k mg \, dx = \Delta K = \frac{1}{2} m(v_2^2 - v_1^2)$$

or

$$v_1^2 - v_2^2 = 2g \int_{x_1}^{x_2} (ax^2 + bx + c) \, dx$$

$$= 2g \left. \frac{a}{3}x^3 + \frac{b}{2} x^2 + cx \right|_{1m}^{2m} = 6.53 \text{ m}^2/\text{s}^2.$$

(The given values of a, b, and c were used.) The minimum speed at the start of the region is the value of v_1 when $v_2 = 0$, or $v_{1,\min} = \sqrt{6.53 \text{ m}^2/\text{s}^2} = 2.56$ m/s.

Problem

61. A 190-g block is launched by compressing a spring of constant $k = 200$ N/m a distance of 15 cm. The spring is mounted horizontally, and the surface directly under it is frictionless. But beyond the equilibrium position of the spring end, the surface has coefficient of friction $\mu = 0.27$. This frictional surface extends of 85 cm, followed by a frictionless curved rise, as shown in Fig. 8-38. After launch, where does the block finally come to rest? Meausure from the left end of the frictional zone.

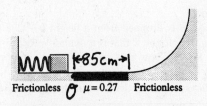

Frictionless $\mu = 0.27$ Frictionless

FIGURE 8-38 Problem 61 Solution.

Solution

The energy of the block when it first encounters friction (at point O) is $K_0 = \frac{1}{2}(200 \text{ N/m})(0.15 \text{ m})^2 = 2.25$ J, if we take the zero of gravitational potential energy at that level. Crossing the frictional zone, the block loses energy $\Delta E = W_{nc} = -\mu_k mg\ell = -(0.27)(0.19 \text{ kg})(9.8 \text{ m/s}^2)(0.85 \text{ m}) = -0.427$ J. Since $K_0/|\Delta E| = 5.27$, five complete crossings are made, leaving the block with energy $K_0 - 5|\Delta E| = 0.113$ J on the curved rise side. This remaining energy is sufficient to move the block a distance $s = 0.113 \text{ J}/\mu_k mg = 22.5$ cm towards point O, so the block comes to rest $85 - 22.5 = 62.5$ cm to the right of point O.

Paired Problems

Problem

65. A block slides down a frictionless incline that terminates in a ramp pointing up at a 45° angle, as shown in Fig. 8-39. Find an expression for the horizontal range x shown in the figure, as a function of the heights h_1 and h_2 shown.

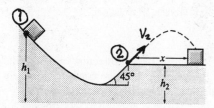

FIGURE 8-39 Problem 65.

Solution

After leaving the ramp (at point 2), with speed v_2 at 45° to the horizontal, the block describes projectile motion with a horizontal range of $x = v_2^2/g$ (see Equation 4-10). Since the track is frictionless (and the normal force does no work), the mechanical energy of the block (kinetic plus gravitational potential) is conserved between point 2 and its start from rest at point 1. Then $\Delta K = \frac{1}{2}mv_2^2 - 0 = -\Delta U = mg(h_1 - h_2)$, and $x = 2(h_1 - h_2)$.

Problem

69. A pendulum consisting of a mass m on a string of length ℓ is pulled back so the string is horizontal, as shown in Fig. 8-42. The pendulum is then released. Find (a) the speed of the mass and (b) the magnitude of string tension when the string makes a 45° angle with the horizontal.

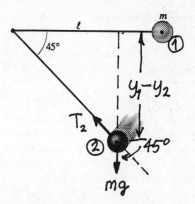

FIGURE 8-42 Problem 69 Solution.

Solution

(a) We assume that the mechanical energy of the pendulum mass is conserved (neglect possible losses), and that this consists of kinetic and gravitational potential energy. At point 1 where the mass is released, $U_1 = mgy_1$ and $K_1 = 0$, while after a 45° swing to point 2, $U_2 = mgy_2$ and $K_2 = \frac{1}{2}mv_2^2$. Energy conservation implies $K_2 = U_1 - U_2$ or $v_2^2 = 2g(y_1 - y_2) = 2g\ell \sin 45° = \sqrt{2}g\ell$, where we used the trigonometry apparent in Fig. 8-42 to express the difference in height in terms of the length of the string and angle of swing. (b) The string tension is in the direction of the centripetal acceleration (also shown in Fig. 8-42 at point 2), so the radial component of Newton's second law gives $T_2 - mg \sin 45° = mv_2^2/\ell$. Using the result of part (a), we find $T_2 = (mv_2^2/\ell) + mg \sin 45° = (m\sqrt{2}g\ell/\ell) + (mg/\sqrt{2}) = 3mg/\sqrt{2}$.

Problem

73. A child sleds down a frictionless hill whose vertical drop is 7.2 m. At the bottom is a level but rough stretch where the coefficient of kinetic friction is 0.51. How far does she slide across the level stretch?

Solution

The child starts near the hilltop with $K_A = 0$ and stops on rough level ground, $K_B = 0$, after falling through a potential energy difference $\Delta U = U_B - U_A = -mg(y_A - y_B)$, where $y_A - y_B = 7.2$ m. The work done by friction (on level ground, $N = mg$) is $W_{nc} = -f_k x = -\mu_k mgx$, where x is the distance slid across the rough level stretch. The energy principle, Equation 8-5, relates these quantities: $W_{nc} = -\mu_k mgx = \Delta K + \Delta U = 0 - mg(y_A - y_B)$. Thus $x = (y_A - y_B)/\mu_k = 7.2 \text{ m}/0.51 = 14.1$ m.

Supplementary Problems

Problem

77. With the brick of Problem 9 standing on end, what is the minimum energy that can be given the brick to make it fall over?

Solution

To fall over, the center of the brick must lie to the right of the vertical through its 8-cm edge, as shown. The potential energy of the brick would have to be increased by $U_b - U_a = mg(y_b - y_a)$. If we assume that energy is conserved between a and b, this is equal to the minimum kinetic energy sought ($K_a - K_b = U_b - U_a = K_a^{min}$, if $K_b = 0$). The numerical value is just the difference between the answers to Problems 9(b) and 9(a), but recalculating, we find: $mg(y_b - y_a) = (1.5 \text{ kg})(9.8 \text{ m/s}^2)\frac{1}{2}[\sqrt{(20 \text{ cm})^2 + (5.5 \text{ cm})^2} - 20 \text{ cm}] = 54.6 \text{ mJ}$.

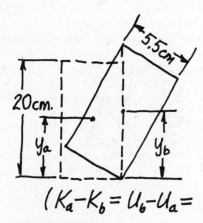

Problem 77 Solution.

Problem

81. An electron with kinetic energy 0.85 fJ enters a region where its potential energy as a function of position is $U = ax^2 - bx$, where $a = 2.7$ fJ/cm^2 and $b = 4.2$ fJ/cm. (a) How far into the region does the electron penetrate? (b) At what position does the electron have its maximum speed? (c) What is this maximum speed?

Solution

(a) The electron's total energy, $E = 0.85$ fJ, equals its initial kinetic energy upon entering the region at $x = 0$. At any other point in the region $x \geq 0$, $E = K + U$, so the point of maximum penetration (the turning point) can be found from the equation $K(x_m) = E - U(x_m) = 0$.

Therefore, $ax_m^2 - bx_m - 0.85 \text{ fJ} = 0$, or $x_m = [4.2 + \sqrt{(4.2)^2 + 4(2.7)(0.85)}]/2(2.7) = 1.74$ cm. (The negative root can be discarded since $x \geq 0$.) (b) The maximum kinetic energy occurs at the point where U is a minimum, that is: $dU/dx = 2ax_0 - b = 0$, or $x_0 = b/2a = 0.778$ cm. (c) $v_{max} = \sqrt{2K_{max}/m_e} = \sqrt{2(E - U_{min})/m_e}$. Now, $U_{min} = x_0(ax_0 - b) = -b^2/4a = -(4.2 \text{ fJ/cm})^2 \div 4(2.7 \text{ fJ/cm}^2) = -1.63$ fJ, so that $v_{max} = \sqrt{2(0.85 + 1.63) \text{ fJ}/(9.11 \times 10^{-31} \text{ kg})} = 7.38 \times 10^7$ m/s.

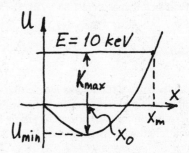

Problem 81 Solution.

Problem

85. A force points in the $-x$ direction with magnitude given by $F = ax^b$, where a and b are constants. Evaluate the potential energy as a function of position, taking $U = 0$ at some point $x_0 > 0$. Use your result to show that an object of mass m released at $x = \infty$ will reach x_0 with finite velocity provided $b < -1$. Find the velocity for this case.

Solution

For one-dimensional motion, Equation 8-2a for the potential energy yields

$$U(x) = -\int_{x_0}^{x} F_x \, dx = -\int_{x_0}^{x} (-ax^b) \, dx$$
$$= a(b + 1)^{-1}(x^{b+1} - x_0^{b+1}),$$

where $U(x_0) = 0$ and $x_0 > 0$. This result applies provided $b \neq -1$; then $U(x)$ is logarithmic. If no other forces act, mechanical energy is conserved, so a particle released at $x = \infty$ with kinetic energy $K(\infty) = 0$, will have kinetic energy at x_0 of $K(x_0) = U(\infty)$. This is finite provided $b + 1 < 0$. (Otherwise, $x^{b+1} \to \infty$, as $x \to \infty$, or if $b = -1$, $\ln x \to \infty$ also.) In case $b + 1 < 0$, $|b + 1| = -(b + 1)$ and $\frac{1}{2}mv_0^2 = (a/|b + 1|)x_0^{-|b+1|}$, or $v_0 = (2a/m|b + 1|)^{1/2}x_0^{-|b+1|/2}$.

● **CHAPTER 9** GRAVITATION

Section 9-2 The Law of Universal Gravitation

Problem

1. Space explorers land on a planet with the same mass as Earth, but they find they weigh twice as much as they would on Earth. What is the radius of the planet?

Solution

At rest on a uniform spherical planet, a body's weight is proportional to the surface gravity, $g = GM/R^2$. Therefore, $(g_p/g_E) = (M_p/M_E)(R_E/R_p)^2 = 2$. Since $M_p/M_E = 1$, $R_p = R_E/\sqrt{2}$.

Problem

5. Two identitcal lead spheres are 14 cm apart and attract each other with a force of 0.25 μN. What is their mass?

Solution

Newton's law of the universal gravitation (Equation 9-1), with $m_1 = m_2 = m$, gives

$$m = \sqrt{\frac{Fr^2}{G}} = \left(\frac{0.25 \times 10^{-6} \text{ N } (0.14 \text{ m})^2}{6.67 \times 10^{-11} \text{ N·m}^2/\text{kg}^2}\right)^{1/2} = 8.57 \text{ kg}.$$

(Newton's law holds as stated, for two uniform spherical bodies, if the distance is taken between their centers.)

Problem

9. A sensitive gravimeter is carried to the top of Chicago's Sears Tower, where its reading for the acceleration of gravity is 0.00136 m/s² lower than at street level. Find the height of the building.

Solution

The difference in the acceleration of gravity between the bottom of the Sears Tower (distance R from the center of the Earth) and the top (distance $R + h$ from the Earth's center, where h is the height of the building) is

$$\Delta g = \frac{GM_E}{R^2} - \frac{GM_E}{(R + h)^2} = \frac{GM_E}{R^2} \frac{h(2R + h)}{(R + h)^2} \approx g_{\text{bot}} \frac{2h}{R}.$$

In the last step, g_{bot} is the value of the acceleration of gravity at the bottom of the building, and since $h \ll R$, we approximated the second fraction by neglecting h compared to R. If we use average values for g_{bot} (about 9.80 m/s² in Chicago) and R (approximately 6370 km) and the given Δg, then $h \approx (\Delta g/g_{\text{bot}})(R/2) = (0.00136/9.80)(6370 \text{ km}/2) = 442$ m. (The actual value of g_{bot}, for example, depends on the shape of the earth, the altitude, the distribution and type of underlying rocks, etc. Present gravimeters can measure differences in g as small as a few tenths of a milligal, where 1 milligal $= 10^{-5}$ m/s² is the unit used to measure gravity anomalies by geologists.)

Section 9-3 Orbital Motion

Problem

13. Find the speed of a satellite in geosynchronous orbit.

Solution

Equation 9-3 and the radius of a geosynchronous orbit from Example 9-4 gives $v = \sqrt{GM_E/r} = [(6.67 \times 10^{-11} \text{ N·m}^2/\text{kg}^2)(5.97 \times 10^{24} \text{ kg})/(42.2 \times 10^6 \text{ m})]^{1/2} = 3.07$ km/s.

Problem

17. During the Apollo moon landings, one astronaut remained with the command module in lunar orbit, about 130 km above the surface. For half of each orbit, this astronaut was completely cut off from the rest of humanity, as the spacecraft rounded the far side of the moon (see Fig. 9-39). How long did this period last?

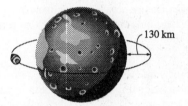

FIGURE 9-39 Problem 17.

Solution

The period of a circular orbit of altitude $h = 130$ km above the moon's surface is (see Equation 9-4) $T = 2\pi \sqrt{(R_m + h)^3/GM_m} = 2\pi[(1.74 + 0.13)^3 \times 10^{18} \text{ m}^3 \div (6.67 \times 10^{-11} \text{ N·m}^2/\text{kg}^2) (7.35 \times 10^{22} \text{ kg})]^{1/2} = 7.26 \times 10^3$ s $= 121$ min. The command module was cut off from Earth communications for roughly half of this period, or about 1 h.

Problem

21. Where should a satellite be placed to orbit the Sun in a circular orbit with a period of 100 days?

Solution

As long as the masses of the orbiting bodies (satellite or planet) are negligible compared to the mass of the Sun, Equation 9-4, as a ratio, can be used to compare the mean orbital radii and periods, $(T_1/T_2)^2 = (r_1/r_2)^3$ (this is, in fact, an approximate statement of Kepler's third law). The mean orbital radius of the Earth is 1 AU (an astronomical unit) $\approx 1.50 \times 10^8$ km and its period is 1 y (a sidereal year) ≈ 365 d, so $(T/1 \text{ y})^2 = (r/1 \text{ AU})^3$ for any satellite of the Sun. If $T = 100$ d, then $(r/1 \text{ AU}) = (100/365)^{2/3} = 0.422$, or $R = 0.422$ AU $= 6.3 \times 10^7$ km. (Of course, Equation 9-4 could be used directly:

$$r = [GM(T/2\pi)^2]^{1/3}$$

$$= \left[\left(6.67 \times 10^{-11} \frac{\text{N} \cdot \text{m}^2}{\text{kg}^2}\right)(1.99 \times 10^{30} \text{ kg})\left(\frac{8.64 \times 10^6 \text{ s}}{2\pi}\right)^2\right]^{1/3}$$

$$= 6.31 \times 10^7 \text{ km.})$$

Problem

25. The asteriod Pasachoff orbits the Sun with a period of 1417 days. What is the semimajor axis of its orbit? Determine using Kepler's third law in comparison with Earth's orbital radius and period.

Solution

As in the solution to Problem 21, $r = (T/1 \text{ y})^{2/3}(1 \text{ AU}) = (1417/365)^{2/3}$ AU $= 2.47$ AU $= 3.71 \times 10^8$ km.

Section 9-4 Gravitational Energy

Problem

29. How much energy does it take to launch a 230-kg instrument package on a vertical trajectory that peaks at an altitude of 1800 km?

Solution

If we neglect any kinetic energy differences associated with the orbital or rotational motion of the Earth or package, the required energy is just the difference in potential energy of the Earth's gravity given by Equation 9-5, $\Delta U = GM_E m[R_E^{-1} - (R_E + h)^{-1}]$. In terms of the more convenient combination of constants $GM_E = gR_E^2$, $\Delta U = mgR_E h/(R_E + h) = (230 \times 9.81 \text{ N})(6370 \times 1800 \text{ km}) \div (8170) = 3.17$ GJ.

Problem

33. Find the energy necessary to put 1 kg, initially at rest on Earth's surface, into geosynchronous orbit.

Solution

The energy of an object at rest on the Earth's surface is $U_0 = -GM_E m/R_E$ (neglect diurnal rotational energy, etc.), while its total mechanical energy in a circular orbit is $E = \frac{1}{2}U = -GM_E m/2r$ (Equation 9-9). The energy necessary to put a mass of $m = 1$ kg into a circular geosynchronous orbit with $r = 4.22 \times 10^7$ m (see Example 9-4) is the difference of these energies, $\Delta E = \frac{1}{2}U - U_0$, or

$$\Delta E = GM_E(1 \text{ kg})\left(\frac{1}{6.37 \times 10^6 \text{m}} - \frac{1}{2(4.22 \times 10^7 \text{m})}\right)$$

$$= 57.8 \text{ MJ}.$$

Problem

37. Drag due to small amounts of residual air causes satellites in low Earth orbit to lose energy and eventually spiral to Earth. What fraction of its orbital energy is lost as a satellite drops from 300 to 100 km, assuming its orbit remains essentially circular?

Solution

The fractional difference in orbital energy is $\Delta E/E = (E - E')/E$, where $E = \frac{1}{2}U = -GM_E m/2r \sim 1/(R_E + h)$ is the orbital energy and h is the altitude of the circular orbit (see Equation 9-9). Thus $\Delta E/E = (r^{-1} - r'^{-1})/r^{-1} = (h' - h)/(R_E + h') = (100 - 300) \text{ km}/(6370 + 100) \text{ km} = 2/64.7 = -3.09 \times 10^{-2} \approx -3\%$. (The energy difference is negative because energy is lost going from a higher to a lower orbit.)

Problem

41. The escape speed from a planet of mass 3.6×10^{24} kg is 9.1 km/s. What is the planet's radius?

Solution

Equation 9-7 implies $R = 2GM/v_{\text{esc}}^2 = 2(6.67 \times 10^{-11}$ N·m²/kg²$)(3.6 \times 10^{24}$ kg$)/(9.1 \times 10^3$ m/s$)^2 = 5.80 \times 10^6$ m.

Problem

45. Neglecting Earth's rotation, show that the energy needed to launch a satellite of mass m into circular orbit at altitude h is

$$\left(\frac{GM_E m}{R_E}\right)\left(\frac{R_E + 2h}{2(R_E + h)}\right).$$

Solution

The energy of a satellite in a circular orbit is $E = \frac{1}{2}U = -GM_E m/2r$, where $r = R_E + h$ (see Equation 9-9). This is the energy in an Earth-centered, non-rotating reference frame, neglecting the gravitational influence of any other body, e.g. the Sun. The energy of a satellite on the Earth's surface depends on its location, because of the Earth's diurnal rotation, $E_0 = U_0 + \frac{1}{2}mv_0^2$, where $U_0 = -GM_E m/R_E$ and v_0 is the speed of the Earth's surface at that location. If we neglect the extra kinetic energy associated with v_0 (since $v_0 \leq 2\pi R_E/1d$, $\frac{1}{2}mv_0^2 \leq 0.34\% |U_0|$), the energy required to launch the satellite into a circular orbit at altitude h is $E - E_0 = \frac{1}{2}U - U_0 = GM_E m[R_E^{-1} - \frac{1}{2}(R_E + h)^{-1}] = GM_E m(R_E + 2h)/2R_E(R_E + h)$ as claimed.

Problem

49. The Pioneer spacecraft left Earth's vicinity moving at about 38 km/s relative to the Sun (this figure combines the effect of rocket boost and Earth's orbital motion). How far out in the solar system could Pioneer get without additional rocket power or use of the "gravitational slingshot" effect?

Solution

If we consider just the gravitational field of the Sun, which is the dominant source of potential energy for the Pioneer spacecraft in this problem, then the conservation of energy in the form $K + U = K_0 + U_0$ (as in Example 9-6) gives $\frac{1}{2}mv^2 - GM_\odot m/r = \frac{1}{2}mv_0^2 - GM_\odot m/r_0$, or $r = [r_0^{-1} - (v_0^2 - v^2)/2GM_\odot]^{-1}$. Here, v_0 is given as 38 km/s and $r_0 = 1$ AU $= 1.50{\times}10^8$ km. We do not know the actual shape of the satellite's orbit, but $v^2 \ll v_0^2$ at the maximum distance from the sun, so $r < r_\odot = [r_0^{-1} - v_0^2/2GM_\odot]^{-1} = [(1.50{\times}10^{11}\text{ m})^{-1} - (38{\times}10^3\text{ m/s})^2/2(6.67{\times}10^{-11}\text{ N·m}^2/\text{kg}^2) \times (1.99{\times}10^{30}\text{ kg})]^{-1} = 8.15{\times}10^{11}$ m $= 5.43$ AU. This is a little farther than the orbit of Jupiter.

Paired Problems

Problem

53. An astronaut hits a golf ball horizontally from the top of a lunar mountain so fast that it goes into circular orbit. What is its orbital period?

Solution

The period of a grazing orbit, $r \approx R$, around a spherical object of mass M and radius R (see Appendix E for lunar values) can be found from Equation 9-4, $T = 2\pi\sqrt{R^3/GM} = 2\pi[(1.74{\times}10^6\text{ m})^3/(6.67{\times}10^{-11}\text{ N·m}^2/\text{kg}^2) \times (7.35{\times}10^{22}\text{ kg})]^{1/2} = 6.51{\times}10^3$ s $= 109$ min.

Problem

57. A satellite is in an elliptical orbit at altitudes ranging from 230 to 890 km. At the high point it's moving at 7.23 km/s. How fast is it moving at the low point?

Solution

The conservation of energy is applied to a satellite in an elliptical Earth orbit in Example 9-7 (where it is a good approximation to neglect the gravitational influence of other bodies, atmospheric drag, etc.) to relate the speed and distance at perigee (the lowest point) to the same quantities at apogee (the highest point): $v_p^2 = v_a^2 + 2GM_E(r_p^{-1} - r_a^{-1})$. The calculation can be simplified by expressing the distances in terms of altitude above the Earth's surface and using a little algebra. Then $r = R_E(1 + h/R_E)$ and $(r_p^{-1} - r_a^{-1}) = (r_a - r_p)/r_a r_p = (h_a - h_p)/R_E^2(1 + h_a/R_E)(1 + h_p/R_E)$. Since $GM_E/R_E^2 = g$, $v_p^2 = (7.23\text{ km/s})^2 + 2(9.81\text{ m/s}^2)(890\text{ km} - 230\text{ km}) \div [(1 + 890/6370)(1 + 230/6370)]$, and $v_p = 7.95$ km/s. (This result also follows from the conservation of angular momentum or Kepler's second law, which implies that $v_a r_a = v_p r_p$.)

Problem

61. A 720-kg spacecraft has total energy $-5.3{\times}10^{11}$ J and is in circular orbit about the Sun. Find (a) its orbital radius, (b) its kinetic energy, and (c) its speed.

Solution

For a small object (the spacecraft) in a circular orbit about a central massive body (the Sun), the total, kinetic, and potential energies are related by Equations 9-8 and 9, $E = -K = \frac{1}{2}U$. (b) Therefore $K = -E = 5.3{\times}10^{11}$ J. (c) Since $K = \frac{1}{2}mv^2$, $v = \sqrt{2K/m} = [2(5.3{\times}10^{11}\text{ J})/720\text{ kg}]^{1/2} = 38.4$ km/s. (a) Since $U = 2E = -GM_\odot m/r$, $r = GM_\odot m/(-2E) = (6.67{\times}10^{-11}\text{ N·m}^2/\text{kg}^2) \times (1.99{\times}10^{30}\text{ kg})(720\text{ kg})/(2 \times 5.3{\times}10^{11}\text{ J}) = 9.02{\times}10^{10}$ m $= 0.601$ AU.

Supplementary Problems

Problem

65. A black hole is an object so dense that its escape speed exceeds the speed of light. Although a full description of black holes requires general relativity, the radius of a black hole can be calculated using Newtonian theory. (a) Show that the radius of a black hole of mass M is $2GM/c^2$, where c is the speed of light. What are the radii of black holes with (b) a the mass of the Earth and (c) the mass of the Sun?

Solution

(a) The so-called Schwarzchild radius of a black hole of mass M turns out to be the same as the radius given by Equation 9-7 with $v_{esc} = c = \sqrt{2GM/r_s}$. Thus $r_s = 2GM/c^2$. (b) $2GM_E/c^2 = 8.85$ mm and (c) $2GM_\odot/c^2 = 2.95$ km.

● **CHAPTER 10** SYSTEMS OF PARTICLES

Section 10-1 Center of Mass

Problem

1. A 28-kg child sits at one end of a 3.5-m long seesaw. Where should her 65-kg father sit so the center of mass will be at the center of the seesaw?

Solution

Take the x axis along the seesaw in the direction of the father, with origin at the center. The center of mass of the child and her father is at the origin, so $x_{cm} = 0 = m_c x_c + m_f x_f$, where the masses are given and $x_c = -(3.5 \text{ m})/2$ (half the length of the seesaw in the negative x direction). Thus, $x_f = -m_c x_c/m_f = (28/65)(1.75 \text{ m}) = 75.4$ cm from the center.

Problem

5. Three equal masses lie at the corners of an equilateral triangle of side ℓ. Where is the center of mass?

Solution

Take x-y coordinates with origin at the center of one side as shown. From the symmetry (for every mass at x, there is an equal mass at $-x$) $x_{cm} = 0$. Since $y = 0$ for the two masses on the x-axis and $y = \ell \sin 60° = \ell\sqrt{3}/2$ for the other mass, Equation 10-2 gives $y_{cm} = m(\ell\sqrt{3}/2)/3m = \ell/2\sqrt{3} = 0.289\,\ell$.

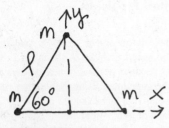

Problem 5 Solution.

Problem

9. Find the center of mass of a pentagon of side a with one triangle missing, as shown in Fig. 10-24. *Hint:* See Example 10-3, and treat the pentagon as a group of triangles.

Solution

Choose coordinates as shown. From symmetry, $x_{cm} = 0$. If the fifth isosceles triangle (with the same assumed uniform density) were present, the center of mass of the whole pentagon would be at the origin, so $0 = (my_5 + 4my_{cm})/5$ m, where y_{cm} gives the position of the center of mass of the figure

we want to find, and y_5 is the position of the center of mass of the fifth triangle. Of course, the mass of the figure is four times the mass of the triangle. In Example 10-3, the center of mass of an isosceles triangle is calculated, so $y_5 = -\frac{2}{3}\ell$, and from the geometry of a pentagon, $\tan 36° = \frac{1}{2}a/\ell$. Therefore, $y_{cm} = -\frac{1}{4}y_5 = \frac{1}{6}\ell = \frac{1}{12}a\cot 36° = 0.115a$.

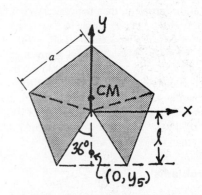

FIGURE 10-24 Problem 9 Solution.

Problem

13. Consider a system of three equal-mass particles moving in a plane; their positions are given by $a_i\hat{\mathbf{i}} + b_i\hat{\mathbf{j}}$, where a_i and b_i are functions of time with the units of position. Particle 1 has $a_1 = 3t^2 + 5$ and $b_1 = 0$; particle 2 has $a_2 = 7t + 2$ and $b_2 = 2$; particle 3 has $a_3 = 3t$ and $b_3 = 2t + 6$. Find the position, velocity, and acceleration of the center of mass as functions of time.

Solution

Since the particles have equal masses, Equation 10-2 gives $x_{cm} = \sum m_i x_i/\sum m_i = \frac{1}{3}\sum a_i$ and $y_{cm} = \frac{1}{3}\sum b_i$. Using the given values of the coefficients (position in meters and time in seconds) we find $x_{cm} = \frac{1}{3}(3t^2 + 5 + 7t + 2 + 3t) = t^2 + (10/3)t + 7/3$ and $y_{cm} = \frac{1}{3}(0 + 2 + 2t + 6) = (2/3)t + 8/3$. Differentiation yields $v_{cm,x} = dx_{cm}/dt = 2t + 10/3$ and $v_{cm,y} = 2/3$ (both in m/s), and $a_{cm,x} = dv_{cm,x}/dt = d^2x_{cm}/dt^2 = 2$ and $a_{cm,y} = 0$ (both in m/s²).

Problem

17. A hemispherical bowl is at rest on a frictionless kitchen counter. A mouse drops onto the rim of the bowl from a cabinet directly overhead. The mouse climbs down the inside of the bowl to eat the crumbs at the bottom. If the bowl moves along the counter a distance equal to one-tenth of its diameter, how does the mouse's mass compare with the bowl's mass?

Solution

When the mouse starts at the rim, the center of mass of the mouse-bowl system has x component:

$$x_{cm} = (m_b x_b + m_m x_m)/(m_b + m_m) = m_m R/(m_b + m_m),$$

since initially, $x_b = 0$ and $x_m = R$. Because there is no external horizontal force (no friction), x_{cm} remains constant as the mouse climbs. When it reaches the center of the bowl, both have x coordinates equal to x_{cm}, which is, therefore, the distance moved by the bowl across the counter, given as $\frac{1}{5}R$. Thus, $x_{cm} = \frac{1}{5}R = m_m R/(m_b + m_m)$, or $m_m = \frac{1}{4}m_b$.

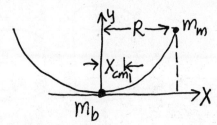

Problem 17 Solution.

Section 10-2 Momentum

Problem

21. A firecracker, initially at rest, explodes into two fragments. The first, of mass 14 g, moves in the positive x direction at 48 m/s. The second moves at 32 m/s. Find its mass and the direction of its motion.

Solution

The momentum of the firecracker/fragments is approximately conserved during a short time interval around the explosion (as for the object in Problem 19) so $0 = m_1 \mathbf{v}_1 + m_2 \mathbf{v}_2$ or $m_2 \mathbf{v}_2 = -(14 \text{ g})(48\hat{\imath} \text{ m/s})$. The direction of the second piece is opposite to that of the first, or $\mathbf{v}_2 = -32\hat{\imath}$ m/s. Then $m_2 = (14 \text{ g})(48/32) = 21$ g.

Problem

25. A runaway toboggan of mass 8.6 kg is moving horizontally at 23 km/h. As it passes under a tree, 15 kg of snow drop onto it. What is its subsequent speed?

Solution

The horizontal component of the momentum of the toboggan and snow is conserved (no net external horizontal force), and the final velocity of each is the same: $m_t v_{ti} + 0 =$ $(m_t + m_s)v_f$, or $v_f = m_t v_{ti}/(m_t + m_s)$ $= (8.6)(23 \text{ km/h})/(8.6 + 15) = 8.38$ km/h.

Problem

29. An 11,000-kg freight car is resting against a spring bumper at the end of a railroad track. The spring has constant $k = 3.2 \times 10^5$ N/m. The car is hit by a second car of 9400 kg mass moving at 8.5 m/s, and the two cars couple together. (a) What is the maximum compression of the spring? (b) What is the speed of the two cars together when they rebound from the spring?

Solution

(a) The motion of the center of mass of the two freight cars (on an assumed horizontal, frictionless track) is determined by the only horizontal external force, that of the spring. If this is conservative, the potential energy of the spring at maximum compression equals the kinetic energy of the center of mass prior to contact with the spring, i.e., $\frac{1}{2}kx_{max}^2 = \frac{1}{2}MV_{cm}^2$. Now $x_{cm} = (m_1 x_1 + m_2 x_2)/M$, so $V_{cm} = m_2 v_2/M$, since the first car is initially at rest. Thus, $V_{cm} = (9,400 \text{ kg})(8.5 \text{ m/s})/(11,000 + 9,400) \text{ kg} = 3.92$ m/s, and $x_{max} = V_{cm}\sqrt{M/k} =$ $(3.92 \text{ m/s})\sqrt{(20,400 \text{ kg})/(3.2 \times 10^5 \text{ N/m})} = 98.9$ cm. (b) When the cars rebound, they are coupled together and both have the same velocity as their center of mass. Since the spring is ideal (by assumption), its maximum potential energy, $\frac{1}{2}kx_{max}^2$, is transformed back into kinetic energy of the cars, $\frac{1}{2}MV_{cm}^2$, so the rebound speed equals the initial V_{cm}, or 3.92 m/s. (Reconsider this problem after reading Chapter 11, especially Example 11-2.)

Problem

33. A 950-kg compact car is moving with velocity $\mathbf{v}_1 = 32\hat{\imath} + 17\hat{\jmath}$ m/s. It skids on a frictionless icy patch, and collides with a 450-kg hay wagon moving with velocity $\mathbf{v}_2 = 12\hat{\imath} + 14\hat{\jmath}$ m/s. If the two stay together, what is their velocity?

Solution

If there are no external horizontal forces acting on the car-wagon system, momentum (in the x-y plane) is conserved: $m_1 \mathbf{v}_1 + m_2 \mathbf{v}_2 = (m_1 + m_2)\mathbf{v}$. Therefore:

$$\mathbf{v} = [(950 \text{ kg})(32\hat{\imath} + 17\hat{\jmath})\text{m/s} + (450 \text{ kg})(12\hat{\imath} + 14\hat{\jmath}) \text{ m/s}]/(950 + 450) \text{ kg}$$
$$= (25.6\hat{\imath} + 16.0\hat{\jmath}) \text{ m/s, or } v = \sqrt{v_x^2 + v_y^2} = 30.2 \text{ m/s,}$$

and

$$\theta = \tan^{-1}(v_y/v_x) = 32.1° \quad \text{(with the } x \text{ axis).}$$

Problem

37. An Ariane rocket ejects 1.0×10^5 kg of fuel in the 90 s after launch. (a) How much thrust is developed if the fuel is ejected at 3.0 km/s with respect to the rocket? (b) What is the maximum total mass of the rocket if it is to get off the ground?

Solution

(a) The thrust, given by Equation 10-10, is:

$$F_{Th} = -v_{ex}\frac{dM}{dt} = -(3 \times 10^3 \text{ m/s})\left(-\frac{10^5 \text{ kg}}{90 \text{ s}}\right)$$

$$= 3.33 \times 10^6 \text{ N}.$$

(b) The thrust must exceed the launch weight of the rocket, $F_{Th} \geq Mg$, or $M \leq F_{Th}/g = 3.40 \times 10^5$ kg.

Section 10-3 Kinetic Energy in Many-Particle Systems

Problem

41. Determine the center-of-mass and internal kinetic energies before and after decay of the lithium nucleus of Example 10-7. Treat the individual nuclei as point particles.

Solution

Before the decay, the system consists of one particle (the ^{5}Li-nucleus), so $K_{int,i} = 0$, and $K_{cm} = K_i = \frac{1}{2}m_{Li}v_{Li}^2 = \frac{1}{2}(5 \times 1.67 \times 10^{-27} \text{ kg})(1.6 \times 10^6 \text{ m/s})^2 = 1.07 \times 10^{-14}$ J $= 66.8$ keV. Afterwards, K_{cm} is the same (since momentum is conserved), while $K_{int,f} = K_f - K_{cm} = \frac{1}{2}m_H v_H^2 + \frac{1}{2}M_{He}v_{He}^2 - K_{cm} = \frac{1}{2}(1.67 \times 10^{-27} \text{ kg})[(4.5 \times 10^6 \text{ m/s})^2 + 4(1.4 \times 10^6 \text{ m/s})^2] - 66.8$ keV $= 79.8$ keV.

Paired Problems

Problem

45. A drinking glass is in the shape of a cylinder whose inside dimensions are 9.0 cm high and 8.0 cm in diameter as shown in Fig. 10-30. Its base has a mass of 140 g, while the mass of the curved, cylindrical sides is 85 g. (a) Where is its center of mass? (b) If the glass is three-quarters filled with juice (density 1.0 g/cm³), where is the center of mass of the glass-juice system? Assume the thickness of the glass is negligible.

Solution

(a) To find the CM of the empty glass, consider it to be composed of two subpieces (as suggested in the text following Example 10-3), namely its base ($m_b = 140$ g, with CM at the center $y_b = 0$) and the curved cylindrical sides ($m_c = 85$ g, with CM at the midpoint of the axis $y_c = 4.5$ cm). Then,

$$y_{CM} = \frac{m_b y_b + m_c y_c}{m_b + m_c} = \frac{85}{225}(4.5 \text{ cm}) = 1.70 \text{ cm}$$

(above the center of the base). (b) Now consider the glass of juice to be composed of two subpieces, namely the empty glass ($m_g = 225$ g, with CM at $y_g = 1.70$ cm) and a cylinder of juice (m_j = density × volume = $\rho\pi r^2 h = (1 \text{ g/cm}^3)\pi(4 \text{ cm})^2(\frac{3}{4} \times 9 \text{ cm}) = 339$ g, with CM at the midpoint on its axis $y_j = \frac{1}{2}(\frac{3}{4} \times 9 \text{ cm}) = 3.38$ cm). Then

$$y_{cm} = \frac{m_g y_g + m_j y_j}{m_g + m_j} = \frac{(225)(1.70 \text{ cm}) + (339)(3.38 \text{ cm})}{225 + 339}$$

$$= 2.71 \text{ cm}$$

is the height of the CM of the glass of juice above the center of the base.

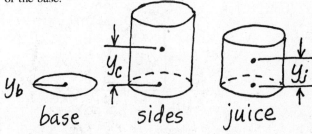

Problem 45 Solution.

Problem

49. A 42-g firecracker is at rest at the origin when it explodes into three pieces. The first, with mass 12 g, moves along the x axis at 35 m/s. The second, with mass 21 g, moves along the y axis at 29 m/s. Find the velocity of the third piece.

Solution

The instant after the explosion (before any external forces have had any time to act appreciably) the total momentum of the pieces of firecracker is still zero, $\mathbf{P}_{tot} = 0 = m_1\mathbf{v}_1 + m_2\mathbf{v}_2 + m_3\mathbf{v}_3$. Therefore

$$\mathbf{v}_3 = \frac{-(12 \text{ g})(35 \text{ m/s})\hat{\imath} - (21 \text{ g})(29 \text{ m/s})\hat{\jmath}}{(42 - 12 - 21) \text{ g}}$$

$$= -(46.7\hat{\imath} + 67.7\hat{\jmath}) \text{ m/s}.$$

Problem

53. Firefighters spray water horizontally at the rate of 41 kg/s from a nozzle mounted on a 12,000-kg fire truck. The water speed is 28 m/s relative to the truck. (a) Neglecting friction, what is the initial acceleration of the truck? (b) If the 12,000-kg truck mass includes 2400 kg of water, how fast will the truck be moving when the water is exhausted? *Hint:* Think about rockets.

Solution

The fire truck behaves like a water rocket with thrust given by Equation 10-10b, $F_{thrust} = -v_{ex}(dM/dt) = -(28 \text{ m/s}) \times (-41 \text{ kg/s}) = 1.15$ kN. (a) If the thrust is the only significant horizontal force, the truck's initial acceleration is $a_i = F_{thrust}/M_i = 1.15$ kN/12000 kg $= 9.57$ cm/s². (b) If the truck starts from rest, Equation 10-11, with $v_i = 0$, gives $v_f = v_{ex} \ln(M_i/M_f) = (28 \text{ m/s}) \ln(12000/(12000 - 2400)) = 6.25$ m/s.

Supplementary Problems

Problem

57. Fig. 10-31 shows a paraboloidal solid of constant density ρ. It extends a height h along the z axis, and is described by $z = ar^2$, where the units of a are m^{-1} and r is the radius in a plane perpendicular to the z axis. Find expressions for (a) the mass of the solid and (b) the z coordinate of its center of mass.

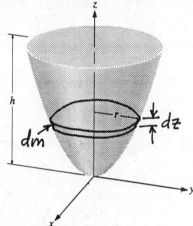

FIGURE 10-31   Problem 57.

Solution

For mass elements, take disks parallel to the x-y plane, of radius $r = \sqrt{z/a}$ and thickness dz, as shown in Fig. 10-28 for Problem 18. Then $dm = \rho \pi r^2 \, dz = \rho \pi (z/a) \, dz$. (a) The total mass is $M = \int_0^h dm = \int_0^h (\rho\pi/a)z \, dz = \rho\pi h^2/2a$. (b) $z_{cm} = \int_0^h z \, dm/M = (2a/\rho\pi h^2)\int_0^h (\rho\pi/a)z^2 \, dz = (2/h^2)(h^3/3) = 2h/3$. (Since the paraboloid is symmetrical about the z axis, $x_{cm} = y_{cm} = 0$.)

Problem

61. While standing on frictionless ice, you (mass 65.0 kg) toss a 4.50-kg rock with initial speed of 12.0 m/s. If the rock is 15.2 m from you when it lands, (a) at what angle did you toss it? (b) How fast are you moving?

Solution

If we assume that the ice surface is horizontal, then the x component of the center of mass of you and the rock stays at rest (at the origin) and the horizontal momentum is conserved:

$$x_{cm} = \frac{m_1 x_1 + m_2 x_2}{m_1 + m_2} = 0, \qquad m_1 v_{1x} + m_2 v_{2x} = 0.$$

(a) The angle of elevation is $\theta_0 = \frac{1}{2}\sin^{-1}(gx_1/v_0^2)$, from Equation 4-10, if we neglect the initial height of the toss. Since $x_1 - x_2 = 15.2$ m and $x_2 = -(4.5 \text{ kg}/65 \text{ kg})x_1$, we find $x_1 = (65/69.5)(15.2 \text{ m}) = 14.2$ m, and $\theta_0 = \frac{1}{2}\sin^{-1}(9.8 \text{ m/s}^2 \times 14.2 \text{ m}/(12.0 \text{ m/s})^2) = 37.7°$. (b) $v_{2x} = -(m_1/m_2)v_{1x} = -(4.5/65)(12.0 \text{ m/s})\cos 37.7° = -65.8$ cm/s.

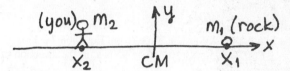

Problem 61 Solution.

Problem

65. (a) Derive an expression for the thrust of a jet aircraft engine. Moving through the air with speed v, the engine takes in air at the rate dM_{in}/dt. It uses the air to burn fuel with a fuel/air ratio f (that is, f kg of fuel burned for each kg of air), and ejects the exhaust gases at speed v_{ex} with respect to the engine. (b) Use your result to find the thrust of a JT-8D engine on a Boeing 727 jetliner, for which $v_{ex} = 1034$ ft/s and $dM_{in}/dt = 323$ lb/s, and which consumes 3760 lb of fuel per hour while cruising at 605 mi/h.

Solution

(a) The operation of a ramjet engine is similar to the rockets described in Section 10-2, except that air, originally at rest, is taken in and exhausted along with the gas from fuel combustion. If we neglect gravity (e.g., in horizontal flight) and air resistance, the momentum of the engine-fuel-air system is conserved. At the begining of a time interval dt, the engine and fuel have mass $m + dm_f$ and horizontal speed v, and the air is at rest, so $P_i = (m + dm_f)v$. After a time dt, the engine has mass m and speed $v + dv$, and combustion gas of mass dm_f and air of mass dM_{in} are ejected with speed $v - v_{ex}$ relative to the ground, so $P_f = m(v + dv) + (dm_f + dM_{in})(v - v_{ex})$. Equate P_i and P_f, expand the products, rearrange terms, and divide by dt: $0 = P_f - P_i = m(v + dv) + (dm_f + dM_{in})(v - v_{ex}) - (m + dm_f)v$, or $m(dv/dt) = v_{ex}(dm_f/dt) + (v_{ex} - v)(dM_{in}/dt) = $ thrust. (Note that if $dM_{in}/dt = 0$, we recover the rocket equation, since $dm_f = -dm$.) In terms of the fuel/air ratio, defined by $dm_f = f \, dM_{in}$, the thrust is $m(dv/dt) = [(1 + f)v_{ex} - v](dM_{in}/dt)$. (b) With the data given for the JT-8D engine, $v_{ex} = 1034$ ft/s, $v = 605$ mi/h $= 887.3$ ft/s, $(dM_{in}/dt)g = 323$ lb/s, $(g = 32.2 \text{ ft/s}^2)$, $f = (3760 \text{ lb}/3600 \text{ s})/(323 \text{ lb/s}) = 3.234 \times 10^{-3}$, the thrust is $[(1 + 3.234 \times 10^{-3})(1034 \text{ ft/s}) - 887.3 \text{ ft/s}] \times (323 \text{ lb/s})/(32.2 \text{ ft/s}^2) = 1504$ lb. (Note that dM_{in}/dt is the rate of air-mass intake, while 323 lb/s is the rate of air-weight intake.)

● CHAPTER 11 COLLISIONS

Section 11-1 Impulse and Collisions

Problem

1. What is the impulse associated with a 950-N force acting for 100 ms?

Solution

The impulse for a constant force (from Equation 11-1) is $I = \int F\, dt = (950\text{ N})(10^{-1}\text{ s}) = 95$ N·s, in the direction of the force.

Problem

5. A 240-g ball is moving with velocity $\mathbf{v}_i = 6.7\hat{\imath}$ m/s when it undergoes a collision lasting 52 ms. After the collision its velocity is $\mathbf{v}_f = -4.3\hat{\imath} + 3.1\hat{\jmath}$ m/s. Find (a) the impulse and (b) the average impulsive force associated with this collision.

Solution

(a) From Equation 11-1, $\mathbf{I} = \Delta\mathbf{p} = m(\mathbf{v}_f - \mathbf{v}_i) =$ $(0.24\text{ kg})(-4.3\hat{\imath} + 3.1\hat{\jmath} - 6.7\hat{\imath})(\text{m/s}) =$ $(-2.64\hat{\imath} + 0.744\hat{\jmath})$ N·s. (b) From Equation 11-2, $\mathbf{F}_{av} = \mathbf{I}/(0.052\text{ s}) = (-50.8\hat{\imath} + 14.3\hat{\jmath})$ N.

Problem

9. A 727 jetliner in level flight with a total mass of 8.6×10^4 kg encounters a downdraft lasting 1.3 s. During this time, the plane acquires a downward velocity component of 85 m/s. Find (a) the impulse and (b) the average impulsive force on the plane.

Solution

(a) The impulse equals the downward component of the change in the plane's momentum (Equation 11-1) $I = \Delta p_y = mv_y = (8.6\times10^4\text{ kg})(-85\text{ m/s}) = -7.31\times10^6$ N·s. (b) The average impulsive force (Equation 11-2) is $F_{av} = I/\Delta t = I/1.3\text{ s} = -5.62$ MN. (Here, minus means downward.)

Section 11-2 Collisions and the Conservation Laws

Problem

13. At the peak of its trajectory, a 1.0-kg projectile moving horizontally at 15 m/s collides with a 2.0-kg projectile at the peak of a vertical trajectory. If the collision takes 0.10 s, how good is the assumption that momentum is conserved during the collision? To find out, compare the change in momentum of the colliding system with the system's total momentum.

Solution

The total momentum before the collision is due to just the first projectile (the second is instantaneously at rest), so $\mathbf{P}_i = m_1\mathbf{v}_{1i} = (1\text{ kg})(15\text{ m/s})\hat{\imath} = (15\text{ kg·m/s})\hat{\imath}$. The change in total momentum is due to the external force (gravity), so $\Delta\mathbf{P} = \mathbf{F}_g\,\Delta t = (1\text{ kg} + 2\text{ kg})(9.8\text{ m/s}^2)(-\hat{\jmath})(0.1\text{ s}) = -(2.94\text{ kg·m/s})\hat{\jmath}$. This is almost 20% of $\mathbf{P}_i$ in magnitude. However, since $\Delta\mathbf{P}$ and $\mathbf{P}_i$ are perpendicular, $P_f = \sqrt{P_i^2 + \Delta P^2} \approx P_i(1 + \frac{1}{2}(\Delta P/P_i)^2 + \cdots)$, which differs from P_i by slightly less than 2% in magnitude, and about 11° in direction.

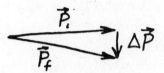

Problem 13 Solution.

Section 11-3 Inelastic Collisions

Problem

17. In a railroad switchyard, a 45-ton freight car is sent at 8.0 mi/h toward a 28-ton car that is moving in the same direction at 3.4 mi/h. (a) What is the speed of the pair after they couple together? (b) What fraction of the initial kinetic energy was lost in the collision?

Solution

(a) If we assume the switchyard track is straight and level, the collision is one-dimensional, totally inelastic, and Equation 11-4 applies: $v_f = [(45\text{ T})(8\text{ mi/h}) + (28\text{ T})(3.4\text{ mi/h})]/(45 + 28)\text{ T} = 6.24$ mi/h. (b) The initial and final kinetic energies are: $K_i = \frac{1}{2}[(45\text{ T})(8\text{ mi/h})^2 + (28\text{ T})(3.4\text{ mi/h})^2] = 1602$ T(mi/h)², $K_f = \frac{1}{2}(45 + 28)\text{ T }(6.24\text{ mi/h})^2 = 1419$ T(mi/h)². The fraction lost is $(K_i - K_f)/K_i = 11.4\%$. (Note: It was not necessary to change to standard units to answer this question.)

Problem

21. A mass m collides totally inelastically with a mass M initially at rest. Show that a fraction $M/(m + M)$ of the initial kinetic energy is lost in the collision.

Solution

In a totally inelastic collision, when one of the two bodies is initially at rest, $v_f = mv_i/(m + M)$ (see Equation 11-4). Then $K_f = \frac{1}{2}(m + M)v_f^2 = \frac{1}{2}m^2v_i^2/(m + M) = K_im/(m + M)$, so the fractional loss of kinetic energy is $(K_i - K_f)/K_i = 1 - K_f/K_i = 1 - m/(m + M) = M/(m + M)$.

Problem

25. A neutron (mass 1 u) strikes a deuteron (mass 2 u), and the two combine to form a tritium nucleus. If the neutron's initial velocity was $28\hat{\imath} + 17\hat{\jmath}$ Mm/s and if the tritium nucleus leaves the reaction with velocity $12\hat{\imath} + 20\hat{\jmath}$ Mm/s, what was the velocity of the deuteron?

Solution

Since the masses and two of the three velocities in Equation 11-3, which describes the conservation of momentum for the totally inelastic collision producing the tritium nucleus, are known, the initial velocity of the deuteron can easily be found:

$$\mathbf{v}_d = \frac{m_t\mathbf{v}_t - m_n\mathbf{v}_n}{m_d} = \frac{3\text{ u}(12\hat{\imath} + 20\hat{\jmath}) - 1\text{ u}(28\hat{\imath} + 17\hat{\jmath})}{2\text{ u}}\frac{\text{Mm}}{\text{s}}$$

$$= (4\hat{\imath} + 21.5\hat{\jmath})\text{ Mm/s}$$

Problem

29. A 400-mg popcorn kernel is skittering across a nonstick frying pan at 8.2 cm/s when it pops and breaks into two equal-mass pieces. If one piece ends up at rest, how much energy was released in the popping?

Solution

In order to conserve momentum, the piece which is not at rest must acquire a velocity equal to twice that of the original kernel, $m\mathbf{v}_i = (\frac{1}{2}m)\mathbf{v}_{1f}$. Therefore, the final energy is twice the initial energy, $K_f = \frac{1}{2}(\frac{1}{2}m)v_{1f}^2 = \frac{1}{2}(\frac{1}{2}m)(2v_i)^2 = mv_i^2 = 2K_i$. The energy released is $K_f - K_i = K_i = \frac{1}{2}(4\times10^{-4}\text{ kg}) \times (0.082\text{ m/s})^2 = 1.34\times10^{-6}$ J.

Section 11-4 Elastic Collisions

Problem

33. While playing ball in the street, a child accidentally tosses a ball at 18 m/s toward the front of a car moving toward him at 14 m/s. What is the speed of the ball after it rebounds elastically from the car?

Solution

In a head-on elastic collision, the relative velocity of separation is equal to the negative of the relative velocity of approach (see Equation 11-8). If $m_2 \gg m_1$ (a car versus a ball), then $v_{2f} \approx v_{2i} = 14$ m/s, and $v_{1f} = v_{2f} + v_{2i} - v_{1i} = 14$ m/s $+ 14$ m/s $- (-18$ m/s$) = 46$ m/s. (We chose positive velocities in the direction of the car.)

Problem

37. Two objects, one initially at rest, undergo a one-dimensional elastic collision. If half the kinetic energy of the initially moving object is transferred to the other object, what is the ratio of their masses?

Solution

If one sets $v_{2i} = 0$ in Equations 11-9a and b, one obtains $v_{1f} = (m_1 - m_2)v_{1i}/(m_1 + m_2)$ and $v_{2f} = 2m_1v_{1i}/(m_1 + m_2)$. (This describes a one-dimensional elastic collision between two objects, one initially at rest.) If half the kinetic energy of the first object is transferred to the second, $\frac{1}{2}K_{1i} = \frac{1}{4}m_1v_{1i}^2 = K_{2f} = \frac{1}{2}m_2[2m_1v_{1i}/(m_1 + m_2)]^2$, or $8m_1m_2 = (m_1 + m_2)^2$. The resulting quadratic equation, $m_1^2 - 6m_1m_2 + m_2^2 = 0$, has two solutions, $m_1 = (3 \pm \sqrt{8})m_2 = 5.83m_2$ or $(5.83)^{-1}m_2$. Since the quadratic is symmetric in m_1 and m_2, one solution equals the other with m_1 and m_2 interchanged. Thus, one object is 5.83 times more massive than the other.

Problem

41. Rework Example 11-5 using a frame of reference in which the carbon nucleus is initially at rest.

Solution

We use Equation 3-10 to express velocities in the rest system of the carbon nucleus. Thus, $v_{2i}' = 0$, $v_{1i}' = (460 - 220)$ km/s $= 240$ km/s, and $v_{2f}' = v_{2f} - v_{2i} = 120$ km/s. From Equation 11-9b, $m_1 = m_2v_{2f}'/(2v_{1i}' - v_{2f}') = (12\text{ u})(120)/(2 \times 240 - 120) = 4$ u (as in Example 11-5). From Equation 11-9a, $v_{1f}' = [(4 - 12)/(4 + 12)]v_{1i}' = -\frac{1}{2}v_{1i}' = -120$ km/s (also consistent with Example 11-5, since in the lab system, $v_{1f} = v_{1f}' + 220$ km/s $= 100$ km/s).

Problem

45. Two pendulums of equal length $\ell = 50$ cm are suspended from the same point. The pendulum bobs are steel spheres with masses of 140 and 390 g. The more massive bob is drawn back to make a 15° angle with the vertical (Fig. 11-18). When it is released the bobs collide elastically. What is the maximum angle made by the less massive pendulum?

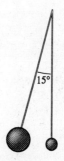

FIGURE 11-18 Problem 45.

Solution

The speed of the larger bob before the collision is (from conservation of energy) $v_{1i} = \sqrt{2g\ell(1 - \cos 15°)}$, (see Example 8-6), while for the smaller bob, $v_{2i} = 0$. After the collision, the smaller bob will reach a maximum angle given by (again from conservation of energy) $v_{2f}^2 = 2g\ell(1 - \cos\theta_2)$, where v_{2f} can be found from Equation 11-9b. Therefore,

$$v_{2f}^2 = \left(\frac{2m_1}{m_1 + m_2}\right)^2 v_{1i}^2 = \left(\frac{2(390)}{530}\right)^2 2g\ell(1 - \cos 15°)$$
$$= 2g\ell(1 - \cos\theta_2), \quad \text{or}$$
$$\cos\theta_2 = 0.926 \quad \text{and} \quad \theta_2 = 22.2°.$$

Problem

49. A tennis ball moving at 18 m/s strikes the 45° hatchback of a car moving away at 12 m/s, as shown in Fig. 11-20. Both speeds are given with respect to the ground. What is the velocity of the ball with respect to the ground after it rebounds elastically from the car? *Hint:* Work in the frame of reference of the car; then transform to the ground frame.

Solution

Let the x-y frame be fixed to the ground, and the x'-y' frame fixed to the car. Then $\mathbf{v}' = \mathbf{v} - \mathbf{v}_{car}$ is the transformation between the two frames, where $\mathbf{v}_{car} = (12 \text{ m/s})\hat{\mathbf{i}}$. In the primed frame, the initial velocity of the tennis ball is $\mathbf{v}'_{1i} = (6 \text{ m/s})\hat{\mathbf{i}}$, and, of course, $\mathbf{v}'_{2i} = 0$ for the car. Since the mass of the car is so much larger than the mass of the ball, in an elastic collision, the speed of the ball is unchanged, and it rebounds at a 45° angle to the hatchback, or $\mathbf{v}'_{1f} = (6 \text{ m/s})\hat{\mathbf{j}}$.
Thus, the ball's final velocity relative to the ground is $\mathbf{v}_{1f} = \mathbf{v}'_{1f} + \mathbf{v}_{car} = (6 \text{ m/s})\hat{\mathbf{j}} + (12 \text{ m/s})\hat{\mathbf{i}}$, or $v_{1f} = \sqrt{6^2 + 12^2}$ m/s $= 13.4$ m/s, at an angle (with the x axis) of $\theta = \tan^{-1}(6/12) = 26.6°$.

FIGURE 11-20 Problem 49 Solution.

Paired Problems

Problem

53. A 590-g basketball is moving at 9.2 m/s when it hits a backboard at 45°. It bounces off at a 45° angle, still moving at 9.2 m/s. If the ball is in contact with the backboard for 22 ms, find the average impulsive force on the ball.

Solution

The normal force of the backboard acts to reverse the perpendicular component of the basketball's momentum. From

Equation 11-2, $F_{av} \Delta t = |\mathbf{p}_f - \mathbf{p}_i| = \sqrt{2}|\mathbf{p}_i|$, since $|\mathbf{p}_f| = |\mathbf{p}_i|$ and their directions are perpendicular. Therefore $F_{av} = \sqrt{2}(0.590 \text{ kg})(9.2 \text{ m/s})/(0.022 \text{ s}) = 349$ N (away from the backboard).

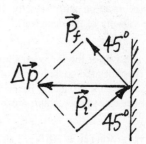

Problem 53 Solution.

Problem

57. In Fig. 11-22 a truck slams into a parked car, and the resulting one-dimensional collision is elastic and transfers 41% of the truck's kinetic energy to the car. Compare the masses of the two vehicles.

Solution

In a two-body one-dimensional elastic collision with one body initially at rest (say $v_{2i} = 0$ for the car), Equations 11-9a and b give $v_{1f}/v_{2f} = (m_1 - m_2)/2m_1$. But 41% of the initial kinetic energy is transfered to m_2 and 59% to m_1, so $m_1 v_{1f}^2 / m_2 v_{2f}^2 = 59/41$. Combining these equations, one obtains $(m_1 - m_2)^2 / 4m_1 m_2 = 59/41$ or $(m_2/m_1)^2 - 2(100 + 59)(41)^{-1}(m_2/m_1) + 1 = 0$. (We wrote the quadratic in this way so that $(100 + 59)/41 = (K_i + K_{1f})/K_{2f}$ can be used in the next problem.) The quadratic formula yields $(m_2/m_1) = (159/41) \pm \sqrt{(159/41)^2 - 1} = 7.62$ or $0.131 = (7.62)^{-1}$. (The two solutions are reciprocals, since the original quadratic was symmetric in m_1 and m_2, see solution to Problem 37.) It is more likely that the truck has the greater mass, so $m_1 = 7.62m_2$ is the best choice here, but the sign of v_{1f}, if known, would determine whether m_1 was greater or less than m_2. Finally, we note that the above solution can be written in various equivalent forms, one of the simplest being $m_2/m_1 = [(\sqrt{K_i} \pm \sqrt{K_{1f}})/\sqrt{K_{2f}}]^2$, where $K_i = K_{1f} + K_{2f}$ represents the sharing of kinetic energy. This follows from the conservation of momentum, $p_{1i} = p_{1f} + p_{2f}$, written in terms of kinetic energies, $\sqrt{2m_1 K_i} = \pm\sqrt{2m_1 K_{1f}} + \sqrt{2m_2 K_{2f}}$, after solving for the mass ratio $\sqrt{m_2/m_1}$.

FIGURE 11-22 Problem 57.

Problem

61. A billiard ball moving at 1.8 m/s strikes an identical ball initially at rest as illustrated in Fig. 11-23. They undergo an elastic collision and the first ball moves off at 23° counterclockwise from its original direction. Find the final speeds of both balls and the direction of the second ball's motion.

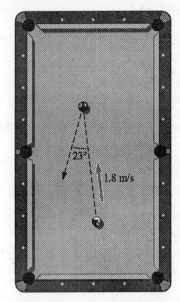

FIGURE 11-23   Problem 61.

Solution

The kinematics of this collision is the same as that in Example 11-6. Therefore $v_{1f} = v_{1i} \cos\theta_1 = (1.8 \text{ m/s})\cos 23° = 1.66$ m/s, $v_{2f} = \sqrt{v_{1i}^2 - v_{1f}^2} = 0.703$ m/s, and $\theta_2 = \sin^{-1}(-v_{1f} \sin\theta_1/v_{2f}) = -67.0°$. (This last result could have been anticipated from the statement of Problem 43.)

Supplementary Problems

Problem

65. A 1400-kg car moving at 75 km/h runs into a 1200-kg car moving in the same direction at 50 km/h (Fig. 11-24). The two cars lock together and both drivers immediately slam on their brakes. If the cars come to rest in a distance of 18 m, what is the coefficient of friction?

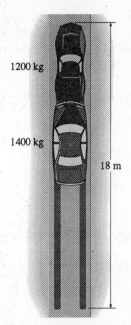

FIGURE 11-24   Problem 65.

Solution

This is a one-dimensional, totally inelastic collision, so the final velocity (positive in the initial direction of motion) is

$$v_f = \frac{(1400 \text{ kg})(75 \text{ km/h}) + (1200 \text{ kg})(50 \text{ km/h})}{1400 \text{ kg} + 1200 \text{ kg}}$$
$$= 63.5 \text{ km/h} = 17.6 \text{ m/s}.$$

On a horizontal road (normal force equals weight and no change in gravitational potential energy), the work-energy theorem implies $W_{\text{nc}} = -\mu_k mgx = \Delta K = -\frac{1}{2}mv_f^2$, therefore

$$\mu_k = \frac{v_f^2}{2gx} = \frac{(17.6 \text{ m/s})^2}{2(9.8 \text{ m/s}^2)(18 \text{ m})} = 0.88.$$

Problem

69. How many head-on collisions must a neutron (mass 1.0087 u) make with stationary ^{12}C nuclei (mass 11.9934 u) in a graphite-moderated nuclear reactor in order to lose as much energy as it would in a water-moderated reactor where it collides with a single proton (mass 1.0073 u)? *Hint*: See hint in Problem 66.

Solution

After n head-on, elastic collisions between particles of mass m_1 and m_2, with m_2 initially at rest, the fraction of the initial kinetic energy retained by m_1 is $[(m_1 - m_2)/(m_1 + m_2)]^{2n}$.

(See Problem 66. For one collision, the fraction follows from Equation 11-9a; for n collisions, the factor is raised to the nth power.) For n collisions of a neutron with stationary C^{12} nuclei to have the same energy loss as one collision with a stationary proton (masses as given):

$$\left(\frac{m_n - m_p}{m_n + m_p}\right)^2 = \left(\frac{1.0087 - 1.0073}{1.0087 + 1.0073}\right)^2 = \left(\frac{m_{C^{12}} - m_n}{m_{C^{12}} + m_n}\right)^{2n}$$

$$= \left(\frac{11.9934 - 1.0087}{11.9934 + 1.0087}\right)^{2n}.$$

Thus, n is the smallest integer greater than or equal to $\ln(2.0160/.0014)/\ln(13.0021/10.9847) = 43.1$, or $n = 44$.

Problem

73. A 1200-kg car moving at 25 km/h undergoes a one-dimensional collision with an 1800-kg car initially at rest. The collision is neither elastic nor totally inelastic; the kinetic energy lost is 5800 J. Find the speeds of both cars after the collision.

Solution

Take positive velocities in the initial direction of motion of the 1200 kg car, so that $m_1/m_2 = 2/3$, $v_{1i} = 25$ km/h and $v_{2i} = 0$. The conservation of momentum, $m_1 v_{1i} = m_1 v_{1f} + m_2 v_{2f}$, and the given change in kinetic energy, $\Delta K = -5800$ J $= \frac{1}{2} m_1 v_{1f}^2 + \frac{1}{2} m_2 v_{2f}^2 - \frac{1}{2} m_1 v_{1i}^2$, provide two equations from which the unknown final velocities can be determined. Solve for v_{2f} in the first equation, $v_{2f} = (m_1/m_2)(v_{1i} - v_{1f})$, and substitute into the second equation, $m_1 v_{1f}^2 + m_2 (m_1/m_2)^2 (v_{1i} - v_{1f})^2 - m_1 v_{1i}^2 - 2\,\Delta K = 0$. This quadratic in v_{1f} can be written as $v_{1f}^2 - 2bv_{1f} - c = 0$, where $b = m_1 v_{1i}/(m_1 + m_2) = (2/5)(25$ km/h$) = 10$ km/h and $c = [(m_2 - m_1)v_{1i}^2 + 2m_2 \,\Delta K/m_1](m_1 + m_2)^{-1} = (1/5)[(25$ km/h$)^2 - (5800$ J$/200$ kg$)(3.6$ km/h$)^2($m/s$)^{-2}] = 49.83$ (km/h)^2. The roots are $v_{1f} = b \pm \sqrt{b^2 + c} = (10 \pm 12.24)$ km/h, so that $v_{2f} = (2/3)(15 \mp 12.24)$ km/h. For a collision between cars, the first car cannot penetrate through the car it strikes, therefore $v_{1f} < v_{2f}$, and one of the roots is unphysical. The other solution, $v_{1f} = -2.24$ km/h and $v_{2f} = 18.2$ km/h, is the appropriate one for this problem.

●

● CHAPTER 12   ROTATIONAL MOTION

Section 12-1   Angular Speed and Acceleration

Problem

1. Determine the angular speed, in rad/s, of (a) Earth about its axis; (b) the minute hand of a clock; (c) the hour hand of a clock; (d) an egg beater turning at 300 rpm.

Solution

The angular speed is $\omega = \Delta\theta/\Delta t$.
(a) $\omega_E = 1$ rev$/1$ d $= 2\pi/86,400$ s $= 7.27 \times 10^{-5}$ s^{-1}.
(b) $\omega_{min} = 1$ rev$/1$ h $= 2\pi/3600$ s $= 1.75 \times 10^{-3}$ s^{-1}.
(c) $\omega_{hr} = 1$ rev$/12$ h $= \frac{1}{12}\omega_{min} = 1.45 \times 10^{-4}$ s^{-1}.
(d) $\omega = 300$ rev/min $= 300 \times 2\pi/60$ s $= 31.4$ s^{-1}.
(Note: Radians are a dimensionless angular measure, i.e., pure numbers, therefore angular speed can be expressed in units of inverse seconds.)

Problem

5. A wheel is turned through 2.0 revolutions while being accelerated from rest at 18 rpm/s. (a) What is the final angular speed? (b) How long does it take to turn the 2.0 revolutions?

Solution

For constant angular acceleration, (a) Equation 12-11 gives $\omega_f = \sqrt{\omega_0^2 + 2\alpha(\theta_f - \theta_0)} = \sqrt{0 + 2(18 \text{ rev} \times 60/\text{min}^2)(2 \text{ rev})} = 65.7$ rpm, and

(b) Equation 12-10 gives $\theta_f - \theta_0 = 0 + \frac{1}{2}\alpha t^2$, or $t = \sqrt{2(2 \text{ rev})/(18 \text{ rev}/60 \text{ s}^2)} = 3.65$ s.

Problem

9. The rotation rate of a compact disc varies from about 200 rpm to 500 rpm (see Problem 7). If the disc plays for 74 min, what is its average angular acceleration in (a) rpm/s and (b) rad/s^2?

Solution

From Equation 12-5 (before the limit is taken), $\alpha_{av} = \Delta\omega/\Delta t = (500 - 200)$ rpm$/(74 \times 60$ s$) = 6.76 \times 10^{-2} \times$ rpm/s $= 7.08 \times 10^{-3}$ s^{-2}. (Recall that 1 rpm $= 2\pi/60$ s.)

Problem

13. A piece of machinery is spinning at 680 rpm. When a brake is applied, its rotation rate drops to 440 rpm while it turns through 180 revolutions. What is the magnitude of the angular deceleration?

Solution

If the angular acceleration is constant, Equation 12-11 gives $\alpha = [(440\pi/30 \text{ s})^2 - (680\pi/30 \text{ s})^2]/2(180 \times 2\pi) = -1.30$ s^{-2}. (A negative acceleration is a positive deceleration.)

Section 12-2 Torque

Problem

17. A torque of 110 N·m is required to start a revolving door rotating. If a child can push with a maximum force of 90 N, how far from the door's rotation axis must she apply this force?

Solution

If the force is applied perpendicular to the door, the radial distance should be $r = \tau/F = 110$ N·m/90 N $= 1.22$ m from the axis. (See Equation 12-12 with $\theta = 90°$.)

Problem

21. A pulley 12 cm in diameter is free to rotate about a horizontal axle. A 220-g mass and a 470-g mass are tied to either end of a massless string, and the string is hung over the pulley. If the string does not slip, what torque must be applied to keep the pulley from rotating?

Solution

If the pulley and string are not moving (no rotation and no slipping) the net torque on the pulley is zero, and the tensions in the string on either side are equal to the weights tied on either end. The tensions are perpendicular to the radii to each point of application, so the torques due to the tensions have magnitude $TR = mgR$, but are in opposite directions. Thus, $0 = \tau_{app} - m_1gR + m_2gR$, where we chose positive torques in the direction of the applied torque, which is opposite to the torque produced by the greater mass. Numerically, $\tau_{app} = (m_1 - m_2)gR = (0.47 - 0.22)$ kg$(9.8$ m/s²$)(0.06$ m$) = 0.147$ N·m. (Note: τ_{app} equals the torque which would be produced by balancing the pulley with 250 g added to the side with lesser mass.)

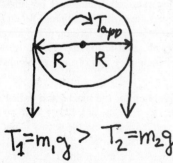

Problem 21 Solution.

Section 12-3 Rotational Inertia and the Analog of Newton's Law

Problem

25. What is the radius of a solid cylinder of mass 6.2 kg and rotational inertia 0.15 kg·m² about its axis?

Solution

The rotational inertia of a solid cylinder or disk about its axis is $I = \frac{1}{2}MR^2$ (see Example 12-8 or Table 12-2), so $R = \sqrt{2I/M} = \sqrt{2(0.15 \text{ kg·m}^2)/(6.2 \text{ kg})} = 22.0$ cm.

Problem

29. The wheel shown in Fig. 12-36 consists of a 120-kg outer rim 55 cm in radius, connected to the center by five 18-kg spokes. Treating the rim as a thin ring and the spokes as thin rods, determine the rotational inertia of the wheel.

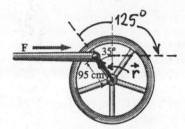

12-36 Problem 29 Solution.

Solution

The sum of the rotational inertia's of the parts is $I = M_{rim}R_{rim}^2 + 5 \times \frac{1}{3}M_{rod}\ell_{rod}^2$, where $R_{rim} = \ell_{rod}$. Thus, $I = (120 \text{ kg} + 5 \times 18 \text{ kg}/3)(0.55 \text{ m})^2 = 45.4$ kg·m² (about an axis perpendicular to the wheel through its center).

Problem

33. (a) Estimate the rotational inertia of the Earth, assuming it to be a uniform solid sphere. (b) What torque would have to be applied to Earth to cause the length of the day to change by one second every century?

Solution

(a) For a uniform solid sphere, and an axis through the center, $I_E = \frac{2}{5}M_ER_E^2 = \frac{2}{5}(5.97 \times 10^{24} \text{ kg})(6.37 \times 10^6 \text{ m})^2 = 9.69 \times 10^{37}$ kg·m². (The Earth has a core of denser material so its actual rotational inertia is less than this.) (b) The angular speed of rotation of the Earth is $\omega = 2\pi/T$, where the period $T = 1$ d $= 86,400$ s. If the period were to change by 1 s per century, $|dT/dt| = 1$ s/$(100 \times 3.16 \times 10^7$ s$)$, this would correspond to an angular acceleration of $\alpha = d\omega/dt = d(2\pi/T)/dt = -(2\pi/T^2)(dT/dt)$. Therefore, a torque of magnitude $I|\alpha| = I(2\pi/T^2)|dT/dt| = (9.69 \times 10^{37}$ kg·m²$)(2\pi/(86,400$ s$)^2)(1/3.16 \times 10^9) = 2.58 \times 10^{19}$ N·m would be required. (Such a torque is actually generated by tidal friction between the moon and the Earth.)

Problem

37. Proof of the parallel-axis theorem: Fig. 12-40 shows an object of mass M with axes through the center of mass and through an arbitrary point A. Both axes are perpendicular to the page. Let $\mathbf{h}$ be a vector from the axis through the center of mass to the axis through point A, $\mathbf{r}_{cm}$ a vector from the axis through the CM to an arbitrary mass element dm, and $\mathbf{r}$ a vector from the axis through point A to the mass element dm, as shown. (a) Use the law of cosines to show that

$$r^2 = r_{cm}^2 + h^2 - 2\mathbf{h}\cdot\mathbf{r}_{cm}.$$

(b) Use this result in the expression $I = \int r^2\, dm$ to calculate the rotational inertia of the object about the axis through A. Each of the three terms in your expression for r^2 leads to a separate integral. Identify one as the rotational inertia about the CM, another as the quantity Mh^2, and show that the third is zero because it involves the position of the center of mass relative to itself. Your result is then a statement of the parallel-axis theorem.

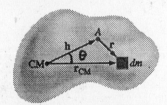

FIGURE 12-40 Problem 37.

Solution

(a) Since $\mathbf{r} = \mathbf{r}_{cm} - \mathbf{h}$, $r^2 = (\mathbf{r}_{cm} - \mathbf{h})\cdot(\mathbf{r}_{cm} - \mathbf{h}) = r_{cm}^2 + h^2 - 2\mathbf{r}_{cm}\cdot\mathbf{h}$ (which is equivalent to the law of cosines; $r^2 = r_{cm}^2 + h^2 - 2r_{cm}h\cos\theta$, where θ is the angle between the sides r_{cm} and h in the triangle in Fig. 12-40).
(b) $I_A = \int r^2\, dm = \int r_{cm}^2\, dm + h^2 \int dm - 2\mathbf{h}\cdot\int \mathbf{r}_{cm}\, dm$. The first integral is I_{cm}, the second is Mh^2, and the third is zero (from the definition of the center of mass—in the CM frame, the origin *is* the CM, so $0 = \int \mathbf{r}_{cm}\, dm/\int dm$).

Problem

41. The crane shown in Fig. 12-42 contains a hollow drum of mass 150 kg and radius 0.80 m that is driven by an engine to wind up a cable. The cable passes over a solid cylindrical 30-kg pulley 0.30 m in radius to lift a 2000-N weight. How much torque must the engine apply to the drum to lift the weight with an acceleration of 1.0 m/s²? Neglect the rotational inertia of the engine and the mass of the cable.

Solution

The equations of motion of the drum, pulley, and weight are: $\tau - T_1 R_1 = I_1\alpha_1$, $(T_1 - T_2)R_2 = I_2\alpha_2$, and $T_2 - mg = ma$, where subscripts 1 and 2 refer to the drum and pulley, respectively, and τ is the engine's torque. We assume $I_1 = M_1 R_1^2$ and $I_2 = \frac{1}{2}M_2 R_2^2$. If the cable does not slip or stretch, $a = \alpha_1 R_1 = \alpha_2 R_2$. When all these results are combined, we may solve for the torque: $\tau = (M_1 a + T_1)R_1$, $T_1 = \frac{1}{2}M_2 a + T_2$, $T_2 = m(g + a)$, therefore $\tau = [M_1 + \frac{1}{2}M_2 + m(1 + g/a)]R_1 a$. Numerically,

$$\tau = \left[150\text{ kg} + \frac{30\text{ kg}}{2} + \frac{2000\text{ kg}}{9.8}(1 + 9.8)\right](0.8\text{ m})\left(1\frac{\text{m}}{\text{s}^2}\right)$$

$$= 1.90\text{ kN·m}.$$

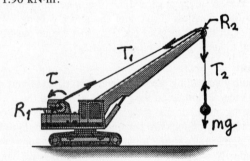

FIGURE 12-42 Problem 41 Solution.

Problem

45. Two blocks of mass m_1 and m_2 are connected by a massless string that passes over a solid cylindrical pulley, as shown in Fig. 12-44. The surface under the block m_2 is frictionless, and the pulley rides on frictionless bearings. The string passes over the pulley without slipping. When released, the masses accelerate at $\frac{1}{3}g$. The tension in the lower half of the string is 2.7 N and that in the upper half, 1.9 N. What are the masses of the pulley and of the two blocks?

Solution

The equations of motion for the blocks and pulley are: $m_1 g - T_1 = m_1 a_1$, $T_2 = m_2 a_2$, and $(T_1 - T_2)R = I\alpha$. (The equation for the vertical forces on block two is not needed, since the horizontal surface on which it slides is frictionless.) Here, R is the radius of the pulley and $I = \frac{1}{2}MR^2$. The given constraints require that $a_1 = a_2$ (inextensible string) $= \alpha R$ (no slipping) $= \frac{1}{3}g$. Substituting these into the equations of motion, we find: $m_1 = 3T_1/2g = 413$ g, $m_2 = 3T_2/g = 582$ g, and $M = 6(T_1 - T_2)/g = 490$ g. ("g" is also the abbreviation for gram.)

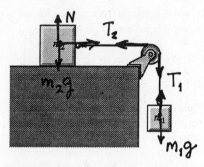

FIGURE 12-44 Problem 45 Solution.

Section 12-4   Rotational Energy

Problem

49. A 150-g baseball is pitched at 33 m/s, spinning at 42 rad/s. What fraction of its kinetic energy is rotational? Treat the baseball as a uniform solid sphere of radius 3.7 cm.

Solution

From Equation 12-31, $K_{rot}/K_{tot} = I_{cm}\omega^2/(Mv^2 + I_{cm}\omega^2)$. For a solid sphere, $I_{cm} = 0.4MR^2$, so the ratio becomes $0.4(R\omega)^2/[v^2 + 0.4(R\omega)^2] = 0.4(0.037 \times 42)^2/[(33)^2 + 0.4(0.037 \times 42)^2] = 8.86 \times 10^{-4} \approx 0.089\%$.

Section 12-5   Rolling Motion

Problem

53. A basketball rolls down a 30° incline. If it starts from rest, what is its speed after it's gone 8.4 m along the incline? (The basketball is hollow.)

Solution

The ball rolls through a vertical height $h = \ell \sin\theta$, where ℓ is the distance along the incline, which makes an angle θ with the horizontal. For a hollow ball, the rotational inertia factor $\alpha = I_{cm}/MR^2$ is $\frac{2}{3}$ (see Table 12-2), so we can use the result of Problem 55 to find $v_{cm} = \sqrt{2gh/(1 + \alpha)} = \sqrt{2(9.8 \text{ m/s})(8.4 \text{ m}) \sin 30°/(1 + \frac{2}{3})} = 7.03$ m/s.

Problem

57. The rotational kinetic energy of a rolling automobile wheel is 40% of its translational kinetic energy. The wheel is then redesigned to have 10% lower rotational inertia and 20% less mass, while keeping its radius the same. By what percentage does its total kinetic energy at a given speed decrease?

Solution

As in Example 12-14, the kinetic energy of the wheel, rolling without slipping, can be written as $K = \frac{1}{2}Mv_{cm}^2 + \frac{1}{2}I_{cm}(v_{cm}/R)^2 = (1 + I_{cm}/MR^2)(\frac{1}{2}Mv_{cm}^2)$. Initially, $I_{cm}/MR^2 = 40\%$, and after redesign, $I'_{cm}/M'R^2 = 0.9I_{cm}/0.8MR^2 =$

$(9/8)\,40\% = 45\%$. The fractional decrease in kinetic energy is

$$\frac{K - K'}{K} = 1 - \frac{(1 + I'_{cm}/M'R^2)M'}{(1 + I_{cm}/MR^2)M} = 1 - \frac{(1.45)0.8M}{(1.40)M}$$
$$= 17.1\%.$$

Paired Problems

Problem

61. A merry-go-round starts from rest and accelerates with angular acceleration 0.010 rad/s² for 14 s. (a) How many revolutions does it make during this time? (b) What is its average angular speed during the spin-up time?

Solution

The kinematics of motion with constant angular acceleration are summarized in Table 12-1, where the initial angular velocity at $t = 0$ is zero if the merry-go-round starts at rest. (a) Equation 12-10 gives the angular displacement: $\Delta\theta = \theta - \theta_0 = \frac{1}{2}\alpha t^2 = \frac{1}{2}(0.01 \text{ s}^2)(14 \text{ s})^2 = (0.98 \text{ radians}) \times (1 \text{ rev}/2\pi) = 0.156 \text{ rev} = 56.1°$. (b) Equations 12-8 and 9 give the average angular speed: $\omega_{av} = \frac{1}{2}\omega_f = \frac{1}{2}\alpha t = \frac{1}{2}(0.01 \text{ s}^{-2})(14 \text{ s}) = 0.07 \text{ s}^{-1} = 0.668$ rpm. (Note that the average angular speed is always equal to $\Delta\theta/\Delta t$, in this case $\frac{1}{2}\alpha t^2/t$, but that Equations 12-8 to 11 apply only to constant α.)

Problem

65. A 50-kg mass is tied to a massless rope wrapped around a solid cylindrical drum. The drum is mounted on a frictionless horizontal axle. When the mass is released, it falls with acceleration $a = 3.7 \text{ m/s}^2$. Find (a) the tension in the rope and (b) the mass of the drum.

Solution

The situation is similar to Example 12-11. (a) The equation of motion (Newton's second law) of the falling mass is $mg - T = ma$ (positive "a" downward), so $T = m(g - a) = (50 \text{ kg})(9.8 - 3.7) \text{ m/s}^2 = 305$ N. (b) The equation of motion of the rotating drum (rotational analog of Newton's second law) is $\tau = RT = I\alpha = (\frac{1}{2}MR^2)(a/R)$, so $M = 2T/a = 2(350 \text{ N})/(3.7 \text{ m/s}^2) = 165 \text{ kg}$ (since $\tau = RT$ and $a = \alpha R$ as explained in Example 12-11).

Supplementary Problems

Problem

69. A solid marble starts from rest and rolls without slipping on the loop-the-loop track shown in Fig. 12-49. Find the minimum starting height of the marble from which it will remain on the track through the loop. Assume the marble radius is small compared with R.

Solution

The CM of the marble travels in a circle of radius $R - r$ inside the loop-the-loop, so at the top, $mg + N = mv^2/(R - r)$. To remain in contact with the track, $N \geq 0$, or $v^2 \geq g(R - r)$. If we assume that energy is conserved between points A (the start) and B (the top of the loop), and use $K_A = 0$ and $K_B = (1 + \frac{2}{5})\frac{1}{2}mv_B^2$, we find $U_A + K_A = mg(h + r) = U_B + K_B = mg(2R - r) + \frac{7}{10}mv_B^2$, or $h = 2(R - r) + \frac{7}{10}(v_B^2/g) \geq 2(R - r) + \frac{7}{10}(R - r) = 2.7(R - r)$.

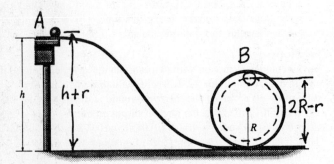

FIGURE 12-49   Problem 69 Solution.

Problem

73. A thin, uniform vane of mass M is in the shape of a right triangle, as shown in Fig. 12-50. Find the rotational inertia about a vertical axis through its apex, as shown in the figure. Express your answer in terms of the triangle's base width b and its mass M.

Solution

The integral in Equation 12-16 for the rotational inertia of a 2-dimensional object is really a double integral over the area of the object. If we choose the x and y axes to coincide with the base of the triangular vane and the axis of rotation, respectively, then the rotational inertia of an infinitessimal mass element $\rho \, dx \, dy$, located at position (x, y), is $x^2\rho \, dx \, dy$, since x is the perpendicular distance to the y axis. ρ is the planar density of the triangle, in this case assumed to be uniform and equal to $M/(\frac{1}{2}bh)$. Therefore Equation 12-16 becomes $I_y = (2M/bh)\iint x^2 \, dx \, dy$, where the integration is over the area of the triangle. If we choose to integrate over y first, the area can be specified by letting y go from 0 to xh/b and x from 0 to b. Then

$$I_y = \left(\frac{2M}{bh}\right)\int_0^b x^2 \, dx \int_0^{xh/b} dy = \left(\frac{2M}{bh}\right)\int_0^b \left(\frac{h}{b}\right)x^3 \, dx$$

$$= \left(\frac{2M}{b^2}\right)\frac{b^4}{4} = \frac{1}{2}Mb^2.$$

Alternatively, if we integrate over x first, the limits of integration are yb/h to b for x, and 0 to h for y, so

$$I_y = \left(\frac{2M}{bh}\right)\int_0^h dy \int_{yb/h}^b x^2 \, dx = \left(\frac{2M}{bh}\right)\int_0^h dy \frac{1}{3}\left[b^3 - \left(\frac{by}{h}\right)^3\right]$$

$$= \left(\frac{2M}{bh}\right)\frac{1}{3}\left[b^3h - \left(\frac{b}{h}\right)^3\frac{h^4}{4}\right] = \frac{1}{2}Mb^2.$$

(Note: The first approach amounts to choosing 1-dimensional mass elements as vertical strips of mass $dm = \rho y \, dx = (2M/bh)(xh/b) \, dx = (2M/b^2)x \, dx$ and rotational inertia $dI_y = x^2 \, dm = (2M/b^2)x^3 \, dx$. Then a 1-dimensional integration from $x = 0$ to $x = b$ gives $I_y = \frac{1}{2}Mb^2$. The second approach corresponds to choosing horizontal strips of length $\ell = b - by/h$ and mass $dm = \rho\ell \, dy$. The rotational inertia of these strips must be calculated from the parallel axis theorem, $dI_y = (x_{cm}^2 + \frac{1}{12}\ell^2) \, dm$, where $x_{cm} = (by/h) + \ell/2$. A little algebra transforms this to the second integrand above, or it may be integrated directly over y, from 0 to h, to get the same $I_y = \frac{1}{2}Mb^2$.)

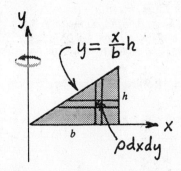

FIGURE 12-50   Problem 73 Solution.

Problem

77. A thin solid rod of length ℓ and mass M is free to pivot about one end. If it makes an angle ϕ with the horizontal, find the torque due to gravity about the pivot point. You'll need to integrate the torques on the individual mass elements comprising the rod.

Solution

Let ℓ' be the distance along the rod from the pivot at O. The force of gravity on each mass element of rod dm makes an angle of $90° - \phi$ with the displacement along ℓ', so the torque of gravity on this element has magnitude

$d\tau_g = \ell'g\,dm\sin(90° - \phi) = \ell'g\cos\phi\,dm$ in a clockwise direction. The total torque of gravity is the sum of these over the whole rod, or $\tau_g = \int_0^\ell \ell'g\cos\phi\,dm$, clockwise. The mass element is the linear density (assumed uniform) times the length element, $dm = (M/\ell)d\ell'$, so $\tau_g = (Mg\cos\phi/\ell)\int_0^\ell \ell'd\ell' = \frac{1}{2}Mg\ell\cos\phi$, clockwise. (Note: The center of gravity of an object is defined as the point at which the total weight of the object would produce the same torque, i.e., $\tau_g = Mg\ell_{CG}\sin(90° - \phi_{cm})$. For the rod, all parts are at the same angle ϕ, so $\ell_{CG} = \int_0^\ell \ell'd(mg)/Mg$, and if g is constant, $\ell_{CG} = \int_0^\ell \ell'\,dm/M = \ell_{cm} = \frac{1}{2}\ell$, the same as the center of mass. See Section 14-2.)

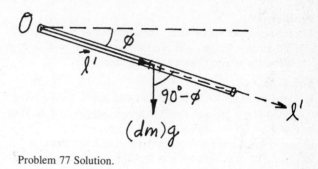

Problem 77 Solution.

CHAPTER 13 ROTATIONAL VECTORS AND ANGULAR MOMENTUM

Section 13-1 Angular Velocity and Acceleration Vectors

Problem

1. A car is headed north at 70 km/h. Give the magnitude and direction of the angular velocity of its 62-cm-diameter wheels.

Solution

If we assume that the wheels are rolling without slipping (see Section 12-5), the magnitude of the angular velocity is $\omega = v_{cm}/r = (70\text{ m}/3.6\text{ s})/(0.31\text{ m}) = 62.7\text{ s}^{-1}$. With the car going north, the axis of rotation of the wheels is east-west. Since the top of a wheel is going in the same direction as the car, the right-hand rule gives the direction of $\boldsymbol{\omega}$ as west.

Problem

5. A wheel is spinning with angular speed $\omega = 5.0$ rad/s, when a constant angular acceleration $\alpha = 0.85$ rad/s^2 is applied at right angles to the initial angular velocity. How long does it take for the angular speed to increase by 10 rad/s?

Solution

For constant angular acceleration, $\boldsymbol{\omega}_f = \boldsymbol{\omega}_i + \boldsymbol{\alpha}t$. If $\boldsymbol{\alpha}$ is perpendicular to $\boldsymbol{\omega}_i$, $\omega_f^2 = \omega_i^2 + (\alpha t)^2$, or $t = \sqrt{\omega_f^2 - \omega_i^2}/\alpha$. Therefore, for $\omega_i = 5\text{ s}^{-1}$ to increase to $\omega_f = 15\text{ s}^{-1}$ under the given conditions requires time $t = \sqrt{(15)^2 - (5)^2}\text{ s}^{-1} \div (0.85\text{ s}^{-2}) = 16.6$ s. (Note: $\boldsymbol{\omega}_f$ is the total angular velocity of the wheel, not just its spin.)

Section 13-2 Torque and the Vector Cross Product

Problem

9. A 12-N force is applied at the point $x = 3$ m, $y = 1$ m. Find the torque about the origin if the force points in (a) the x direction, (b) the y direction, and (c) the z direction.

Solution

Equation 13-2 and the definition of the cross product give:
(a) $\boldsymbol{\tau} = \mathbf{r} \times \mathbf{F} = (3\hat{\mathbf{i}} + \hat{\mathbf{j}}) \times 12\hat{\mathbf{i}}\text{ N·m} = -12\hat{\mathbf{k}}\text{ N·m}$,
(b) $(3\hat{\mathbf{i}} + \hat{\mathbf{j}}) \times 12\hat{\mathbf{j}}\text{ N·m} = 36\hat{\mathbf{k}}\text{ N·m}$,
(c) $(3\hat{\mathbf{i}} + \hat{\mathbf{j}}) \times 12\hat{\mathbf{k}}\text{ N·m} = (12\hat{\mathbf{i}} - 36\hat{\mathbf{j}})\text{ N·m}$, (magnitude 37.9 N·m in the x-y plane, 71.6° clockwise from the x axis).

Problem

13. Vector **A** points 30° counterclockwise from the x axis. Vector **B** is twice as long as **A**. Their product $\mathbf{A} \times \mathbf{B}$ has length A^2, and points in the negative z direction. What is the direction of vector **B**?

Solution

The cross product, $\mathbf{A} \times \mathbf{B} = -A^2\hat{\mathbf{k}}$, is perpendicular to the plane of **A** and **B**, so these vectors lie in the x-y plane. The right-hand rule implies that the angle between **A** and **B**, measured clockwise from **A**, is less than 180°, i.e., $\theta_A - \theta_B < 180°$, or $-150° < \theta_B < 30° = \theta_A$, where θ_A and θ_B are measured counterclockwise from the x axis. The magnitude of $\mathbf{A} \times \mathbf{B}$ is $AB\sin(\theta_A - \theta_B) = 2A^2\sin(\theta_A - \theta_B) = A^2$ (as given, with $B = 2A$), so $\sin(\theta_A - \theta_B) = \frac{1}{2}$ or $\theta_A - \theta_B = 30°$ or 150°. When this is combined with the given value of θ_A and the range of θ_B, one finds that $\theta_B = 0°$ or $-120°$ (i.e., along the x axis or 120° clockwise from the x axis).

Problem

17. Find a force vector that, when applied at the point $x = 2.0$ m, $y = 1.5$ m, will produce a torque $\tau = 4.7\hat{k}$ N·m about the origin. (There are many answers.)

Solution

$\mathbf{F}$ is perpendicular to the torque and thus lies in the x-y plane (i.e., $\mathbf{F} = \hat{i}F_x + \hat{j}F_y$ and $F_z = 0$). Specifically, $\tau = 4.7\hat{k}$ N·m $= \mathbf{r} \times \mathbf{F} = (2\hat{i} + 1.5\hat{j}) \times (F_x\hat{i} + F_y\hat{j})$ m $= (2F_y - 1.5F_x)\hat{k}$ m. Therefore, any force with components $2F_y - 1.5F_x = 4.7$ N and $F_z = 0$ will produce the given torque about the origin, when applied at the given point.

Section 13-3 Angular Momentum

Problem

21. A gymnast of rotational inertia 62 kg·m² is tumbling head over heels. If her angular momentum is 470 kg·m²/s, what is her angular speed?

Solution

If we regard the gymnast as a rigid body rotating about a fixed (instantaneous) axis, Equation 13-4 gives $\omega = L/I = (470 \text{ kg·m}^2/\text{s})/(62 \text{ kg·m}^2) = 7.58 \text{ s}^{-1} = 1.21$ rev/s as her angular speed about that axis.

Problem

25. A weightlifter's barbell consists of two 25-kg masses on the ends of a 15-kg rod 1.6 m long. The weightlifter holds the rod at its center and spins it at 10 rpm about an axis perpendicular to the rod. What is the magnitude of the barbell's angular momentum?

Solution

The rotational inertia of the weights and bar about the specified axis is $I = 2m_{wt}(\ell/2)^2 + m_{bar}\ell^2/12$ (see Table 12-2) so the angular momentum about this axis is $L = I\omega = [2(25 \text{ kg})(0.8 \text{ m})^2 + (15 \text{ kg})(1.6 \text{ m})^2/12] \times (10\pi/30 \text{ s}) = 36.9$ J·s.

Problem

29. A 1.0-kg particle is moving at a constant 3.5 m/s along the line $y = 0.62x + 1.4$, where x and y are in meters and where the motion is toward the positive x and y directions. Find its angular momentum (a) about the origin (b) about the point $x = 0$, $y = 1.4$, and (c) about the point $x = 2.0$, $y = 4.8$.

Solution

The line of motion makes an angle of $\tan^{-1} 0.62 = 31.8°$ with the x axis, so the momentum of the particle is $\mathbf{p} = (3.5 \text{ kg·m/s})(\hat{i}\cos 31.8° + \hat{j}\sin 31.8°)$. The displacement from a point P in the x-y plane to the particle is $\mathbf{r} - \mathbf{r}_p = \hat{i}(x - x_p) + \hat{j}(y - y_p)$, where x and $y = (\tan 31.8°)x + 1.4$ are the coordinates of the particle and x_p and y_p those

of the point, in meters. The angular momentum about point P is simply $\mathbf{L}_p = (\mathbf{r} - \mathbf{r}_p) \times \mathbf{p} = (3.5 \text{ kg·m}^2/\text{s}) \times ((x - x_p)\hat{i} + (y - y_p)\hat{j})(\hat{i}\cos 31.8° + \hat{j}\sin 31.8°) = (3.5 \text{ J·s})\hat{k} [(x - x_p)\sin 31.8° - (y - y_p)\cos 31.8°] = (3.5 \text{ J·s})\hat{k} [(x - x_p)\sin 31.8° - (x\tan 31.8° + 1.4 - y_p)\cos 31.8°] = (3.5 \text{ J·s})\hat{k} [(y_p - 1.4)\cos 31.8° - x_p\sin 31.8°]$. (Note: The quantity in square brackets is just the perpendicular distance from P to the line of motion, see Problems 26 and 27 and the sketch.) (a) If P is the origin, $x_p = y_p = 0$, and $\mathbf{L}_0 = -(3.5 \text{ J·s})\hat{k} (1.4\cos 31.8°) = -4.16\hat{k}$ J·s. (b) If P is on the line of motion, $\mathbf{L}_p = 0$. (c) If $x_p = 2$ and $y_p = 4.8$, then $\mathbf{L}_p = (3.5 \text{ J·s})\hat{k} [(4.8 - 1.4) \times \cos 31.8° - 2\sin 31.8°] = 6.43\hat{k}$ J·s.

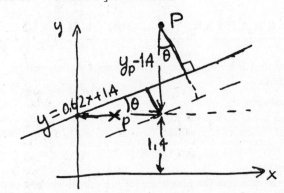

Problem 29 Solution.

Section 13-4 Conservation of Angular Momentum

Problem

33. In Fig. 13-33 the lower disk, of mass 440 g and radius 3.5 cm, is rotating at 180 rpm on a frictionless shaft of negligible radius. The upper disk, of mass 270 g and radius 2.3 cm, is initially not rotating. It drops freely down the shaft onto the lower disk, and frictional forces act to bring

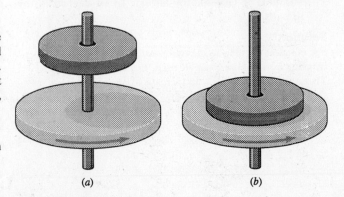

(a)   (b)

FIGURE 13-33 Problem 33.

the two disks to a common rotational speed. (a) What is that speed? (b) What fraction of the initial kinetic energy is lost to friction?

Solution

(a) There are no external torques acting about the frictionless shaft, so the total angular momentum of the two-disk system in this direction is conserved, or $\mathbf{L}_1 + \mathbf{L}_2 = $ const. (The frictional torques between the disks are internal to the system.) Initially, $L_1 = I_1\omega_{1i}$ and $L_2 = 0$, and finally $L_1 + L_2 = (I_1 + I_2)\omega_f$, therefore $\omega_f = \omega_{1i}I_1/(I_1 + I_2) = $ (180 rpm)$[440 \times 3.5^2/(440 \times 3.5^2 + 270 \times 2.3^2)] = $ 142 rpm. (We canceled common factors of $\frac{1}{2}$g·cm^2 from the rotational inertias.) (b) The fraction of initial kinetic energy lost is

$$\frac{K_i - K_f}{K_i} = 1 - \frac{K_f}{K_i} = 1 - \frac{I_1}{I_1 + I_2} = 1 - \frac{\omega_f}{\omega_{1i}}$$
$$= 1 - 0.791 = 20.9\%.$$

(We used $K = \frac{1}{2}I\omega^2 = L^2/2I$ for the kinetic energy of a rigid body rotating about a fixed axis, and the result of part (a).)

Problem

37. A turntable of radius 25 cm and rotational inertia 0.0154 kg·m^2 is spinning freely at 22.0 rpm about its central axis, with a 19.5-g mouse on its outer edge. The mouse walks from the edge to the center. Find (a) the new rotation speed and (b) the work done by the mouse.

Solution

(a) We suppose that "spinning freely" means that there are no external torques acting about the axis of rotation, so that the total angular momentum of the turntable and mouse in this direction is constant, or $I_i\omega_i = I_f\omega_f$. Initially, the rotational inertia of the mouse, mr^2, must be added to that of the turntable, I_0, but we assume the mouse contributes nothing when at the center, so $\omega_f = \omega_i(I_i/I_f) = $ (22 rpm) $\times$ [0.0154 kg·m^2 + 0.0195 kg (0.25 m)^2]/(0.0154 kg·m^2) = 23.7 rpm. (b) The work done by the mouse (when it exerts reaction forces to friction between its feet and the turntable) can be found from the work-energy theorem, $W_{nc} = K_f - K_i = \frac{1}{2}I_f\omega_f^2 - \frac{1}{2}I_i\omega_i^2 = \frac{1}{2}I_f\omega_f^2(1 - \omega_i/\omega_f)$, where we used the conservation of angular momentum from part (a). Numerically, $W_{nc} = \frac{1}{2}(0.0154)(23.7\pi/30)^2$ J$(1 - 22/23.7) = $ 3.49 mJ.

Problem

41. Eight 60-kg skaters join hands and skate down an ice rink at 4.6 m/s. Side by side, they form a line 12 m long. The skater at one end stops abruptly, and the line proceeds to rotate rigidly about that skater. Find (a) the angular speed, (b) the linear speed of the outermost skater, and (c) the force that must be exerted on the outermost skater.

Solution

The force that abruptly stops the skater at one end exerts no torque about that skater (point P), so the total angular momentum about a vertical axis through P is conserved. Initially, the angular momentum of each of the other seven skaters about P is mv_0b, where b is the perpendicular distance from the original straight-line motion to the point P (see Problems 26 or 27). For these seven skaters, $b_n = n\ell/7$, where $n = 1, 2, \ldots, 7$ and $\ell = 12$ m, so $L = \Sigma_1^7 mv_0(n\ell/7) = (mv_0\ell/7)\Sigma_1^7 n = (mv_0\ell/7)(7 \times 8/2) = 4mv_0\ell$. (a) When the skaters rotate rigidly about P with angular speed ω, their angular momentum is $L = I\omega = (\Sigma_1^7 mb_n^2)\omega = (m\ell^2\omega/49)\Sigma_1^7 n^2 = (m\ell^2\omega/49)(7 \times 8 \times 15/6) = 20m\ell^2\omega/7$. Since L is constant, $4mv_0\ell = 20m\ell^2\omega/7$ implies $\omega = (7v_0/5\ell) = (1.4)(4.6 \text{ m/s})/(12 \text{ m}) = 0.537 \text{ s}^{-1}$. (b) The linear speed of the outermost skater is $v_7 = \omega b_7 = \omega\ell = (0.537 \text{ s}^{-1})(12 \text{ m}) = 6.44 \text{ m/s}$. (c) The force is the centripetal force, $mv_7^2/\ell = m\ell\omega^2 = (60 \text{ kg})(12 \text{ m})(0.537 \text{ s}^{-1})^2 = 207$ N. (Note: the sums of integers and squares of integers can be found in mathematical tables.)

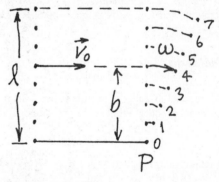

Problem 41 Solution.

Section 13-5 Rotational Dynamics in Three Dimensions

Problem

45. A gyroscope consists of a disk and shaft mounted on frictionless bearings in a frame of diameter d, as shown in Fig. 13-36. Initially the gyroscope is spinning with angular speed ω and is perfectly balanced so it's not precessing. When a mass m is hung from the frame at one of the shaft bearings, the gyroscope precesses about a vertical axis with angular speed Ω. Find an expression for the rotational inertia of the gyroscope.

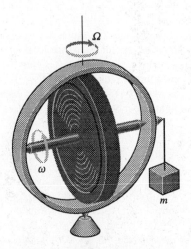

FIGURE 13-36   Problem 45.

Solution

The spin axis is horizontal and the gravitational torque in Fig. 13-36 is $mg(d/2)$, so Equation 13-7, for a gyroscope under these circumstances, gives $\Omega = \tau/L = (mgd/2)/I\omega$ or $I = mgd/2\omega\Omega$.

Paired Problems

Problem

49. The dot product of a pair of vectors is twice the magnitude of their cross product. What is the angle between the vectors?

Solution

The given relationship means that $\mathbf{A} \cdot \mathbf{B} = AB\cos\theta = 2|\mathbf{A} \times \mathbf{B}| = 2AB\sin\theta$, so $\theta = \tan^{-1}(\frac{1}{2}) = 26.6°$.

Problem

53. The turntable in Fig. 13-37 has rotational inertia 0.021 kg·m², and is rotating at 0.29 rad/s about a frictionless vertical axis. A wad of clay is tossed onto the turntable and sticks 15 cm from the rotation axis. The clay

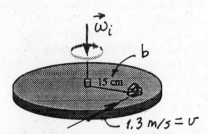

FIGURE 13-37   Problem 53.

impacts with horizontal velocity component 1.3 m/s, at right angles to the turntable's radius, and in a direction that opposes the rotation, as suggested in Fig. 13-37. After the clay lands the turntable has slowed to 0.085 rad/s. Find the mass of the clay.

Solution

There are no external torques in the direction of the turntable's axis, so the vertical angular momentum of the turntable/clay system is conserved. (The horizontal forces, which cause the clay to stick to the turntable, are internal forces.) If we take the sense of rotation of the turntable to define the positive direction of vertical angular momentum, then the system's initial angular momentum is $I\omega_i - mvb$, where I is the rotational inertia of the turntable, v the horizontal component of the velocity of the mass, m, of clay, and b the perpendicular distance to the axis of rotation. After the clay lands, this angular momentum equals $(I + mb^2)\omega_f$, so solving for the mass, we find $m = I(\omega_i - \omega_f)/(vb + b^2\omega_f) = (0.021 \text{ kg·m}^2)(0.29 \text{ s}^{-1} - 0.085 \text{ s}^{-1})/[(1.3 \text{ m/s})(0.15 \text{ m}) + (0.15 \text{ m})^2(0.085 \text{ s}^{-1})] = 21.9$ g. (We assumed ω_f and ω_i have the same sense. If the sense of rotation of the turntable were reversed after impact of the clay, i.e. $\omega_f = -0.085 \text{ s}^{-1}$, then $m = 40.8$ g.)

Supplementary Problems

Problem

57. An advanced civilization lives on a solid spherical planet of uniform density. Running out of room for their expanding population, the civilization's government calls an engineering firm specializing in planetary reconfiguration. Without adding any material or angular momentum, the engineers reshape the planet into a hollow shell whose thickness is one-fifth of its outer radius. Find the ratio of the new to the old (a) surface area and (b) length of day.

Solution

Since the planetary mass (M) and density (ρ) are constant, the volume of the original sphere (radius R_0) is the same as the volume of the new spherical shell (outer radius R, inner radius $R - \frac{1}{5}R = \frac{4}{5}R$). Thus, $\frac{4}{3}\pi R_0^3 = \frac{4}{3}\pi R^3(1 - 0.8^3)$, or $R/R_0 = (1 - 0.512)^{-1/3} = 1.27$. (a) The ratio of the outer surface area of the new to old planet is $4\pi R^2/4\pi R_0^2 = (R/R_0)^2 = (1.27)^2 = 1.61$. (b) Since the angular momentum is also constant, $I_0\omega_0 = I_0(2\pi/T_0) = I\omega = I(2\pi/T)$, or $T/T_0 = I/I_0$, where T is the planetary rotation period (length of day) and I the rotational inertia about the axis of rotation. For a solid uniform sphere, $I_0 = \frac{2}{5}MR_0^2 = \frac{2}{5}(\frac{4}{3}\pi\rho)R_0^5$, while for the uniform shell, $I = \frac{2}{5}(\frac{4}{3}\pi\rho)[R^5 - (0.8R)^5]$, so the ratio is $T/T_0 = (R/R_0)^5(1 - 0.8^5) = (1.27)^5(1 - 0.8^5) = 2.22$.

Problem

61. About 99.9% of the solar system's total mass lies in the Sun. Using data from Appendix E, estimate what fraction of the solar system's angular momentum about its center is associated with the Sun. Where is most of the rest of the angular momentum?

Solution

The planets orbit the Sun in planes approximately perpendicular to the Sun's rotation axis, so most of the angular momentum in the solar system is in this direction. We can estimate the orbital angular momentum of a planet by mvr, where m is its mass, v its average orbital speed, and r its mean distance from the Sun (see the solution to Problem 24, where one also finds that $L_{\text{orb}} \gg L_{\text{rot}}$). Compared to the orbital angular momentum of the four giant planets, everything else is negligible, except for the rotational angular momentum of the Sun itself. The latter can be estimated by assuming the Sun to be a uniform sphere rotating with an average period of $\frac{1}{2}(27 + 36)$ d. (This is an overestimate since most of the Sun's mass is in its core, with radius about $\frac{1}{4}R_\odot$). The numerical data in Appendix E results in the following estimates:

ORBITAL ANGULAR MOMENTUM (mvr)		%
Jupiter	19.2×10^{42} J·s	59.7
Saturn	7.85×10^{42} J·s	24.4
Uranus	1.69×10^{42} J·s	5.2
Neptune	2.52×10^{42} J·s	7.8
ROTATIONAL ANGULAR MOMENTUM ($\frac{2}{5}MR^2\omega$)		
Sun	0.89×10^{42} J·s	2.8
Total	32.2×10^{42} J·s	99.9

Problem

65. The contraption shown in Fig. 13-41 consists of two solid rubber wheels each of mass M and radius R mounted on an axle of negligible mass in a rigid square frame made of thin rods of mass m and length ℓ. The axle bearings are frictionless, and the wheels are mounted symmetrically about the center line of the frame, just far enough apart that they don't touch. The whole contraption is floating freely in space, and the frame is not rotating. The wheels are rotating with angular speed ω in the same direction, as shown. A mechanism built into the frame moves the axles very slightly so the wheels touch and frictional forces act between them. Describe quantitatively the motion of the system after the wheels have stopped.

Solution

Since the contraption is isolated in free space, its angular momentum (calculated about the CM) is constant. Initially, just the wheels are spinning, so $L_i = 2(\frac{1}{2}MR^2)\omega_i = MR^2\omega_i$, in the direction of the wheel axes (the z axis). After the wheels have stopped, the whole contraption rotates in the same direction about the z axis such that $L_f = I_f\omega_f = L_i$. Here, I_f is the total rotational inertia, including four rods and two wheels. With the aid of Table 12-2 and the parallel axis theorem, we have $I_f = 2(\frac{1}{12}m\ell^2) + 2m(\frac{1}{2}\ell)^2 + 2(\frac{1}{2}MR^2 + MR^2) = \frac{2}{3}m\ell^2 + 3MR^2$. Thus, $\omega_f = L_i/I_f = MR^2\omega_i/(\frac{2}{3}m\ell^2 + 3MR^2)$. (When they touch, the wheels exert reaction forces on their axles, causing the contraption to rotate.)

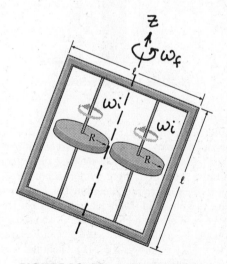

FIGURE 13-41 Problem 65 Solution.

● CHAPTER 14 STATIC EQUILIBRIUM

Section 14-1 Conditions for Equilibrium

Problem

1. Five forces act on a rod, as shown in Fig. 14-26. Write the torque equations that must be satisfied for the rod to be in static equilibrium taking the torques (a) about the top of the rod and (b) about the center of the rod.

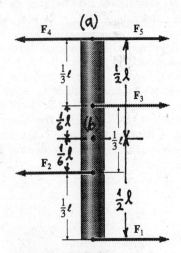

FIGURE 14-26 Problem 1.

Solution

All of the forces lie in the plane of Fig. 14-26, so all of the torques about any point on the rod are into or out of the page. Suppose the latter direction, out of the page or counterclockwise, is positive. Moreover, all of the forces are perpendicular to the rod, so their lever arms about any point on the rod (recall that the magnitude of the torque is force times lever arm) can easily be read-off from Fig. 14-26. (a) About the top of the rod, F_4 and F_5 contribute zero torque, and Equation 14-2 becomes $0 = \frac{1}{3}\ell F_3 - \frac{2}{3}\ell F_2 + \ell F_1$. (b) About the center of the rod, the perpendicular distances to F_2 and F_3 are $\frac{1}{6}\ell$, and to F_1, F_4 and F_5 are $\frac{1}{2}\ell$, so $0 = \frac{1}{2}\ell(F_1 + F_4 - F_5) - \frac{1}{6}\ell(F_2 + F_3)$.

Problem

5. In Fig. 14-27 the forces shown all have the same magnitude F. For each of the cases shown, is it possible to place a third force so the three forces meet both conditions for static equilibrium? If so, specify the force and a suitable application point; if not, why not?

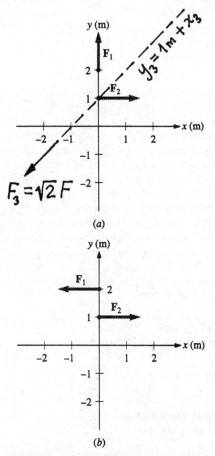

FIGURE 14-27 Problem 5.

Solution

The conditions for static equilibrium, under the action of three forces, can be written as: $F_3 = -(F_1 + F_2)$ and $r_3 \times F_3 = -(r_1 \times F_1 + r_2 \times F_2)$. (a) In this case, $F_1 = F\hat{j}$, $r_1 = (2 \text{ m})\hat{j}$, $F_2 = F\hat{i}$, and $r_2 = (1 \text{ m})\hat{j}$. Thus, $F_3 = -F(\hat{i} + \hat{j})$,

which is a force of magnitude $\sqrt{2}F$, 45° down into the third quadrant ($\theta_x = 225°$ or $-135°$ CCW from the x axis). The point of application, r_3, can be found from the second condition, $r_3 \times F_3 = (x_3\hat{i} + y_3\hat{j}) \times (-F\hat{i} - F\hat{j}) = (-x_3 + y_3)F\hat{k} = -r_1 \times F_1 - r_2 \times F_2 = 0 - (1 \text{ m})\hat{j} \times F\hat{i} = (1 \text{ m})F\hat{k}$. Thus, $-x_3 + y_3 = 1$ m, or the line of action of F_3 passes through the point of application of F_2 (the point (0, 1 m)). Any point on this line is a suitable point of application for F_3 (e.g. the point (0, 1m)). (b) In this case, $F_1 = -F_2$ so $F_3 = 0$, but $r_1 \times F_1 + r_2 \times F_2 = (r_2 - r_1) \times F_2 \neq 0$ so $r_3 \times F_3 \neq 0$. Thus there is no single force that can be added to produce static equilibrium.

Section 14-2 Center of Gravity

Problem

9. Figure 14-29a shows a thin, uniform square plate of mass m and side ℓ. The plate is in a vertical plane. Find the magnitude of the gravitational torque on the plate about each of the three points shown.

Solution

The center of gravity is at the center of a uniform plate. In calculating the gravitational torque, one may consider the entire weight as acting at the center of gravity. (a) $\mathbf{r}_A = \sqrt{2}\ell/2$ at 135° from the weight of the plate, so $\tau_A = (\sqrt{2}\ell/2)mg \sin 135° = \frac{1}{2}mg\ell$. (b) $\mathbf{r}_B$ is colinear with the weight, so $\tau_B = 0$. (c) $\tau_C = \frac{1}{2}\ell \, mg \sin 90° = \frac{1}{2}mg\ell$ (but note that $\tau_C = -\tau_A$). (We also assumed that B and C are at the centers of their respective sides. Alternatively, the torques can be found from the lever arms shown.)

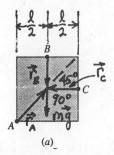

FIGURE 14-29(a) Problem 9 Solution.

Section 14-3 Examples of Static Equilibrium

Problem

13. Where should the child in Fig. 14-31 sit if the scale is to read (a) 100 N and (b) 300 N?

Solution

If we consider torques about the pivot point (so that the force exerted by the pivot does not contribute) then Equation 14-2 is sufficient to determine the position of the child. As shown on Fig. 14-31, the weight of the tabletop (acting at its center of gravity), the weight of the child (acting a distance x from the left end), and the scale force, F_s, produce zero torque about the pivot:

$$\left(1/g\right)\left(\sum \tau\right)_P = 0 =$$

$$\left(F_s/g\right)(160 \text{ cm}) - (60 \text{ kg})(40 \text{ cm}) + (40 \text{ kg})(80 \text{ cm} - x).$$

Therefore, $x = 20$ cm $+ (F_s/9.8 \text{ N})4$ cm. If (a) $F_s = 100$ N, then $x = 20$ cm $+ (400/9.8)$ cm $= 60.8$ cm, and if (b) $F_s =$

300 N, $x = 142$ cm. (Note that the child is on opposite sides of the pivot in parts (a) and (b), since without the child, $F_s = 147$ N.)

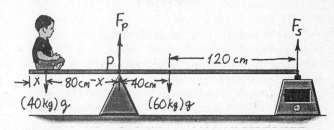

FIGURE 14-31 Problem 13 Solution.

Problem

17. Fig. 14-34 shows a traffic signal, with masses and positions of its various members indicated. The structure is mounted with two bolts, located symmetrically about the vertical member's centerline, as indicated. What tension force must the left-hand bolt be capable of withstanding?

Solution

The forces on the traffic signal structure, and their lever arms about point 0 (on the vertical member's centerline between the bolts) as shown on Fig. 14-34. The normal forces exerted by the bolts and the ground on the vertical member are designated by N_ℓ and N_r, measured positive upward. (Of course, the ground can only make a positive contribution, and the bolts only a negative contribution, to these normal forces.) The two conditions of static equilibrium needed to determine N_ℓ and N_r are: $\Sigma F_y = 0 = N_\ell + N_r - (9.8)(320 + 170 + 65)$ N (the vertical component of

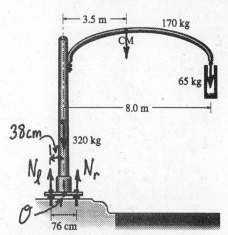

FIGURE 14-34 Problem 17 Solution.

Equation 14-1, positive up) and $(\Sigma \tau_z)_0 = 0 =$
$(N_r - N_\ell)(0.38 \text{ m}) - (170 \times 3.5 + 65 \times 8)(9.8 \text{ N·m})$ (the
out-of-the-page-component of Equation 14-2, positive CCW).
These can be written as $N_r + N_\ell = 5.44$ kN, and $N_r -$
$N_\ell = 28.8$ kN. Thus $N_\ell = -11.7$ kN, which is downward
and must be exerted by the bolt. The reaction force on the
bolt is upward and is a tensile force. (Really, N_ℓ is the difference between the downward force exerted by the bolt and the
upward force exerted by the ground. Tightening the bolt increases the tensile force it must withstand beyond the minimum value calculated above, under the assumption that the
ground exerts no force.)

Problem

21. A 15.0-kg door measures 2.00 m high by 75.0 cm wide. It
hangs from hinges mounted 18.0 cm from top and bottom.
Assuming that each hinge carries half the door's weight,
determine the horizontal and vertical forces that the door
exerts on each hinge.

Solution

If the door is properly hung, all the forces *on* the door are
coplanar. We assume that the CM is at the geometrical center
of the door. The conditions for equilibrium are:

$$\Sigma F_x = 0 = F_{Ax} + F_{Bx},$$
$$\Sigma F_y = 0 = F_{Ay} + F_{By} - Mg,$$
$$(\Sigma \tau)_B = 0 = \mathbf{r}_A \times \mathbf{F}_A + \mathbf{r}_{cm} \times M\mathbf{g}$$
$$= (164 \text{ cm}\hat{\mathbf{j}}) \times (F_{Ax}\hat{\mathbf{i}} + F_{Ay}\hat{\mathbf{j}})$$
$$+ (37.5 \text{ cm}\hat{\mathbf{i}} + 82 \text{ cm}\hat{\mathbf{j}}) \times (-Mg\hat{\mathbf{j}}),$$

where we chose to calculate torques about the lower hinge at
B. Expanding the cross products, $0 = 164F_{Ax}(-\hat{\mathbf{k}}) -$
$37.5Mg(\hat{\mathbf{k}})$, we find $F_{Ax} = -(37.5)(15 \text{ kg})(9.8 \text{ m/s}^2)/(164) =$
-33.6 N. From the x equation, $F_{Bx} = -F_{Ax} = 33.6$ N, and
by assumption, $F_{Ay} = F_{By} = \frac{1}{2}Mg = 73.5$ N. Of course, the

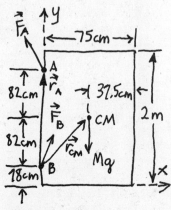

Problem 21 Solution.

forces exerted *by* the door on the hinges (by Newton's third
law) are the reactions to the forces, $\mathbf{F}_A$ and $\mathbf{F}_B$, just calculated.

Problem

25. A uniform sphere of radius R is supported by a rope attached
to a vertical wall, as shown in Fig. 14-40. The point where
the rope is attached to the sphere is located so a continuation
of the rope would intersect a horizontal line through the
sphere's center a distance $R/2$ beyond the center, as shown.
What is the smallest possible value for the coefficient of
friction between wall and sphere?

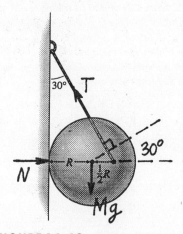

FIGURE 14-40 Problem 25 Solution.

Solution

In equilibrium, the sum of the torques about the center of the
sphere must be zero, so the frictional force is up, as shown.
The lever arm of the tension in the rope is $\frac{1}{2}R\cos 30° =$
$\sqrt{3}R/4$, and the weight and normal force exert no torque
about the center. Thus, $fR = T\sqrt{3}R/4$. The sum of the horizontal components of the forces is zero also, so $0 = N -$
$T\sin 30°$, or $T = 2N$. Therefore $f = \frac{1}{2}\sqrt{3} N$. Since $f \leq$
$\mu_s N$, this implies $\mu_s \geq f/N = \frac{1}{2}\sqrt{3} = 0.87$.

Problem

29. The boom in the crane of Fig. 14-42 is free to pivot about
point P and is supported by the cable that joins halfway
along its 18-m total length. The cable passes over a pulley
and is anchored at the back of the crane. The boom has mass
1700 kg, distributed uniformly along its length, and the mass
hanging from the end of the boom is 2200 kg. The boom
makes a 50° angle with the horizontal. What is the tension
in the cable that supports the boom?

Solution

The forces on the boom are shown superposed on the figure.
By assumption, T is horizontal and acts at the CM of the

boom. To find T, we compute the torques about P, $(\Sigma \tau)_P = 0$, obtaining:

$$T\tfrac{1}{2}\ell \sin 50° - m_b g \tfrac{1}{2}\ell \cos 50° - mg\ell \cos 50° = 0,$$

or

$$T = (2m + m_b)g \cot 50° = (4400 + 1700)(9.8 \text{ N})\cot 50°$$
$$= 50.2 \text{ kN}.$$

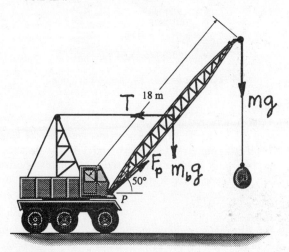

FIGURE 14-42 Problem 29 Solution.

Problem

33. Fig. 14-45 shows a uniform board dangling over a *frictionless* edge, secured by a *horizontal* rope. If the angle θ in Fig. 14-45 were 30°, what fraction would the distance d shown in the figure be of the board length ℓ?

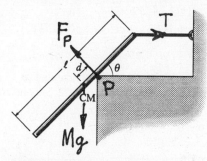

FIGURE 14-45 Problem 33 Solution.

Solution

The three forces acting on the board are in the same configuration as those in Problem 31, so Equation 14-2 about the edge gives $Mgd \cos \theta = T(\tfrac{1}{2}\ell - d)\sin \theta$. If the edge is frictionless, then $\mathbf{F}_P$ is perpendicular to the board, so Equation 14-1 requires $T = F_P \sin \theta$ and $Mg = F_P \cos \theta$. Substitut-

ing above, we find $F_P d \cos^2 \theta = F_P(\tfrac{1}{2}\ell - d)\sin^2 \theta$, or $d(\cos^2 \theta + \sin^2 \theta) = \tfrac{1}{2}\ell \sin^2 \theta$ and $\theta = \sin^{-1}\sqrt{2d/\ell}$. For $\theta = 30°$, $d/\ell = \tfrac{1}{2}\sin^2 30° = \tfrac{1}{8}$.

Problem

37. The potential energy as a function of position for a certain particle is given by

$$U(x) = U_0\left(\frac{x^3}{x_0^3} + a\frac{x^2}{x_0^2} + 4\frac{x}{x_0}\right),$$

where U_0, x_0, and a are constants. For what values of a will there be two static equilibria? Comment on the stability of these equilibria.

Solution

The equilibrium condition, $dU/dx = 0$ (Equation 14-5), requires $3(x/x_0)^2 + 2a(x/x_0) + 4 = 0$. This quadratic has two real roots if the discriminant is positive, i.e. $a^2 - 12 > 0$, or $|a| > 2\sqrt{3}$. The roots are $(x/x_0)_\pm = \tfrac{1}{3}(-a \pm \sqrt{a^2 - 12})$. The second derivative of the potential energy, evaluated at these roots, is

$$\left(\frac{d^2U}{dx^2}\right)_\pm = \frac{U_0}{x_0^2}\left[6\left(\frac{x}{x_0}\right)_\pm + 2a\right] = \pm 2\sqrt{a^2 - 12}\left(\frac{U_0}{x_0^2}\right).$$

Thus, the "plus" root is a position of metastable equilibrium (Equation 14-6), while the "minus" root represents unstable equilibrium, (Equation 14-7). A plot of the potential energy, which is a cubic, will clarify these remarks. For $|a| \leq 2\sqrt{3}$, $U(x)$ has no wiggles, as shown. (U passes through the origin, but its position depends on the value of "a", and is not shown.)

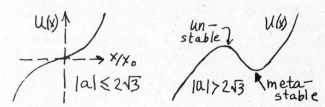

Problem 37 Solution.

Problem

41. A 4.2-kg plant hangs from the bracket shown in Fig. 14-49. The bracket has a mass of 0.85 kg, and its center of mass lies 9.0 cm from the wall. A single screw holds the bracket to the wall, as shown. Find the horizontal tension force in the screw. *Hint:* Imagine that the bracket is slightly loose and pivoting about its bottom end. Assume the wall is frictionless.

Solution

We assume that the screw provides the total support for the bracket, exerting a force with horizontal component F_x (the reaction to which is a tensile force on the screw) and vertical component F_y (the reaction to which is a shearing force on the screw equal to the total weight) as shown. A normal contact force exerted by the wall could be distributed along the bracket (e.g. by tightening the screw), but if we only wish to estimate the minimum F_x, we may consider all the normal force to act at the lowest point of contact of the bracket, point O. Then Equation 14-2 about O gives $F_x(7.2 \text{ cm}) =$ $[(4.2 \text{ kg})(28 \text{ cm}) + (0.85 \text{ kg})(9.0 \text{ cm})](9.8 \text{ m/s}^2)$ or $F_x =$ 170 N.

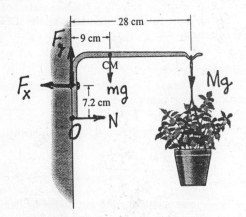

FIGURE 14-49 Problem 41 Solution.

Problem
45. A uniform, solid cube of mass m and side s is in stable equilibrium when sitting on a level tabletop. How much energy is required to bring it to an unstable equilibrium where it's resting on its corner?

Solution
When balancing on a corner, the CM of a uniform cube (i.e. its center) is a distance $\sqrt{(s/2)^2 + (s/2)^2 + (s/2)^2} = \sqrt{3}s/2$ above the corner resting on the tabletop. When in stable equilibrium, the CM is $s/2$ above the tabletop. Thus, the potential energy difference is $\Delta U = mg \, \Delta y_{cm} = mgs(\sqrt{3} - 1)/2$.

Problem
49. One end of a board of negligible mass is attached to a spring of spring constant k, while its other end rests on a frictionless surface, as shown in Fig. 14-51. If a mass m is placed on the middle of the board, by how much will the spring compress?

Solution
If the frictionless surface is horizontal, the three forces acting on the board are vertical, as shown. For equilibrium, $N + F_s = mg$ and $(\Sigma \, \tau)_{cm} = 0$. The latter implies $N = F_s$, so $F_s = \frac{1}{2}mg = k \, \Delta x$, and the compression of the spring is $\Delta x = mg/2k$.

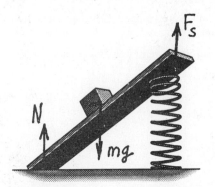

FIGURE 14-51 Problem 49 Solution.

Problem
53. A 2.0-m-long rod has a density described by $\lambda = a + bx$, where λ is the density in kilograms per meter of length, $a = 1.0 \text{ kg/m}$, $b = 1.0 \text{ kg/m}^2$, and x is the distance in meters from the left end of the rod. The rod rests horizontally with its ends each supported by a scale. What do the two scales read?

Solution
The rod is in static equilibrium under the three vertical forces shown in the sketch, so $\Sigma \, F_y = 0$ implies $F_{s\ell} + F_{sr} = Mg$, and $(\Sigma \, \tau)_{cm} = 0$ implies $F_{s\ell}x_{cm} = F_{sr}(\ell - x_{cm})$. The solution for the left and right scale forces is $F_{s\ell} = Mg - F_{sr} = Mg(1 - x_{cm}/\ell)$. Equation 10-5 gives

$$x_{cm} = \int_0^\ell \lambda x \, dx \bigg/ \int_0^\ell \lambda \, dx$$
$$= \int_0^\ell (ax + bx^2) \, dx \bigg/ \int_0^\ell (a + bx) \, dx$$
$$= \left(a\tfrac{1}{2}\ell^2 + b\tfrac{1}{3}\ell^3\right) \bigg/ \left(a\ell + b\tfrac{1}{2}\ell^2\right)$$
$$= \ell(3a + 2b\ell)/(6a + 3b\ell).$$

For the values given, $x_{cm}/\ell = \frac{7}{12}$ and note that $M = a\ell + \frac{1}{2}b\ell^2 = 4 \text{ kg}$. Thus, $F_{sr} = Mgx_{cm}/\ell = 22.9 \text{ N}$ and $F_{s\ell} = 16.3 \text{ N}$.

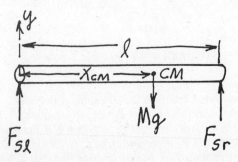

Problem 53 Solution.

Problem

57. A uniform solid cone of height h and base diameter $\frac{1}{3}h$ is placed on the board of Fig. 14-53. The coefficient of static friction between the cone and incline is 0.63. As the slope of the board is increased, will the cone first tip over or begin sliding? *Hint:* Begin with an integration to find the center of mass.

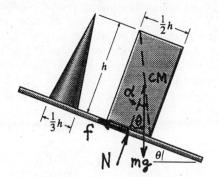

FIGURE 14-53 Problem 55 Solution.

Solution

The analysis for Problem 55 applies to the cone, where α is the angle between the symmetry axis and a line from the CM to the edge of the base. The integration to find the CM is

fastest when the cone is oriented like the aircraft wing in Example 10-3, for then, the equation of the sloping side is simple, as shown in the sketch. For mass elements, take thin disks parallel to the base, so $dm = \rho\pi y^2\,dx = (3M/h^3)x^2\,dx$, where $\rho = M/\frac{1}{3}\pi R^2 h$ is the density (assumed constant) and M is the mass of the cone. Then $x_{cm} = M^{-1}\int x\,dm = (3/h^3)\int_0^h x^3\,dx = \frac{3}{4}h$, or the CM is $\frac{1}{4}h$ above the base. Since $\tan\alpha = (\frac{1}{6}h)/(\frac{1}{4}h) = \frac{2}{3} > 0.63 = \mu_s$, this cone will slide before tipping over.

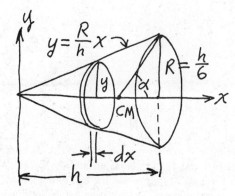

Problem 57 Solution.

● PART 1 CUMULATIVE PROBLEMS

Problem

1. A 170-g apple sits atop a 2.8-m-high post. A 45-g arrow moving horizontally at 130 m/s passes horizontally through the apple and strikes the ground 36 m from the base of the post, as shown in Fig. 1. Where does the apple hit the ground? Neglect the effect of air resistance on either object as well as any friction between apple and post.

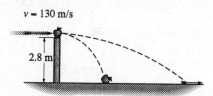

FIGURE 1 Cumulative Problem 1.

Solution

We can assume that momentum is conserved during the inelastic collision (in a brief interval at $t = 0$) between the arrow (m_1) and the apple (m_2). The velocities of the arrow before and after are specified to be horizontal (in the

x direction), therefore the velocity of the apple (which was at rest before) is also horizontal after the collision. Thus, $m_1 v_{1i,x} = m_1 v_{1f,x} + m_2 v_{2f,x}$. Since both are moving horizontally after the collision, the arrow and the apple will each fall to the ground through the same vertical distance y (equal to the height of the post), in the same time $t = \sqrt{2y/g}$. However, they strike the ground at different horizontal positions, which (in the absence of air resistance) are $x_1 = v_{1f,x}t$ and $x_2 = v_{2f,x}t$, relative to the base of the post. Since $x_1 = 36$ m, $y = 2.8$ m, and $v_{1i,x} = 130$ m/s are given, $v_{1f,x}$ and $v_{2f,x}$ can be eliminated and a solution for x_2 obtained: $x_2 = (m_1/m_2)(v_{1i,x} - v_{1f,x})t = (m_1/m_2)(v_{1i,x}\sqrt{2y/g} - x_1) = (45/170)[(130 \text{ m/s})\sqrt{2(2.8 \text{ m})/(9.8 \text{ m/s}^2)} - 36 \text{ m}] = 16.5$ m. (Alternatively, since external horizontal forces are neglected, the center of mass of the arrow/apple system moves horizontally at constant speed until it reaches ground level, $v_{cm,x} = \text{constant} = m_1 v_{1i,x} \div (m_1 + m_2)$ (its value before the collision). Then at ground level, $m_2 x_2 = (m_1 + m_2)x_{cm} - m_1 x_1 = (m_1 + m_2)v_{cm,x}t - m_1 x_1 = m_1(v_{1i,x}t - x_1)$, as before.) Refer to relevant material in Chapters 4, 10, and 11 if necessary.

Problem

5. A solid ball of radius R is set spinning with angular speed ω about a horizontal axis. The ball is then lowered vertically with negligible speed until it just touches a horizontal surface and is released (Fig. 5). If the coefficient of kinetic friction betwen the ball and the surface is μ, find (a) the linear speed of the ball once it achieves pure rolling motion and (b) the distance it travels before its motion is pure rolling.

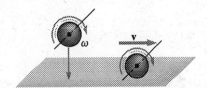

FIGURE 5 Cumulative Problem 5.

Solution

(a) While there is relative motion at the point of contact between the ball and the horizontal surface, the force of sliding friction ($f = \mu N = \mu mg$) slows the ball's rotation and accelerates its center of mass. The equation for the latter is $f = \mu mg = ma_{cm}$, or $a_{cm} = \mu g$ (positive to the right in the sketch and Fig. 5). The equation for the former is $\tau = -fR = -\mu mgR = I\alpha = (2mR^2/5)\alpha$, where

$\alpha = -5\mu g/2R$ is the angular acceleration about the horizontal axis through the center of the ball (positive clockwise in the sketch and Fig. 5). The accelerations are constant, so the velocities are given by Equations 2-17 and 12-9 as $v_{cm} = a_{cm}t = \mu gt$ and $\omega = \omega_0 + \alpha t = \omega_0 - 5\mu gt/2R$, where the ball is released at $t = 0$ and the initial velocities, $v_0 = 0$ and ω_0, are given. The accelerated motion continues until the point of contact is instantaneously at rest (no more sliding friction). The ball rolls without slipping thereafter, at a constant velocity given by $v_{cm} = \omega R$. This occurs at a time t, when $(\omega_0 - 5\mu gt/2R)R = \mu gt$, or $t = 2\omega_0 R/7\mu g$. Thus, the final velocity is $v_{cm} = \mu gt = 2\omega_0 R/7$. (b) The distance traveled during this time is $\Delta x = \frac{1}{2}a_{cm}t^2 = \frac{1}{2}(\mu g)(2\omega_0 R/7\mu g)^2 = 2\omega_0^2 R^2/49\mu g$ (or $\Delta x = v_{cm}^2/2a_{cm}$ with the same result).

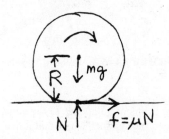

Cumulative Problem 5 Solution.

PART 2 OSCILLATIONS, WAVES, AND FLUIDS

● CHAPTER 15 OSCILLATORY MOTION

Sections 15-1 and 15-2 Oscillations and Simple Harmonic Motion

Problem

1. A doctor counts 77 heartbeats in one minute. What are the period and frequency of the heart's oscillations?

Solution

If 77 heartbeats take 1 min., then one heartbeat (one cycle) takes $T = 1 \text{ min}/77 = 60 \text{ s}/77 = 0.77$ s, which is the period. The frequency is $f = 77/\text{min} = 77/60 \text{ s} = 1.28$ Hz. (One can see that $T = 1/f$.)

Problem

5. Determine the amplitude, angular frequency, and phase constant for each of the simple harmonic motions shown in Fig. 15–33.

Solution

The amplitude is the maximum displacement, read along the x axis (ordinate) in Fig. 15-33. The angular frequency is 2π

times the reciprocal of the period, which is the time interval between corresponding points read along the t axis (abscissa). The phase constant can be determined from the intercept and slope (displacement and velocity) at $t = 0$. One sees that
(a) $A \simeq 20$ cm, $\omega \simeq 2\pi/4 \text{ s} \simeq \frac{1}{2}\pi\text{s}^{-1}$, and $\phi \simeq 0$;
(b) $A \simeq 30$ cm, $\omega \simeq 2\pi/3.2 \text{ s} \simeq 2 \text{ s}^{-1}$, and $\phi \simeq -90° \simeq -\frac{1}{2}\pi$;
(c) $A \simeq 40$ cm, $\omega \simeq 2\pi/(2\times2 \text{ s}) \simeq \frac{1}{2}\pi\text{s}^{-1}$, and $\phi \simeq \cos^{-1}(27/40) \simeq 48° \simeq \frac{1}{4}\pi$.

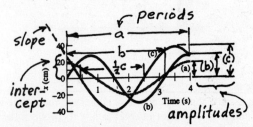

FIGURE 15-33 Problem 5 Solution.

Problem

9. A simple model of a carbon dioxide (CO_2) molecule consists of three mass points (the atoms) connected by two springs (electrical forces), as suggested in Fig. 15-34. One way this system can oscillate is if the carbon atom stays fixed and the two oxygens move symmetrically on either side of it. If the frequency of this oscillation is 4.0×10^{13} Hz, what is the effective spring constant? The mass of an oxygen atom is 16 u.

FIGURE 15-34 Problem 9.

Solution

With the carbon atom end of either "spring" fixed, the frequency of either oxygen atom is $\omega = 2\pi f = \sqrt{k/m}$. Therefore $k = (2\pi \times 4 \times 10^{13}$ Hz$)^2 (16 \times 1.66 \times 10^{-27}$ kg$) = 1.68 \times 10^3$ N/m.

Problem

13. A mass m slides along a frictionless horizontal surface at speed v_0. It strikes a spring of constant k attached to a rigid wall, as shown in Fig. 15-35. After a completely elastic encounter with the spring, the mass heads back in the direction it came from. In terms of k, m, and v_0, determine (a) how long the mass is in contact with the spring and (b) the maximum compression of the spring.

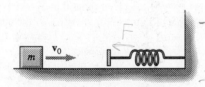

FIGURE 15-35 Problem 13.

Solution

(a) While the mass is in contact with the spring, the net horizontal force on it is just the spring force, so it undergoes half a cycle of simple harmonic motion before leaving the spring with speed v_0 to the left. This takes time equal to half a period $\frac{1}{2}T = \pi/\omega = \pi \sqrt{m/k}$. (b) v_0 is the maximum speed, which is related to the maximum compression of the spring (the amplitude) by $v_0 = \omega A$. Thus $A = v_0/\omega = v_0 \sqrt{m/k}$.

Section 15-3 Applications of Simple Harmonic Motion

Problem

17. How long should you make a simple pendulum so its period is (a) 200 ms; (b) 5.0 s; (c) 2.0 min?

Solution

The period and length of a simple pendulum (at the surface of the Earth) are related by Equation 15-21, therefore $\ell = (T/2\pi)^2 g = (0.248$ m/s$^2)T^2 = 9.93$ mm, 6.21 m, and 3.57 km respectively, for the three values given.

Problem

21. A pendulum of length ℓ is mounted in a rocket. What is its period if the rocket is (a) at rest on its launch pad; (b) accelerating upward with asceleration $a = \frac{1}{2}g$; (c) accelerating downward with acceleration $a = \frac{1}{2}g$; (d) in free fall?

Solution

(It may be helpful to think of an elevator instead of a rocket in this problem.) Let **a** be the acceleration of the pendulum relative to the rocket, and let $\mathbf{a_0}$ be the acceleration of the rocket relative to the ground (assumed to be an inertial system). Then Newton's second law is $\Sigma \mathbf{F} = m(\mathbf{a} + \mathbf{a_0})$, or $\Sigma \mathbf{F} - m\mathbf{a_0} = m\mathbf{a}$. The rotational analog of this equation is the appropriate generalization of Equation 15-18 for a simple pendulum in an accelerating frame. The "fictitious" torque (about the point of suspension), $\mathbf{r} \times (-m\mathbf{a_0})$, can be combined with the torque of gravity, $\mathbf{r} \times m\mathbf{g}$, if we replace g by $|\mathbf{g} - \mathbf{a_0}|$, while the right-hand side, $|\mathbf{r} \times m\mathbf{a}| = I\alpha$, is the same as Equation 15-18. For small oscillations about the equilibrium position (which depends on $\mathbf{a_0}$), the period is $T = 2\pi \sqrt{\ell/|\mathbf{g} - \mathbf{a_0}|}$. (a) If $\mathbf{a_0} = 0$, $T = 2\pi \sqrt{\ell/g}$, as before. (b) Take the y axis positive up. Then $\mathbf{a_0} = \frac{1}{2}g\hat{\mathbf{j}}$ and $\mathbf{g} = -g\hat{\mathbf{j}}$, so $T = 2\pi \sqrt{\ell/(g + \frac{1}{2}g)} = 2\pi \sqrt{2\ell/3g}$. (c) If $\mathbf{a_0} = -\frac{1}{2}g\hat{\mathbf{j}}$, $T = 2\pi \sqrt{\ell/(g - \frac{1}{2}g)} = 2\pi \sqrt{2\ell/g}$. (d) If $\mathbf{a_0} = \mathbf{g}$, $T = \infty$ (there is no restoring torque and the pendulum does not oscillate).

Problem

25. A solid disk of radius R is suspended from a spring of linear spring constant k and torsional constant κ, as shown in Fig. 15-36. In terms of k and κ, what value of R will give the same period for the vertical and torsional oscillations of this system?

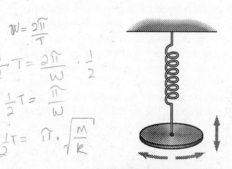

FIGURE 15-36 Problem 25.

Solution

Equating the angular frequencies for vertical and torsional oscillations (Equations 15-11 and 15-17), we find $k/M = \kappa/I = \kappa/(\frac{1}{2}MR^2)$, or $R = \sqrt{2\kappa/k}$.

Problem

29. A thin, uniform hoop of mass M and radius R is suspended from a thin horizontal rod and set oscillating with small amplitude, as shown in Fig. 15-38. Show that the period of the oscillations is $2\pi\sqrt{2R/g}$. *Hint:* You may find the parallel-axis theorem useful.

Solution

Equation 15-19 gives $T = 2\pi\sqrt{I/Mg\ell} = 2\pi\sqrt{(MR^2 + MR^2)/MgR} = 2\pi\sqrt{2/Rg}$, where we used the parallel axis theorem for I, with $\ell = h = R$.

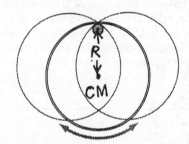

FIGURE 15-38 Problem 29 Solution.

Problem

33. A cyclist turns her bicycle upside down to tinker with it. After she gets it upside down, she notices the front wheel executing a slow, small-amplitude back-and-forth rotational motion with a period of 12 s. Considering the wheel to be a thin ring of mass 600 g and radius 30 cm, whose only irregularity is the presence of the tire valve stem, determine the mass of the valve stem.

Solution

The bicycle wheel may be regarded as a physical pendulum, with rotational inertia $I = MR^2 + mR^2$ about its central axle, where $M = 600$ g is the mass of the wheel (thin ring) and m is the mass of the valve stem (a circumferential point mass). The distance of the CM from the axle is given by $(M + m)\ell = mR$ (this is just Equation 10-4, with origin at the center of the wheel so $x_1 = 0$ for M, $x_2 = R$ for m, and $x_{cm} = \ell$). For small oscillations, the period is given by Equation 15-19 (where $M + m$ is the total mass), therefore $T = 2\pi/\omega = 2\pi\sqrt{I/(M + m)g\ell} = 2\pi\sqrt{(M + m)R^2/mgR}$. Solving for m, we find $m = M[(g/R)(T/2\pi)^2 - 1]^{-1} = (600 \text{ g})[(9.8/0.3)(12/2\pi)^2 - 1]^{-1} = 5.04$ g.

Problem

37. The equation for an ellipse is

$$\frac{x^2}{a^2} + \frac{y^2}{b^2} = 1.$$

Show that two-dimensional simple harmonic motion whose two components have different amplitudes and are $\pi/2$ out of phase gives rise to elliptical motion. How are a and b related to the amplitudes?

Solution

Simple harmonic motions in the x and y directions, with different amplitudes and $\pi/2$ out of phase, are $x = a\cos(\omega t + \phi)$ and $y = b\cos(\omega t + \phi \pm \pi/2) = \mp b\sin(\omega t + \phi)$. These describe an elliptical path with semi-major or minor axis equal to the amplitudes, a or b, since $(x/a)^2 + (y/b)^2 = \cos^2(\omega t + \phi) + \sin^2(\omega t + \phi) = 1$.

Problem

41. The motion of a particle is described by

$$x = (45 \text{ cm})[\sin(\pi t + \pi/6)],$$

with x in cm and t in seconds. At what time is the potential energy twice the kinetic energy? What is the position of the particle at this time?

Solution

The condition that the potential energy equal twice the kinetic energy implies that $U(t) = \frac{1}{2}kx(t)^2 = 2K(t) = mv(t)^2$, or $\omega x(t)/v(t) = \pm\sqrt{2}$, where $\omega = \sqrt{k/m}$. For $x(t) = (45 \text{ cm})\sin(\pi t + \pi/6)$ as given (note that $\omega = \pi(\text{s}^{-1})$), $v(t) = dx/dt = \omega(45 \text{ cm})\cos(\pi t + \pi/6)$, so the above condition becomes $\tan(\pi t + \pi/6) = \pm\sqrt{2}$. There are four angles in each cycle which satisfy this (since $\tan\theta = -\tan(\pi - \theta) = \tan(\pi + \theta) = -\tan(2\pi - \theta)$), which are $\pi(t + \frac{1}{6}) = 0.955$, 2.19, 4.10, and 5.33 radians. (We chose the cycle with phases between 0 and 2π radians; for any other cycle, an integer multiple of 2π can be added to these angles.) The times corresponding the these phases are $t = (0.955/\pi) - \frac{1}{6} = 0.137$ s, and 0.529 s, 1.14 s, and 1.53 s, respectively. (An integer multiple of 2 can be added to get the times for any other cycle.) The positions of the particle corresponding to these phases are $x(0.137 \text{ s}) = (45 \text{ cm})\sin(0.955) = 36.7 \text{ cm} = x(0.529 \text{ s}) = -x(1.14 \text{ s}) = -x(1.53 \text{ s})$, since $\sin\theta = \sin(\pi - \theta) = -\sin(\pi + \theta) = -\sin(2\pi - \theta)$. (During each cycle, the particle passes each of the points ± 36.7 cm twice, traveling with the same speed, but in opposite directions.)

Problem

4$. A solid cylinder of mass M and radius R is mounted on an axle through its center. The axle is attached to a horizontal spring of constant k, and the cylinder rolls back and forth without slipping (Fig. 15-43). Write the statement of energy conservation for this system, and differentiate it to obtain an equation analogous to Equation 15-4 (see previous problem). Comparing your result with Equation 15-4, determine the angular frequency of the motion.

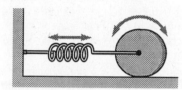

FIGURE 15-43 Problem 45.

Solution

With reference to Equation 12-30 (and the condition $v = \omega R$ for rolling without slipping) $K = \frac{1}{2}Mv^2 + \frac{1}{2}I_{cm}\omega^2 = \frac{1}{2}Mv^2 + \frac{1}{2}(\frac{1}{2}MR^2)(v/R)^2 = \frac{3}{4}Mv^2$. The potential energy of the spring is $U = \frac{1}{2}kx^2$, where $v = dx/dt$, so $E = K + U = \frac{3}{4}M(dx/dt)^2 + \frac{1}{2}kx^2$. Differentiating, we find:

$$\frac{dE}{dt} = 0 = \frac{3}{4}M\cdot2\left(\frac{dx}{dt}\right)\left(\frac{d^2x}{dt^2}\right) + \frac{1}{2}k\cdot2x\left(\frac{dx}{dt}\right), \quad \text{or}$$

$$\frac{d^2x}{dt^2} = -\frac{2k}{3M}x \equiv -\omega^2x.$$

(The energy method is particularly convenient for analyzing small oscillations, since complicated details of the forces can be avoided.)

Section 15-7 Driven Oscillations and Resonance

Problem

49. A mass-spring system has $b/m = \omega_0/5$, where b is the damping constant and ω_0 the natural frequency. How does its amplitude when driven at frequencies 10% above and below ω_0 compare with its amplitude at ω_0?

Solution

The amplitude at resonance ($\omega_d = \omega_0$) is $A_{res} = F_0/b\omega_0$, so that Equation 15-26 can be rewritten as:

$$\frac{A}{A_{res}} = \frac{A}{(F_0/b\omega_0)} = \frac{(b\omega_0/m)}{\sqrt{(\omega_d^2 - \omega_0^2)^2 + b^2\omega_d^2/m^2}}$$

$$= \left[\left(\frac{m\omega_0}{b}\right)^2\left(\frac{\omega_d^2}{\omega_0^2} - 1\right)^2 + \frac{\omega_d^2}{\omega_0^2}\right]^{-1/2}$$

If $(m\omega_0/b) = 5$, and $(\omega_d/\omega_0) = 1.1$ (10% above resonance), then $A/A_{res} = 1/\sqrt{25(1.21 - 1)^2 + 1.21} = 65.8\%$, while for $\omega_d/\omega_0 = 0.9$ (10% below resonance), $A/A_{res} = 1/\sqrt{25(0.81 - 1)^2 + 0.81} = 76.4\%$.

Paired Problems

Problem

53. A particle undergoes simple harmonic motion with amplitude 25 cm and maximum speed 4.8 m/s. Find (a) the angular frequency, (b) the period, and (c) the maximum acceleration.

Solution

(a) Since $v_{max} = \omega A$, $\omega = (4.8 \text{ m/s})/(0.25 \text{ m}) = 19.2 \text{ s}^{-1}$.
(b) $T = 2\pi/\omega = 0.327 \text{ s}$. (c) $a_{max} = \omega v_{max} = 92.2 \text{ m/s}^2$.

Problem

57. A meter stick is suspended from one end and set swinging. What is the period of the resulting oscillations, assuming they have small amplitude?

Solution

The meter stick is a physical pendulum whose CM is $\ell = 0.5$ m below the point of suspension through one end. The rotational inertia of the stick about one end is $\frac{1}{3}M(1 \text{ m})^2$, so Equation 15-19 gives the period as $T = 2\pi\sqrt{I/Mg\ell} = 2\pi\sqrt{2(1 \text{ m})/3g} = 1.64 \text{ s}$.

Problem

61. Two mass-spring systems with the same mass are undergoing oscillatory motion with the same amplitudes. System 1 has twice the frequency of system 2. How do (a) their energies and (b) their maximum accelerations compare?

Solution

(a) The energy of a mass-spring system is $E = \frac{1}{2}m\omega^2A^2$ (see Section. 15.5). If m and A are the same, but $\omega_1 = 2\omega_2$, then $E_1 = 4E_2$. (b) The maximum acceleration is $a_{max} = \omega^2A$, so $a_{1,max} = 4a_{2,max}$ as well.

Problem

65. A 500-g block on a frictionless surface is connected to a rather limp spring of constant $k = 8.7$ N/m. A second block rests on the first, and the whole system executes simple harmonic motion with a period of 1.8 s. When the amplitude of the motion is increased to 35 cm, the upper block just begins to slip. What is the coefficient of static friction between the blocks?

Solution

If the surfaces of contact are horizontal, it is the frictional force which accelerates the upper block, hence $f_s = m_2 a(t) \le \mu_s N = \mu_s m_2 g$, or $a(t) \le \mu_s g$. In simple harmonic motion, $a_{max} = \omega^2 A$, so when the upper block begins to slip, $\omega^2 A = \mu_s g$, or $\mu_s = (2\pi/1.8\text{ s})^2(0.35\text{ m})/(9.8\text{ m/s}^2) = 0.44$. (Note: the data given in the problem which were not used to find μ_s (i.e., k and m_1) can be used to calculate that $m_2 = 214$ g, since $\omega = 2\pi/T = \sqrt{k/(m_1 + m_2)}$.)

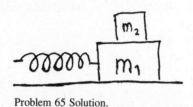

Problem 65 Solution.

Problem

69. A 1.2-kg block rests on a frictionless surface and is attached to a horizontal spring of constant $k = 23$ N/m (Fig. 15-46). The block is oscillating with amplitude 10 cm and with phase constant $\phi = -\pi/2$. A block of mass 0.80 kg is moving from the right at 1.7 m/s. It strikes the first block when the latter is at the rightmost point in its oscillation. The collision is completely inelastic, and the two blocks stick together. Determine the frequency, amplitude, and phase constant (relative to the *original* $t = 0$) of the resulting motion.

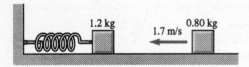

FIGURE 15-46 Problem 69.

Solution

The simple harmonic motion with just the first block on the spring can be described by Equation 15-14 and the given amplitude and phase constant; $x(t) = (10\text{ cm})\cos(\omega_1 t - \pi/2) = (10\text{ cm})\sin\omega_1 t$, where $\omega_1 = \sqrt{k/m_1} = \sqrt{(23\text{ N/m})/(1.2\text{ kg})} = 4.38\text{ s}^{-1}$. This equation holds up to the time of the collision, i.e. for $t < t_c$, where $t_c = \pi/2\omega_1$, since for the rightmost point of oscillation, $\sin\omega_1 t_c = 1$, or $\omega_1 t_c = \pi/2$. (This specifies the original zero of time appropriate to the given phase constant of $-\frac{\pi}{2}$.)

Equation 15-14 also describes the simple harmonic motion after the collision; $x(t) = A\cos(\omega_2 t + \phi)$ for $t > t_c$, where $\omega_2 = \sqrt{k/(m_1 + m_2)} = 3.39\text{ s}^{-1}$ is the angular frequency when both blocks oscillate on the spring $f_2 = \omega_2/2\pi = $

0.540 Hz, as asked in the problem.) It follows from this that $v(t) = -\omega_2 A\sin(\omega_2 t + \phi)$. The amplitude A and phase constant ϕ can be determined from these two equations evaluated just after the collision, essentially at t_c, if we assume that the collision takes place almost instantaneously; then conservation of momentum during the collision can be applied (see Equation 11-4). Just after the collision, $x(t_c) = 10$ cm (given) and $v(t_c) = (m_1 v_1 + m_2 v_2)/(m_1 + m_2)$, where just before the collision, $v_1 = 0$ (given m_1 at rightmost point of its original motion) and $v_2 = -1.7$ m/s (also given). Numerically, $v(t_c) = (-1.7\text{ m/s})(0.8)/(0.8 + 1.2) = -68$ cm/s. Thus, the two equations become $x(t_c) = 10\text{ cm} = A\cos(\omega_2 t_c + \phi)$, and $v(t_c) = -68\text{ cm/s} = -\omega_2 A\sin(\omega_2 t_c + \phi)$, where ω_2 and t_c are known. (In fact, $\omega_2 t_c = \omega_2 \pi/2\omega_1 = (\pi/2)\sqrt{m_1/(m_1 + m_2)} = \sqrt{0.6}\,(\pi/2)$ radians $= 69.7°$.)

Solving for A (using $\sin^2 + \cos^2 = 1$), we find $A = \sqrt{x(t_c)^2 + [-v(t_c)/\omega_2]^2} = \sqrt{(10\text{ cm})^2 + (68\text{ cm}/3.39)^2} = 22.4$ cm. Solving for ϕ (using $\sin/\cos = \tan$), we find $\phi = \tan^{-1}[-v(t_c)/\omega_2 x(t_c)] - \omega_2 t_c = \tan^{-1}(68/3.39 \times 10) - 69.7° = -6.22° = -0.109$ radians.

(Note: The solution for A is equivalent to calculating the various energies in the second simple harmonic motion, since just after the collision, $K(t_c) = \frac{1}{2}(m_1 + m_2)v(t_c)^2 = \frac{1}{2}(2\text{ kg})(-0.68\text{ m/s})^2 = 0.462$ J, $U(t_c) = \frac{1}{2}kx(t_c)^2 = \frac{1}{2}(23\text{ N/m})(0.1\text{ m})^2 = 0.115$ J, $E = K(t_c) + U(t_c) = 0.577$ J $= \frac{1}{2}kA^2$, or $A = \sqrt{2(0.577\text{ J})/(23\text{ N/m})}$. Once A is known, ϕ can also be found from either expression for $x(t_c)$ or $v(t_c)$, e.g., $\omega_2 t_c + \phi = \cos^{-1}(10/22.4) = \sin^{-1}(68/3.39 \times 22.4)$.)

Problem

73. A mass m is connected between two springs of length L, as shown in Fig. 15-47. At equilibrium, the tension force in each spring is F_0. Find the period of oscillations *perpendicular* to the springs, assuming sufficiently small amplitude that the magnitude of the spring tension is essentially unchanged.

FIGURE 15-47 Problem 73.

Solution

Suppose that no forces with components in the direction of motion act on the mass other than the spring forces. If m is given a small displacement perpendicular to the springs (as sketched), the net force is $F_y = -2F_0 \sin\theta = -2F_0 y/\sqrt{L^2 + y^2} \approx -2F_0 y/L$, for $y \ll L$. Newton's second law gives $md^2y/dt^2 = F_y$, or $d^2y/dt^2 \approx -(2F_0/mL)y$. This is the equation for simple harmonic motion with angular frequency $\omega = \sqrt{2F_0/mL}$ and period $T = 2\pi/\omega = 2\pi\sqrt{mL/2F_0}$.

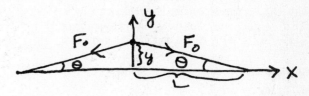

Problem 73 Solution.

● CHAPTER 16 WAVE MOTION

Section 16-1 Wave Properties

Problem

1. Ocean waves with 18-m wavelength travel at 5.3 m/s. What is the time interval between wave crests passing under a boat moored at a fixed location?

Solution

Wave crests (adjacent wavefronts) take a time of one period to pass a fixed point, traveling at the wave speed (or phase velocity) for a distance of one wavelength. Thus $T = \lambda/v = 18$ m/(5.3 m/s) = 3.40 s.

Problem

5. A 145-MHz radio signal propagates along a cable. Measurement shows that the wave crests are spaced 1.25 m apart. What is the speed of the waves on the cable? Compare with the speed of light in vacuum.

Solution

The distance between adjacent wave crests is one wavelength, so the wave speed in the cable (Equation 16-1) is $v = f\lambda = (145\times10^6$ Hz$)(1.25$ m$) = 1.81\times10^8$ m/s $= 0.604c$, where $c = 3\times10^8$ m/s is the wave speed in vacuum.

Problem

9. In Fig. 16-28 two boats are anchored offshore and are bobbing up and down on the waves at the rate of six complete cycles each minute. When one boat is up the other is down. If the waves propagate at 2.2 m/s, what is the minimum distance between the boats?

Solution

The boats are $180° = \pi$ rad out of phase, so the minimum distance separating them is half a wavelength. (In general, they could be an odd number of half-wavelengths apart.) The frequency is 6/60 s $= 0.1$ Hz, so $\frac{1}{2}\lambda = \frac{1}{2}v/f = \frac{1}{2}(2.2$ m/s$)/(0.1/$s$) = 11$ m. (Fig. 16-28 shows the answer, not the question.)

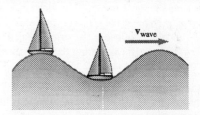

FIGURE 16-28 Problem 9 Solution.

Section 16-2 Mathematical Description of Wave Motion

Problem

13. A simple harmonic wave of wavelength 16 cm and amplitude 2.5 cm is propagating along a string in the negative x direction at 35 cm/s. Find (a) the angular frequency and (b) the wave number. (c) Write a mathematical expression describing the displacement y of this wave (in centimeters) as a function of position and time. Assume the displacement at $x = 0$ is a maximum when $t = 0$.

Solution

(b) Equation 16-4 gives $k = 2\pi/16$ cm $= 0.393$ cm^{-1}, and (a) Equation 16-6 gives $\omega = kv = (0.393$ cm$^{-1})(35$ cm/s$) = 13.7$ s^{-1}. (c) Equation 16-5, for a wave moving in the negative x direction, becomes $y(x, t) = (2.5$ cm$)\cos[(0.393$ cm$^{-1})x + (13.7$ s$^{-1})t]$.

Problem

17. At time $t = 0$, the displacement in a transverse wave pulse is described by $y = 2(x^4 + 1)^{-1}$, with both x and y in cm. Write an expression for the pulse as a function of position x and time t if it is propagating in the positive x direction at 3 cm/s.

Solution

From the shape of the pulse at $t = 0$, $y(x, 0) = f(x)$, a pulse with the same waveform, traveling in the positive x direction with speed v, can be obtained by replacing x by $x - vt$, $y(x, t) = f(x - vt)$. For the given $f(x)$ and v, $y(x, t) = 2[(x - 3t)^4 + 1]^{-1}$, with x and y in cm and t in s.

Section 16-3   Waves on a String

Problem

21. The main cables supporting New York's George Washington Bridge have a mass per unit length of 4100 kg/m and are under tension of 250 MN. At what speed would a transverse wave propagate on these cables?

Solution

$v = \sqrt{F/\mu} = \sqrt{(2.5 \times 10^8 \text{ N})/(4100 \text{ kg/m})} = 247$ m/s (from Equation 16-7).

Problem

25. A 3.1-kg mass hangs from a 2.7-m-long string whose total mass is 0.62 g. What is the speed of transverse waves on the string? *Hint:* You can ignore the string mass in calculating the tension but not in calculating the wave speed. Why?

Solution

The tension in the string is approximately equal to the weight of the 3.1 kg mass (since the weight of the string is only 2% of this). Thus, $v = \sqrt{F/\mu} = \sqrt{(3.1 \text{ kg})(9.8 \text{ m/s}^2)(2.7 \text{ m})/(0.62 \text{ g})} = 364$ m/s. (0.62 g is small compared to 3.1 kg, but not small compared to zero!)

Problem

29. A 25-m-long piece of 1.0-mm-diameter wire is put under 85 N tension. If a transverse wave takes 0.21 s to travel the length of the wire, what is the density of the material comprising the wire?

Solution

From the length of wire, travel time and Equation 16-7, $v = 25$ m/0.21 s $= \sqrt{85 \text{ N}/\mu}$, so $\mu = 6.00 \times 10^{-3}$ kg/m. But for a uniform wire of length ℓ and diameter d, $\rho = \mu/\frac{1}{4}\pi d^2 = (6.00 \times 10^{-3} \text{ kg/m})/\frac{1}{4}\pi(1 \text{ mm})^2 = 7.64$ g/cm³ (see solution to Problem 27).

Section 16-4   Wave Power and Intensity

Problem

33. A rope with 280 g of mass per meter is under 550 N tension. A wave with frequency 3.3 Hz and amplitude 6.1 cm is propagating on the rope. What is the average power carried by the wave?

Solution

The average power transmitted by transverse traveling waves in a string is given by Equation 16-8, $\bar{P} = \frac{1}{2}\mu\omega^2 A^2 v = \frac{1}{2}(0.28 \text{ kg/m})(2\pi \times 3.3 \text{ Hz})^2(0.061 \text{ m})^2 \sqrt{550 \text{ N}/(0.28 \text{ kg/m})} = 9.93$ W. (We used Equation 16-7 for v.)

Problem

37. Figure 16-32 shows a wave train consisting of two cycles of a sine wave propagating along a string. Obtain an expression for the total energy in this wave train, in terms of the string tension F, the wave amplitude A, and the wavelength λ.

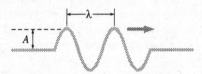

FIGURE 16-32   Problem 37.

Solution

The average wave energy, $d\bar{E}$, in a small element of string of length dx, is transmitted in time, dt, at the same speed as the waves, $v = dx/dt$. From Equation 16-8, $d\bar{E} = \bar{P}\,dt = \frac{1}{2}\mu\omega^2 A^2 v\,dt = \frac{1}{2}\mu\omega^2 A^2\,dx$, so the average linear energy density is $d\bar{E}/dx = \frac{1}{2}\mu\omega^2 A^2$. The total average energy in a wave train of length $\ell = 2\lambda$ is $\bar{E} = (d\bar{E}/dx)\ell = \frac{1}{2}\mu\omega^2 A^2(2\lambda)$. In terms of the quantities specified in this problem (see Equations 16-1 and 7) $\bar{E} = \frac{1}{2}(F/v^2)(2\pi v/\lambda)^2 A^2(2\lambda) = 4\pi^2 FA^2/\lambda$. (Note: The relation derived can be written as $\bar{P} = (d\bar{E}/dx)v$. For a one-dimensional wave, $\bar{P}$ is the intensity, so the average intensity equals the average energy density times the speed of wave energy propagation. This is a general wave property, e.g., see the first unnumbered equation for S in Section 34-10.)

Problem

41. Use data from Appendix E to determine the intensity of sunlight at (a) Mercury and (b) Pluto.

Solution

Equation 16-8 gives the ratio of intensities at two distances from an isotropic source of spherical waves as $I_2/I_1 = (r_1/r_2)^2$. If we use the average intensity of sunlight given in Table 16-1 and mean orbital distances to the sun from Appendix E, we obtain (a) $I_{\text{Merc}} = I_E(r_E/r_{\text{Merc}})^2 = (1368 \text{ W/m}^2)(150/57.9)^2 = 9.18$ kW/m², and (b) $I_{\text{Pluto}} = (1368 \text{ W/m}^2)(150/5.91 \times 10^3)^2 = 0.881$ W/m². (Alternatively, the luminosity of the sun, $\bar{P} = 3.85 \times 10^{26}$ W, from Appendix E, could be used directly in Equation 16-8, with only slightly different numerical results.)

Problem

45. Use Table 16-1 to determine how close to a rock band you should stand for it to sound as loud as a jet plane at 200 m. Treat the band and the plane as point sources. Is this assumption reasonable?

Solution

To have the same loudness, the soundwave intensities should be equal, i.e., $I_{\text{band}}(r) = I_{\text{jet}}(200 \text{ m})$. Regarded as isotropic point sources, use of Equation 16-8 gives $\bar{P}_{\text{band}}/r^2 =$

$\bar{P}_{jet}/(200 \text{ m})^2$. The average power of each source can be found from Table 16-1 and a second application of Equation 16-8, $\bar{P}_{band} = 4\pi(4 \text{ m})^2(1 \text{ W/m}^2)$ and $\bar{P}_{jet} = 4\pi(50 \text{ m})^2(10 \text{ W/m}^2)$. Then $r^2 = (\bar{P}_{band}/\bar{P}_{jet})(200 \text{ m})^2 = (200 \text{ m})^2(4 \text{ m})^2(1 \text{ W/m}^2)/(50 \text{ m})^2(10 \text{ W/m}^2)$, or $r = 5.06$ m. The size of a rock band is several meters, nearly equal to this distance, so a point source is not a good approximation. Besides, the acoustical output of a rock band usually emanates from an array of speakers, which is not point-like. Moreover, the size of a jet plane is also not very small compared to 50 m.

Section 16-5 The Superposition Principle and Wave Interference

Problem

49. You're in an airplane whose two engines are running at 560 rpm and 570 rpm. How often do you hear the sound intensity increase as a result of wave interference?

Solution

As mentioned in the text, pilots of twin-engine airplanes use the beat frequency to synchronize the rpm's of their engines. The beat frequency is simply the difference of the two interfering frequencies, $f_{beat} = (570 - 560)/60 \text{ s} = \frac{1}{6} \text{ s}^{-1}$, so you would hear one beat every six seconds.

Section 16-6 The Wave Equation

Problem

53. The following equation arises in analyzing the behavior of shallow water:

$$\frac{\partial^2 y}{dx^2} - \frac{1}{gh}\frac{\partial^2 y}{dt^2} = 0,$$

where h is the equilibrium depth and y the displacement from equilibrium. Give an expression for the speed of waves in shallow water. (Here shallow means the water depth is much less than the wavelength.)

Solution

The equation given is in the standard form for the one-dimensional linear wave equation (Equation 16-11), so the wave speed is the reciprocal of the square root of the quantity multiplying $\partial^2 y/\partial t^2$. Thus $v = \sqrt{gh}$.

Paired Problems

Problem

57. A spring of mass m and spring constant k has an unstretched length ℓ_0. Find an expression for the speed of transverse waves on this spring when it has been stretched to a length ℓ.

Solution

The spring may be regarded as a stretched string with tension, $F = k(\ell - \ell_0)$, and linear mass density $\mu = m/\ell$. Equation 16-7 gives the speed of transverse waves as $v = \sqrt{k\ell(\ell - \ell_0)/m}$.

Problem

61. Two motors in a factory produce sound waves with the same frequency as their rotation rates. If one motor is running at 3600 rpm and the other at 3602 rpm, how often will workers hear a peak in the sound intensity?

Solution

The beat frequency equals the difference in the motors' rpm's, so the period of the beats is $T_{beat} = 1/f_{beat} = 1/(3602 - 3600) \text{ min}^{-1} = 30$ s. (See also Problem 49.)

Supplementary Problems

Problem

65. An ideal spring is compressed until its total length is ℓ_1, and the speed of transverse waves on the spring is measured. When it's compressed further to a total length ℓ_2, waves propagate at the *same* speed. Show that the uncompressed spring length is just $\ell_1 + \ell_2$.

Solution

The tension in a compressed spring has magnitude $k(\ell_0 - \ell)$ while its linear mass-density is $\mu = m/\ell$. Therefore, the speed of transverse waves is $v = \sqrt{F/\mu} = \sqrt{k\ell(\ell_0 - \ell)/m}$ (as in Problem 57 for a stretched spring). If $v_1 = v_2$ for two different compressed lengths, then $\ell_1(\ell_0 - \ell_1) = \ell_2(\ell_0 - \ell_2)$ or $(\ell_1 - \ell_2)\ell_0 = \ell_1^2 - \ell_2^2 = (\ell_1 - \ell_2)(\ell_1 + \ell_2)$. Since $\ell_1 \neq \ell_2$, division by $\ell_1 - \ell_2$ gives $\ell_0 = \ell_1 + \ell_2$.

Problem

69. In Example 16-5, how much farther would you have to row to reach a region of maximum wave amplitude?

Solution

In general, the interference condition for waves in the geometry of Example 16-5 is $AP - BP = n\lambda/2$, where n is an odd integer for destructive interference (a node) and n is an even integer for constructive interference (a maximum amplitude). (In Example 16-5, $n = 1$ gave the first node and in Problem 51, $n = 3$ gave the second node.) If $d = 20$ m is the distance between the openings, $\ell = 75$ m is the perpendicular distance from the breakwater, and x is the distance parallel to the breakwater measured from the midpoint of the openings, the interference condition is $\sqrt{\ell^2 + (x + \frac{1}{2}d)^2} - \sqrt{\ell^2 + (x - \frac{1}{2}d)^2} = n\lambda/2$ (see Fig. 16-26). In this problem, we wish to find x for the first maximum, $n = 2$, and the wavelength calculated in Example 16-5, $\lambda = 16.01$ m. Solving for x, we find:

$$x^2 = \frac{[\ell^2 + (\frac{1}{2}d)^2 - (\frac{1}{4}n\lambda)^2]}{(2d/n\lambda)^2 - 1}$$

$$= \frac{[(75)^2 + (10)^2 - (8.005)^2] \text{ m}^2}{(40/32.02)^2 - 1} = (100.5 \text{ m})^2.$$

This is 100.5 m $-$ 33 m $=$ 67.5 m farther than the first node in Example 16-5. (Note: We rounded off to three figures; if you round off to two figures, the answer is 67 m. Also, if $x = 33$ m is substituted into the general interference condition, one can recapture the wavelengths of the first and second nodes, for $n = 1$ and 3, calculated in Example 16-5 and Problem 51, respectively.) ●

● **CHAPTER 17**   SOUND AND OTHER WAVE PHENOMENA

Section 17-1 and 17-2   Sound Waves and the Speed of Sound in Gases

Problem

1. Show that the quantity $\sqrt{P/\rho}$ has the units of speed.

Solution

The units of pressure (force per unit area) divided by density (mass per unit volume) are $(\text{N/m}^2)/(\text{kg/m}^3) = (\text{N/kg})(\text{m}^3/\text{m}^2) = (\text{m/s}^2)\,\text{m} = (\text{m/s})^2$, or those of speed squared.

Problem

5. Timers in sprint races start their watches when they see smoke from the starting gun, not when they hear the sound (Fig. 17-24). Why? How much error would be introduced by timing a 100-m race from the sound of the shot?

Solution

The sound of the starting gun takes $(100 \text{ m})/(340 \text{ m/s}) = 0.294$ s to reach the finish line. An error of this magnitude is significant in short races, where world records are measured in hundredths of a second. (This problem is almost the same as Problem 2, Chapter 2.)

Problem

9. A gas with density 1.0 kg/m^3 and pressure 8.0×10^4 N/m^2 has sound speed 365 m/s. Are the gas molecules monatomic or diatomic?

Solution

Solving for γ in Equation 17-1, we find $\gamma = \rho v^2/P = (1.0 \text{ kg/m}^3)(365 \text{ m/s})^2/(8.0\times10^4 \text{ N/m}^2) = 1.67$, very close to the value for an ideal monatomic gas. (Actually, $\gamma - 5/3 = -1.35\times10^{-3}$ for this gas.)

Problem

13. You see an airplane straight overhead at an altitude of 5.2 km. Sound from the plane, however, seems to be coming from a point back along the plane's path at a 35° angle to the vertical (Fig. 17-25). What is the plane's speed, assuming an average 330 m/s sound speed?

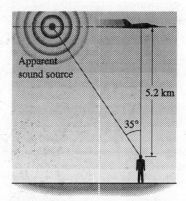

FIGURE 17-25   Problem 13.

Solution

The travel time of the sound from the airplane, reaching you along a line making an angle of 35° with the vertical (from the apparent sound source), is $\Delta t = d/v$. During this time, the airplane moved a horizontal distance $\Delta x = d\sin 35°$, so its speed is $u = \Delta x/\Delta t = d\sin 35°/(d/v) = v\sin 35° = (330 \text{ m/s})\sin 35° = 189$ m/s. (The airplane's altitude, 5.2 km $= d\cos 35°$, was not needed in this calculation.)

Section 17-3   Sound Intensity

Problem

17. A speaker produces 440-Hz sound with total power 1.2 W, radiating equally in all directions. At a distance of 5.0 m, what are (a) the average intensity, (b) the decibel level, (c) the pressure amplitude, and (d) the displacement amplitude?

Solution

(a) Equation 16-9 gives the average intensity at a given distance from an isotropic point source, $\bar{I} = \mathcal{P}/4\pi r^2 = (1.2 \text{ W})/4\pi(5 \text{ m})^2 = 3.82$ mW/m^2. (b) From Equation 17-4, this corresponds to a sound level intensity of $\beta = (10 \text{ dB})\log(I/I_0) = (10 \text{ dB})\log(3.82\times10^{-3}/10^{-12}) = 95.8$ dB $\simeq 96$ dB. (c) The pressure amplitude, for "normal air" of Example 17-1, follows from Equation 17-3b, $\Delta P_0 = \sqrt{2\rho v \bar{I}} = \sqrt{2(1.20 \text{ kg/m}^3)(343 \text{ m/s})(3.82 \times 10^{-3} \text{ W/m}^2)} = 1.77 \text{ N/m}^2$. (d) The corresponding displacement amplitude is $s_0 = \Delta P_0/\rho\omega v = (1.77 \text{ } P_a)/(1.20 \text{ kg/m}^3)(343 \text{ m/s})(2\pi \times 440 \text{ Hz}) = 1.56 \text{ } \mu\text{m}$.

Problem

21. What are the intensity and pressure amplitudes in sound waves with intensity levels of (a) 65 dB and (b) −5 dB?

Solution

The exponentiation of Equation 17-4, to the base ten, relates the intensity to the decibel level, $I/I_0 = 10^{\beta/10 \text{ dB}}$ while Equation 17-3b gives the pressure amplitude, $\Delta P_0 = \sqrt{2\rho v I}$. Here, I is the average intensity, I_0 the threshold level, and we use values of ρ and v for air under the "normal" conditions in Example 17-1. (a) For $\beta = 65$ dB, $I = (10^{-12} \text{ W/m}^2)10^{6.5} = 3.16 \times 10^{-6} \text{ W/m}^2$, and $\Delta P_0 = \sqrt{2(1.2 \text{ kg/m}^3)(343 \text{ m/s})(3.16 \times 10^{-6} \text{ W/m}^2)} = 5.10 \times 10^{-2} \text{ N/m}^2$. (b) For $\beta = -5$ dB, $I = I_0 10^{-5/10} = 3.16 \times 10^{-13} \text{ W/m}^2$ and $\Delta P_0 = 1.61 \times 10^{-5} \text{ N/m}^2$.

Problem

25. Show that a doubling of sound intensity corresponds to very nearly a 3 dB increase in the decibel level.

Solution

If the sound intensity is doubled, $I' = 2I$, Equation 17-4 shows that $\beta' = (10 \text{ dB})\log(I'/I_0) = (10 \text{ dB})\log(2I/I_0) = (10 \text{ dB})\log(I/I_0) + (10 \text{ dB})\log 2 = \beta + 3.01$ dB, or the decibel level increases by about 3 dB.

Problem

29. Sound intensity from a certain extended source drops as $1/r^n$, where r is the distance from the source. If the intensity level drops by 3 dB every time the distance is doubled, what is n?

Solution

A 3 dB drop corresponds to a drop in intensity by a factor of one half (see Problem 25). Thus, if $I' = \frac{1}{2}I$ when $r' = 2r$, $I'/I = \frac{1}{2} = (1/2r)^n/(1/r)^n = 1/2^n$ implies $n = 1$.

Section 17-4 Sound Waves in Liquids and Solids

Problem

33. The speed of sound in body tissues is essentially the same as in water. Find the wavelength of 2.0 MHz ultrasound used in medical diagnostics.

Solution

From Table 17-2, the speed of sound in water is 1497 m/s, so Equation 16-1 gives the wavelength of 2.0 MHz ultrasound as $\lambda = v/f = (1497 \text{ m/s})/(2.0 \text{ MHz}) = 0.749$ mm.

Section 17-5 Wave Reflection

Problem

37. A string with mass per unit length μ is joined to another with 4μ. If a wave of amplitude A propagates from the lighter toward the heavier string, find the amplitudes of (a) the transmitted and (b) the reflected waves.

Solution

Equations 17-7 and 8 give the reflected and transmitted wave amplitudes at the boundary of two strings, in terms of the incident amplitude:
(a) $A_R/A_I = (\sqrt{\mu} - \sqrt{4\mu})/(\sqrt{\mu} + \sqrt{4\mu}) = -\frac{1}{3}$, and
(b) $A_T/A_I = 2\sqrt{\mu}/(\sqrt{\mu} + \sqrt{4\mu}) = \frac{2}{3}$. (Note that consideration of the wave power transmitted across the boundary leads to the relation $\sqrt{\mu} A_I^2 = \sqrt{\mu} A_R^2 + \sqrt{4\mu} A_T^2$, which is satisfied by Equations 17-7 and 8 and the numbers in this particular problem. See Problem 81.)

Problem

41. A rope is made from a number of identical strands twisted together. The rope is frayed, with only a single strand continuing, as shown in Fig. 17-26. The rope is held under tension, and a 2.0-cm-high pulse is sent from the single strand. The first pulse reflected back along the string has 0.90 cm amplitude. How many strands are in the rope?

FIGURE 17-26 Problem 41.

Solution

Let the mass per unit length of a single strand be μ_1, and that for the entire rope be $\mu_2 = n\mu_1$, where n is the number of strands. Then Equation 17-7 and the given amplitudes yield $A_R/A_I = -0.90 \text{ cm}/2.0 \text{ cm} = (1 - \sqrt{n})/(1 + \sqrt{n})$, or $n = (1 + 0.45)^2/(1 - 0.45)^2 = 6.95$. The number of strands is probably seven. (Note that when $\mu_2 > \mu_1$, the reflected wave is 180° out of phase with the incident wave, and A_R/A_I is negative.)

Section 17-6 Standing Waves

Problem

45. Show that only odd harmonics are allowed on a taut string with one end tight and the other free.

Solution

For a string free at one end, the amplitude factor in Equation 17-9 is a maximum for $x = L$, i.e., $2A \sin kL = \pm 2A$. Therefore, $kL = (2m - 1)\pi/2$, where $2m - 1$ is an odd integer for $m = 1, 2, \ldots$ In terms of standing-wave wavelengths, $kL = (2\pi/\lambda_m)L = (2m - 1)\pi/2$, or $L = (2m - 1)\lambda_m/4$, as stated in the next paragraph after Equation 17-10. In terms of frequency, $f_m = v/\lambda_m = (2m - 1)f_1$, where $f_1 = v/4L$ is the frequency of the fundamental. Thus, only odd harmonics occur.

Problem

49. "Vibrato" in a violin is produced by sliding the finger back and forth along the vibrating string. The G-string on a particular violin measures 30 cm between the bridge and its far end and is clamped rigidly at both points. Its fundamental frequency is 197 Hz. (a) How far from the end should the violinist place a finger so that the G-string plays the note A (440 Hz)? (b) If the violinist executes vibrato by moving the finger 0.50 cm to either side of the position in part (a), what range of frequencies results?

Solution

(a) The fundamental frequency of a string fixed at both ends is $f = v/2L$. Since fingering does not change the tension (and hence v) in a violin string appreciably, $f'/f = L/L'$, or $L' = (197 \text{ Hz}/440 \text{ Hz})(30 \text{ cm}) = 13.4$ cm. This is the sounding length of the string, so the finger must be placed a distance $(30 - 13.4)$ cm $= 16.6$ cm from the ("nut") end.

(b) Alteration of L' by ± 0.5 cm yields frequencies between:

$$f'' = (440 \text{ Hz})(13.4)/(13.4 \pm 0.5) = 424 \text{ to } 457 \text{ Hz}.$$

Problem

53. What would be the fundamental frequency of the double bassoon of Example 17-4, if it were played in helium under conditions of Example 17-1?

Solution

The wavelength of the fundamental mode depends on the dimensions of the instrument, so the difference in fundamental frequency, for the bassoon played in helium versus air, is due to the change in the velocity of sound only, $f = v/\lambda$. Thus, if the speed of sound in helium from Example 17-1 is used in place of that in air in Example 17-4, one finds $f = (1000 \text{ m/s})/(11 \text{ m}) = 90.9$ Hz.

Section 17-7 The Doppler Effect

Problem

57. A fire truck's siren at rest wails at 1400 Hz; standing by the roadside as the truck approaches, you hear it at 1600 Hz. How fast is the truck going?

Solution

One can solve Equation 17-12 for u (with the minus sign appropriate to an approaching source) with the result: $u = v(1 - f/f') = (343 \text{ m/s})(1 - 1400/1600) = 42.9 \text{ m/s} = 154$ km/h. (We used the speed of sound in air from Example 17-1.)

Problem

61. You're standing by the roadside as a truck approaches, and you measure the dominant frequency in the truck noise at 1100 Hz. As the truck passes the frequency drops to 950 Hz. What is the truck's speed?

Solution

The result of part (a) of the preceding problem gives $u/v = (1100 - 950)/(1100 + 950) = 0.0732$. For sound waves in "normal" air (Example 17-1), this implies a truck speed of $u = 0.0732(343 \text{ m/s}) = 25.1 \text{ m/s} = 90.4$ km/h. (From Equation 17-12, the frequency emitted by the truck is $f = f_1(1 - u/v) = f_2(1 + u/v)$, where f_1 and f_2 are the observed frequencies when the truck is approaching or receding, respectively. The solution of this equation for the source's speed is $u/v = (f_1 - f_2)/(f_1 + f_2)$.)

Paired Problems

Problem

65. A 1.0-W sound source emits uniformly in all directions. Find (a) the intensity and (b) the decibel level 12 m from the source.

Solution

(a) The average intensity at any distance from an isotropic sound source is given by Equation 16-9, $\bar{I} = \bar{P}/4\pi r^2 = (1 \text{ W})/4\pi(12 \text{ m})^2 = 5.53 \times 10^{-4}$ W/m^2. (b) The corresponding decibel level (Equation 17-4) is $(10 \text{ dB})\log(5.53 \times 10^{-4}/10^{-12}) = 87.4$ dB.

Problem

69. Find the wave speed in a medium where a 28 m/s source speed causes a 3% increase in frequency measured by a stationary observer.

Solution

To cause an increase in frequency, the source must be approaching the stationary observer. Solving Equation 17-12 for the wave speed, we find $v = u/(1 - f/f')$. The given increase is 3%, so $f/f' = 1/1.03$ and $v = (28 \text{ m/s})/(1 - 1/1.03) = 961$ m/s.

Supplementary Problems

Problem

73. A rectangular trough is 2.5 m long and is much deeper than its length, so Equation 16-10 applies. Determine the wavelength and frequency of (a) the longest and (b) the next longest standing waves possible in this trough. Why isn't the higher frequency twice the lower?

Solution

Since the volume of water in the trough is constant, there must be as many "hills" as there are "valleys" in the standing-wave patterns of sinusoidal surface waves. Therefore, there are antinodes at each end, as shown. Since the distance between two antinodes is a multiple of half-wavelengths, $L = m\lambda/2$, so the two longest standing-wave wavelengths are $\lambda_1 = 2L = 5$ m and $\lambda_2 = L = 2.5$ m. If we use Equation 16-10 for the speed of deep water surface waves, the corresponding standing-wave frequencies are $f_1 = v/\lambda_1 = \sqrt{g/2\pi\lambda_1} = \sqrt{(9.8 \text{ m/s}^2)/2\pi \times 5 \text{ m}} = 0.559$ Hz, and $f_2 = \sqrt{2}\,f_1 = 0.790$ Hz. These are not multiples of one another because of the way the wave speed depends on the wavelength, i.e., the dispersion relation for these waves is *not* $\omega = (\text{constant})k$.

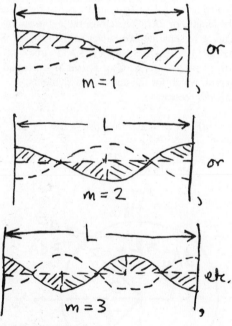

Problem 73 Solution.

Problem

77. Consider an object moving at speed u through a medium, and reflecting sound waves from a stationary source back toward the source. The object receives the waves at the shifted frequency given in the preceding problem, and when it re-emits them they are shifted once again, this time according to Equation 17-12. Find an expression for the overall frequency shift that results, and show that, for $u \ll v$, this shift is approximately $2fu/v$.

Solution

The object receives waves at the frequency of an observer moving toward a stationary source, so $f' = f(1 + u/v)$. (See the solution to the preceding problem. Briefly, in a time t, all the waves in a distance $vt + ut$ are received, their number being $(v + u)t/\lambda$, where λ is the wavelength emitted by the source. The observed frequency, f', is this number divided by t, whereas the source's frequency is $f = v/\lambda$, therefore $f' = (u + v)t/\lambda t = (u + v)/(v/f) = f(1 + u/v)$.) The reflected waves are reemitted by the moving object at this frequency, f', and so are received by the original stationary source at frequency $f'' = f'/(1 - u/v) = f(1 + u/v)/(1 - u/v)$. (See Equation 17-12.) The overall frequency shift is $f'' - f = \Delta f = f[(v + u)(v - u)^{-1} - 1] = 2uf/(v - u)$. If $u \ll v$, then $\Delta f \approx 2uf/v$. (Note: If the object is moving away from the stationary source, one replaces u with $-u$ in the above treatment.)

Problem

81. A wave pulse of total energy E is propagating on a string with mass per unit length μ_1 toward the junction with a second string whose mass per unit length is μ_2. Find an expression for the fraction of the pulse energy transmitted into the second string.

Solution

The frequency of the incident, reflected, and transmitted waves is the same, and so is the string tension on either side of the junction. The two quantities that change across the junction are the linear mass density of the strings and the respective wave amplitudes. The average wave energy flow, or power, written in terms of these quantities, is $(d\bar{E}/dt) = \frac{1}{2}\mu\omega^2 A^2\sqrt{F/\mu} = \frac{1}{2}\omega^2\sqrt{F}\sqrt{\mu}A^2 = \text{const.}\sqrt{\mu}A^2$ (See Equations 16-7 and 8), so $(d\bar{E}/dt)_I \sim \sqrt{\mu_1}\,A_I^2$, $(d\bar{E}/dt)_R \sim \sqrt{\mu_1}\,A_R^2$, and $(d\bar{E}/dt)_T \sim \sqrt{\mu_2}\,A_T^2$ for pulse energy in the respective waves. The fraction of the incident pulse energy which is transmitted is $\sqrt{\mu_2}\,A_T^2/\sqrt{\mu_1}\,A_I^2 = \sqrt{\mu_2/\mu_1}(2\sqrt{\mu_1}/(\sqrt{\mu_1} + \sqrt{\mu_2}))^2 = 4\sqrt{\mu_1\mu_2}/(\sqrt{\mu_1} + \sqrt{\mu_2})^2$, where we used Equation 17-8 for the amplitude ratio. (The fraction of energy reflected is $(A_R/A_I)^2 = (\sqrt{\mu_1} - \sqrt{\mu_2})^2/(\sqrt{\mu_1} + \sqrt{\mu_2})^2$; note that energy conservation is satisfied, $(A_R/A_I)^2 + \sqrt{\mu_2/\mu_1}\,(A_T/A_I)^2 = 1$, or $(d\bar{E}/dt)_R + (d\bar{E}/dt)_T = (d\bar{E}/dt)_I$.)

● CHAPTER 18 FLUID MOTION

Section 18-1 Describing Fluids: Density and Pressure

Problem

1. The density of molasses is 1600 kg/m³. Find the mass of the molasses in a 0.75-liter jar.

Solution

The mass of molasses, which occupies a volume equal to the capacity of the jar, is $\Delta m = \rho \, \Delta V = (1600 \text{ kg/m}^3) \times (0.75 \times 10^{-3} \text{ m}^3) = 1.2 \text{ kg}$.

Problem

5. A plant hangs from a 3.2-cm diameter suction cup affixed to a smooth horizontal surface (Fig. 18-41). What is the maximum weight that can be suspended (a) at sea level and (b) in Denver, where atmospheric pressure is about 0.80 atm?

Suction cup

FIGURE 18-41 Problem 5.

Solution

(a) The force exerted on the suction cup by the atmosphere is $F = PA = P_{\text{atm}}(\pi d^2 / 4) = (1.013 \times 10^5 \text{ Pa})\pi (0.016 \text{ m})^2 = 81.5 \text{ N}$ (perfect vacuum inside cup assumed). This is equal to the maximum weight. (b) At Denver, $P = 0.8 P_{\text{atm}}$, so the maximum weight is 80% of that in part (a), or 65.2 N (a slight variation in g with altitude is neglected).

Problem

9. The fuselage of a 747 jumbo jet is roughly a cylinder 60 m long and 6 m in diameter. If the interior of the plane is pressurized to 0.75 atm, what is the net pressure force tending to separate half the cylinder from the other half when the plane is flying at 10 km, where air pressure is about 0.25 atm? (The earliest commercial jets suffered structural failure from just such forces; modern planes are better engineered.)

Solution

Consider the skin of the fuselage to be divided into infinitesimal strips, parallel to the cylinder's axis, of area dA (shown in cross-section in the sketch). Because of the pressure difference between the cabin interior and the outside, $\Delta P = (0.75 - 0.25)$ atm at 10 km altitude, there is a net force radially outward on dA of magnitude $dF = \Delta P \, dA$. These forces produce stresses in the skin, i.e., forces of one part of the cylinder on another part, that this problem asks us to estimate. The pressure force on one half of the cylinder is balanced by the stress force exerted by the other half. By symmetry, for every dA located at angle θ shown, there is a dA' at angle $-\theta$ with opposite y component of pressure force, $dF_y + dF_y' = 0$, so only the x component of dF contributes to the net pressure force on the half cylinder. But $dF_x = \Delta P \, dA \cos \theta = \Delta P \, dA_y$, where dA_y is the projection of the area dA onto an axial plane parallel to the y axis, and the total projected area of the half-cylinder is just the diameter times the length, or $2RL$. Therefore, the net pressure force tending to separate two halves of the fuselage is $F_x = \Delta P(2RL) = (0.75 - 0.25)(101.3 \text{ kPa})(6 \times 60 \text{ m}^2) = 1.82 \times 10^7 \text{ N} \approx 2050$ tons. Note: F_x can be expressed as a surface integral, which for the above area elements, $dA = \Delta R \, d\theta$, reduces to

$$\int_{\text{half-cylinder}} dF_x = \int_{-\pi/2}^{\pi/2} \Delta P \cos \theta \, dA = \Delta P \, LR \int_{-\pi/2}^{\pi/2} \cos \theta \, d\theta$$

$$= \Delta P L R \sin \theta \Big|_{-\pi/2}^{\pi/2} = \Delta P \cdot 2RL.$$

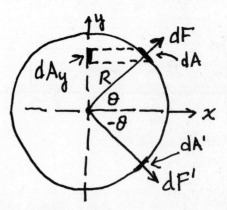

Problem 9 Solution.

Problem

13. When a couple with a total mass of 120 kg lies on a waterbed, the pressure in the bed increases by 4700 Pa. What surface area of the two bodies is in contact with the bed?

Solution

The pressure increase times the average horizontal contact surface area equals the weight of the couple, or $A_{av} = mg/\Delta P = (120 \times 9.8 \text{ N})/(4700 \text{ Pa}) = 0.250 \text{ m}^2$.

Section 18-2 Fluids at Rest: Hydrostatic Equilibrium

Problem

17. What is the density of a fluid whose pressure increases at the rate of 100 kPA for every 6.0 m of depth?

Solution

The increase in pressure with depth, for an incompressible fluid, is given by Equation 18-3. Thus, $\rho = \Delta P/gh = (100 \text{ kPa})/(9.8 \times 6 \text{ m}^2/\text{s}^2) = 1.70 \times 10^3 \text{ kg/m}^3$

Problem

21. A vertical tube open at the top contains 5.0 cm of oil (density 0.82 g/cm^3) floating on 5.0 cm of water. Find the *gauge* pressure at the bottom of the tube.

Solution

The pressure at the top of the tube is atmospheric pressure, P_a. The absolute pressure at the interface of the oil and water is $P_i = P_a + \rho_{oil}gh_{oil}$, and at the bottom is $P = P_i + \rho_{water}gh_{water} = P_a + \rho_{oil}gh_{oil} + \rho_{water}gh_{water}$ (see Equation 18-3). Therefore, the gauge pressure at the bottom is $P - P_a = (\rho_{oil}h_{oil} + \rho_{water}h_{water})g = (0.82 + 1.00)(10^3 \text{ kg/m}^3)(0.05 \text{ m})(9.8 \text{ m/s}^2) = 892$ Pa (gauge).

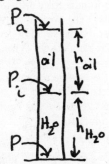

Problem 21 Solution.

Problem

25. A U-shaped tube open at both ends contains water and a quantity of oil occupying a 2.0-cm length of the tube, as shown in Fig. 18-44. If the oil's density is 0.82 times that of water, what is the height difference h?

Solution

From Equation 18-3, the pressure at points at the same level in the water is the same, $P_1 = P_2$. Now, $P_1 = P_{atm} + \rho_{H_2O}g(2 \text{ cm} - h)$ and $P_2 = P_{atm} + \rho_{oil}g(2 \text{ cm})$, so $h = (2 \text{ cm})(1 - \rho_{oil}/\rho_{H_2O}) = (2 \text{ cm})(1 - .82) = 3.6$ mm. (h is positive as shown).

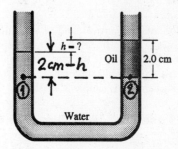

FIGURE 18-44 Problem 25 Solution.

Problem

29. A garage lift has a 45-cm diameter piston supporting the load. Compressed air with a maximum pressure of 500 kPa is applied to a small piston at the other end of the hydraulic system. What is the maximum mass the lift can support?

Solution

If we neglect the variation of pressure with height in the hydraulic system (which is usually small compared to the applied pressure), the fluid pressure is the same throughout, or $P_{appl} = F/A$ (for either the small or large cylinders). Thus, $F_{max} = (500 \text{ kPa})\frac{1}{4}\pi(0.45 \text{ m})^2 = 79.5$ kN, which corresponds to a mass-load of $F_{max}/g = 8.11$ tonnes (metric tons).

Section 18-3 Archimedes's Principle and Buoyancy

Problem

33. The density of styrofoam is 160 kg/m^3. What per cent error is introduced by weighing a styrofoam block in air, which exerts an upward buoyancy force, rather than in vacuum? The density of air is 1.2 kg/m^3.

Solution

The fractional error is $(W - W_{app})/W = \rho_{air}/\rho_{styro} = 1.2/160 = 0.75\%$ (see Example 18-4).

Problem

37. A typical supertanker has mass 2.0×10^6 kg and carries twice that much oil. If 9.0 m of the ship is submerged when it's empty, what is the minimum water depth needed for it to navigate when full? Assume the sides of the ship are vertical.

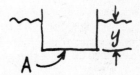

Problem 37 Solution.

Solution

If the sides of the hull are vertical, and its bottom flat, the volume it displaces is proportional to its draft (depth in the water), i.e., $V = Ay$, where A is the cross-sectional area. Since the total mass of the full supertanker is 3× that when empty, the draft when full is simply $3 \times (9 \text{ m}) = 27$ m.

Sections 18-4 and 18-5 Fluid Dynamics and Applications

Problem

41. A fluid is flowing steadily, roughly from left to right. At left it is flowing rapidly; it then slows down, and finally speeds up again. Its final speed at right is not as great as its initial speed at left. Sketch a streamline pattern that could represent this flow.

Solution

In order to maintain a constant volume rate of flow, in an incompressible fluid, streamlines must be closer together (smaller cross-section of tube of flow) where the velocity is greater, as sketched.

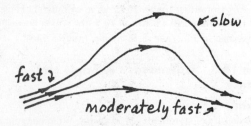

Problem 41 Solution.

Problem

45. A typical human aorta, or main artery from the heart, is 1.8 cm in diameter and carries blood at a speed of 35 cm/s. What will be the flow speed around a clot that reduces the flow area by 80%?

Solution

The continuity equation (Equation 18-5) is a reasonable approximation for blood circulation in an artery, so $v' = v(A/A')$. If the cross-sectional area is reduced by 80%, then $A/A' = 100\%/20\% = 5$, so $v' = 5(35 \text{ cm/s}) = 1.75$ m/s.

Problem

49. The water in a garden hose is at a gauge pressure of 140 kPa and is moving at negligible speed. The hose terminates in a sprinkler consisting of many small holes. What is the maximum height reached by the water emerging from the holes?

Solution

The pressure, velocity, and height of the water in the hose (point ①) are $P_1 = P_{atm} + 140$ kPa, $v_1 \approx 0$, and $y_1 = 0$, while at the highest point of a jet of water from a hole

(point ②), $P_2 = P_{atm}$, $v_2 \approx 0$, and $y_2 = h$. (We assume that the jets from the holes are the same.) Then Bernoulli's equation (Equation 18-6a) yields $P_{atm} + 140$ kPa $= P_{atm} + \rho gh$, or $h = 140 \text{ kPa}/(9800 \text{ N/m}^3) = 14.3$ m.

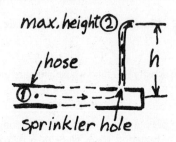

Problem 49 Solution.

Paired Problems

Problem

53. A steel drum has volume 0.23 m³ and mass 16 kg. Will it float in water when filled with (a) water or (b) gasoline (density 860 kg/m³)? Neglect the thickness of the steel.

Solution

An object will float in water if its average density is less than the density of water. (This follows from Archimedes' principle, since the volume of water displaced by an object floating on the surface is less than its total volume, i.e., $V_{dis} < V$. Because the buoyant force equals the weight of a floating object, $F_b = \rho_{H_2O} g V_{dis} = W = \rho_{av} gV$, this implies $\rho_{av} < \rho_{H_2O}$.)
(a) Since $\rho_{steel} > \rho_{H_2O}$, when the drum is filled with water, $\rho_{av} > \rho_{H_2O}$ and the drum will sink. (The average density of a composite object is always greater the the smallest density of its components; see solution to Problem 3.) (b) When the drum is filled with gasoline, its average density is $\rho_{av} = (M_{steel} + M_{gas})/V$. If we neglect the volume occupied by the steel compared to the volume of the drum, then $M_{gas} = \rho_{gas}V$ and $\rho_{av} = (16 \text{ kg}/0.23 \text{ m}^3) + 860 \text{ kg/m}^3 = 930 \text{ kg/m}^3$, which is less than ρ_{H_2O} so the drum floats.

Problem

57. Water at a pressure of 230 kPa is flowing at 1.5 m/s through a pipe, when it encounters an obstruction where the pressure drops by 5%. What fraction of the pipe's area is obstructed?

Solution

Assume horizontal flow in a narrow pipe (so there is no dependance on height). Then $v_1 A_1 = v_2 A_2$, and $P_1 + \frac{1}{2}\rho v_1^2 = P_2 + \frac{1}{2}\rho v_2^2$ (Equations 18-5 and 18-6a for steady incompressible fluid flow), where subscript 2 refers to the obstruction. Since the pressure at the obstruction is 5% less, $P_1 - P_2 = 0.05P_1 = \frac{1}{2}\rho v_1^2[(A_1^2/A_2^2) - 1]$, where we eliminated v_2. Then $(A_1/A_2)^2 = 1 + (0.1 \times 230 \text{ kPa})/(10^3 \text{ kg/m}^3)(1.5 \text{ m/s})^2 = (3.35)^2$. The fraction of area obstructed is $(A_1 - A_2)/A_1 = 1 - (1/3.35) = 70.1\%$.

Supplementary Problems

Problem

61. A 1.0-m-diameter tank is filled with water to a depth of 2.0 m and is open to the atmosphere at the top. The water drains through a 1.0-cm-diameter pipe at the bottom; that pipe then joins a 1.5-cm-diameter pipe open to the atmosphere, as shown in Fig. 18-52. Find (a) the flow speed in the narrow section and (b) the water height in the *sealed* vertical tube shown.

Solution

(a) If we assume a steady incompressible flow (Equation 18-5 and 6), an argument similar to Example 18-8, comparing point ① at the top of the tank with point ② at the opening of the 1.5 cm pipe, gives $P_a + \rho g y_1 \approx P_a + \rho g y_2 + \frac{1}{2}\rho v_2^2$ or $v_2 = \sqrt{2g(y_1 - y_2)}$. Here, we neglected the flow speed at the top, $v_1 \approx 0$, and $y_1 - y_2 = 2$ m. The continuity equation gives the speed in the narrower section of pipe, $v_3 = v_2(A_2/A_3) = (1.5/1.0)^2\sqrt{2(9.8 \text{ m/s}^2)(2 \text{ m})} = 14.1$ m/s.
(b) The pressure at the bottom of the stagnant column of water over the narrow section of pipe is $P_3 = \rho g h$, because there is no pressure exerted by a vacuum, and we assume this pressure is uniform over the cross-section of the narrow pipe. Another application of Bernoulli's equation gives $P_a + \rho g y_1 \approx P_3 + \rho g y_2 + \frac{1}{2}\rho v_3^2 = \rho g h + \rho g y_2 + (1.5)^4\rho g(y_1 - y_2)$, where we have again neglected the flow speed at the top of the tank, and we used the expression for v_3 from part (a). Therefore, $h = (P_a/\rho g) - (y_1 - y_2)((1.5)^4 - 1) = (101.3 \text{ kPa}/9800 \text{ N/m}^3) - (2 \text{ m})(65/16) = 2.21$ m.

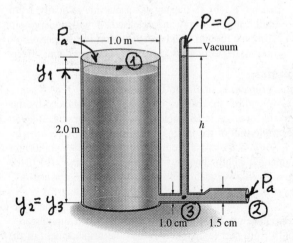

P_a y_1 1.0 m $P=0$ Vacuum   ①   2.0 m h $y_2 = y_3$ 1.0 cm   ③   1.5 cm   ② P_a

FIGURE 18-52 Problem 61 Solution.

Problem

65. With its throttle valve wide ope has a throat diameter of 2.4 cr engine draws 0.50 L of air th engine speed of 3000 rpm, what are
(b) the airflow speed, and (c) the difference bet
spheric pressure and air pressure in the carburetor throa
The density of air is 1.2 kg/m³.

Solution

(a) The volume rate of flow of intake air is $R_V = (3000 \text{ rpm})(1 \text{ min}/60 \text{ s})(0.5 \text{ L/rev}) = 25$ L/s $= 0.025$ m³/s.
(b) This rate of flow, assumed constant over the cross-sectional area of the carburetor throat, implies a flow speed of $v = R_V/A = (0.025 \text{ m}^3/\text{s})/\frac{1}{4}\pi(0.024 \text{ m})^2 = 55.3$ m/s.
(c) Since the flow speed is much smaller than the speed of sound in air at this density, we can use Equation 18-6 to calculate the pressure difference. We suppose that air enters the carburetor intake at a speed which is negligible compared to v, at essentially the same height as the throat. Then $\Delta P = P_a - P_{\text{throat}} = \frac{1}{2}\rho v^2 = \frac{1}{2}(1.2 \text{ kg/m}^3)(55.3 \text{ m/s})^2 = 1.83$ kPa.

Problem

69. A circular pan of liquid (density ρ) is centered on a horizontal turntable rotating with angular speed ω. Its axis coincides with the rotation axis, as shown in Fig. 18-55. Atmospheric pressure is P_a. Find expressions for (a) the pressure at the bottom of the pan and (b) the height of the liquid surface as functions of the distance r from the axis, given that the height at the center is h_0.

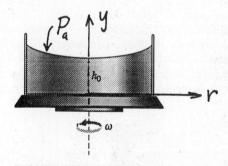

FIGURE 18-55 Problem 69.

Solution

When the water is in equilibrium at constant angular velocity, the vertical change in pressure balances the weight of the water, the radial change in pressure supplies the centripetal acceleration, and there is no change in pressure in the direction tangent to the rotation, Introduce vertical, radial and tangential coordinates, y, r, and φ respectively, with origin at the bottom center of the pan and y axis positive upward.

are cylindrical coordinates.) Consider a fluid element $\rho\, dV = \rho\, dr(rd\varphi)\, dy$ as shown. (ρ is the density and is the volume element.)

The vertical pressure difference balances the gravitational force, as in Equation 18-2, $dF_{\text{press}} + dF_{\text{grav}} = 0 = -dP_y A_y - \rho g\, dV$, or $\partial P/\partial y = -\rho g$. Here, $A_y = r\, dr\, d\varphi$ is the area of the faces perpendicular to the y direction, and we wrote a partial derivative because the pressure varies with both y and r. Note that $\partial P/\partial y$ is negative because the pressure increases with depth (decreasing y).

Similarly, the pressure force in the radial direction equals the mass element times the centripetal acceleration, $dF_{\text{press}} = -dm\,\omega^2 r = -dP_r A_r = -\rho\, dV\,\omega^2 r$. (Recall that $a_r = -v^2/r = -\omega^2 r$.) In this equation, $A_r = r\, d\varphi\, dy$ is the area of the faces perpendicular to the radial direction. Since $dV = A_r\, dr$, after canceling $-A_r$, we find $dP_r = \rho\omega^2 r\, dr$, or $\partial P/\partial r = \rho\omega^2 r$. Here, $\partial P/\partial r$ is positive because the pressure increases with r.

For an incompressible fluid, ρ is a constant (not a function of r and y), thus $\partial P/\partial y = -\rho g$ and $\partial P/\partial r = \rho\omega^2 r$ require the presence of terms equal to $-\rho g y$ and $\frac{1}{2}\rho\omega^2 r^2$ in the expression for the pressure (then the partial derivatives have their specified values). Thus $P(r, y) = -\rho g y + \frac{1}{2}\rho\omega^2 r^2 +$ constant. The constant term can be evaluated, since the pressure is atmospheric pressure at the surface above the center,

i.e., $P(0, h_0) = P_a = -\rho g h_0 +$ constant, or the constant $= P_a + \rho g h_0$. Then $P(r, y) = P_a - \rho g(y - h_0) + \frac{1}{2}\rho\omega^2 r^2$.
(a) Along the bottom of the pan ($y = 0$), $P(r, 0) = P_a + \rho g h_0 + \frac{1}{2}\rho\omega^2 r^2$. (b) The pressure at the water's surface is P_a for all values of r, so the height of the surface, $y = h(r)$, is given by the equation $P(r, h(r)) = P_a$, or $-\rho g[h(r) - h_0] + \frac{1}{2}\rho\omega^2 r^2 = 0$. Thus, $h(r) = h_0 + \omega^2 r^2/2g$, i.e., parabolic. (Such a technique is used to shape large mirrors for astronomical telescopes by a process called spin casting.)

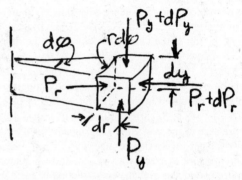

Problem 69 Solution.

● PART 2 CUMULATIVE PROBLEMS

Problem

1. A cylindrical log of total mass M and uniform diameter d has an uneven mass distribution that causes it to float in a vertical position, as shown in Figure 1. (a) Find an expression for the length ℓ of the submerged portion of the log when it is floating in equilibrium, in terms of M, d, and the water density ρ. (b) If the log is displaced vertically from its equilibrium position and released, it will undergo simple harmonic motion. Find an expression for the period of this motion, neglecting viscosity and other frictional effects.

Solution

(This problem is similar to Problem 18-68.)
(a) At equilibrium, the weight of the log is balanced by the buoyant force, as in Example 18-6. The former has magnitude Mg, while that of the latter is $F_b = \rho g V_{\text{sub}} = \rho g A\ell$, where $A = \frac{1}{4}\pi d^2$ is the cross-sectional area and ℓ the equilibrium submerged length. Thus $\ell = M/\rho A = 4M/\rho\pi d^2$. (b) If the log is given a vertical displacement y (positive upwards as shown), the net force (neglecting frictional effects) is $F_b - Mg = \rho g A(\ell - y) - \rho g A\ell = -(Mg/\ell)y$, where we expressed the weight in terms of the equilibrium submerged length from part (a). Since y is also the displacement of the log's center of mass from its equilibrium position, the net force equals $M\, d^2y/dt^2$, or $d^2y/dt^2 = -(g/\ell)y$. This is the equation for simple harmonic motion with frequency $\omega^2 = g/\ell$ and period $T = 2\pi/\omega = 2\pi\sqrt{\ell/g} = 4\sqrt{\pi M/\rho g d^2}$.

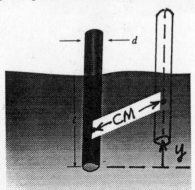

FIGURE 1 Cumulative Problem 1 Solution.

Problem

5. A U-shaped tube containing liquid is mounted on a table that tilts back and forth through a slight angle, as shown in Fig. 4. The diameter of the tube is much less than either the height of it's arms or their separation. When the table is rocked very slowly or very rapidly, nothing particularly dramatic happens. But when the rocking takes place at a few times per second, the liquid level in the tube oscillates violently, with maximum amplitude at a rocking frequency of 1.7 Hz. Explain what is going on, and find the total length of the liquid including both vertical and horizontal portions.

Solution

When the tube is rocked back and forth, the liquid in it is dragged along by viscous forces. We suppose that the dimensions of the tube (its small diameter compared to the height of the tube arms or their separation) allow us to treat the column of liquid, of total length ℓ, as a one-dimensional system which undergoes underdamped oscillations with weak damping (as in Section 15-7). The maximum amplitude occurs at a driving frequency very close to the undamped natural frequency, $\omega_d = 2\pi(1.7 \text{ Hz}) \approx \omega_0$. To find ω_0 in terms of ℓ, suppose one end of the liquid column is depressed a distance x from equilibrium, as shown. The net restoring force is the weight of a length $2x$ of liquid, or $F = -(\Delta m)g = -\rho gA\cdot 2x$, where A is the cross-sectional area of the tube and ρ is the density of the liquid. The mass of the entire column is $\rho A\ell$, so Newton's second law gives $-2\rho gAx = \rho A\ell \; d^2x/dt^2$. This is the equation for simple harmonic motion with natural frequency $\omega_0^2 = 2g/\ell$. In this case, $\ell = 2g/\omega_0^2 = 2(9.8 \text{ m/s}^2)/(2\pi \times 1.7 \text{ Hz})^2 = 17.2 \text{ cm}$.

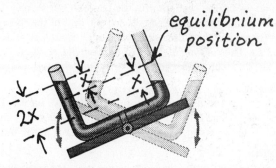

FIGURE 4 Cumulative Problem 5 Solution.

PART 3 THERMODYNAMICS

● CHAPTER 19 TEMPERATURE AND HEAT

Section 19-1 Macroscopic and Microscopic Descriptions

Problem

1. The macroscopic state of a carton capable of holding a half-dozen eggs is specified by giving the number of eggs in the carton. The microscopic state is specified by telling where each egg is in the carton. How many microscopic states correspond to the macroscopic state of a full carton?

Solution

If we number the eggs (so that they are distinguishable), we could put the first egg in any one of six places, the second in any one of the remaining five places, etc. The total number of microscopic states is $6\cdot5\cdot4\cdot3\cdot2\cdot1 = 6! = 720$. (If the eggs are indistinguishable, there is only one microscopic state for a full carton.)

Section 19-3 Measuring Temperature

Problem

5. At what temperature do the Fahrenheit and Celsius scales coincide?

Solution

In Equation 19-3, T_F and T_C are numerically equal when $T_F = (\frac{9}{5})T_C + 32 = T_C$, or $T_C = -(\frac{5}{4})(32) = -40 = T_F$.

Problem

9. A constant-volume gas thermometer is filled with air whose pressure is 101 kPa at the normal melting point of ice. What would its pressure be at (a) the normal boiling point of water, (b) the normal boiling point of oxygen (90.2 K), and (c) the normal boiling point of mercury (630 K)?

Solution

The thermometric equation for an ideal constant-volume gas thermometer is $P/T = P_{ref}/T_{ref}$. (This is Equation 19-1 written for a reference point not necessarily equal to the triple point of water.) If we use the given values at the normal melting point of ice, $P = T(101 \text{ kPa})/(273.15 \text{ K})$. When the temperatures of the normal boiling points of water (100°C = 373.15 K), oxygen (90.2 K), and mercury (630 K) are substituted, pressures of (a) 138 kPa, (b) 33.4 kPa, and (c) 233 kPa are calculated.

Problem

13. A constant-volume gas thermometer supports a 72.5-mm-high mercury column when it's immersed in liquid nitrogen at −196°C. What will be the column height when the thermometer is in molten lead at 350°C?

Solution

For a constant-volume gas thermometer, P/T is constant (see Equation 19-1). Since pressure can be measured in mm of mercury ($P = \rho g h$) it is also true that h/T is constant. Under the conditions specified for liquid nitrogen and molten load, $h/T = h'/T'$ implies $h' = (T'/T)h = (7.25 \text{ mm})(350 + 273)/(-196 + 273) = 587 \text{ mm}$. (Note: the temperature in Equation 19-1 is the absolute Kelvin temperature.)

Sections 19-4 and 19-5 Temperature and Heat, Heat Capacity and Specific Heat

Problem

17. Typical fats contain about 9 kcal per gram. If the energy in body fat could be utilized with 100% efficiency, how much mass could a 78-kg person lose running a 26.2-mile marathon? The energy expenditure rate for that mass is 125 kcal/mile.

Solution

The energy expended in running a marathon for a person with the given mass is $(125 \text{ kcal/mi})(26.2 \text{ mi}) = 3.28 \times 10^3$ kcal. This is equivalent to the energy content of $3.28 \times 10^3 \text{ kcal}/(9 \text{ kcal/g}) = 364$ g, or about 13 oz, of fat.

Problem

21. How much heat is required to raise an 800-g copper pan from 15°C to 90°C if (a) the pan is empty; (b) the pan contains 1.0 kg of water; (c) the pan contains 4.0 kg of mercury?

Solution

(a) When just the pan is heated, $\Delta Q = m_{Cu} c_{Cu} \Delta T = (0.8 \text{ kg})(386 \text{ J/kg·K})(90 - 15)\text{K} = 23.2 \text{ kJ} = 5.54$ kcal.
(b) If the pan contains water and both are heated between the same temperatures, $\Delta Q = (m_{Cu} c_{Cu} + m_w c_w) \Delta T = 23.2 \text{ kJ} + (1 \text{ kg})(4184 \text{ J/kg·K})(75 \text{ K}) = 337 \text{ kJ} = 80.5$ kcal.

(c) With 4 kg of mercury replacing the water, $\Delta Q = 23.2 \text{ kJ} + (4 \text{ kg})(140 \text{ J/kg·K})(75 \text{ K}) = 65.2 \text{ kJ} = 15.6$ kcal. (See Table 19-1 for the specific heats.)

Problem

25. You insert your microwave oven's temperature probe in a roast and start it cooking. You notice that the temperature goes up 1°C every 20 s. If the roast has the same specific heat as water, and if the oven power is 500 W, what is the mass of the roast? Neglect heat loss.

Solution

With no losses, the heat absorbed by the roast per second (power) equals the mass times the specific heat of the roast times the temperature rise per second, or $\mathcal{P} = \Delta Q/\Delta t = mc\,\Delta T/\Delta t$. Thus $m = (500 \text{ W})/(4184 \text{ J/kg·K})(1 \text{ K}/20 \text{ s}) = 2.39$ kg, or about $5\frac{1}{4}$ lb.

Problem

29. A 1.2-kg iron tea kettle sits on a 2.0-kW stove burner. If it takes 5.4 min to bring the kettle and the water in it from 20°C to the boiling point, how much water is in the kettle?

Solution

The energy supplied by the stove burner heats the kettle and water in it from 20°C to 100°C. If we neglect any losses of heat and the heat capacity of the burner, this energy is just the burner's power output times the time, so $\Delta Q = \mathcal{P}\,\Delta t = (m_w c_w + m_K c_K)\,\Delta T$ (Equation 19-5 for water and kettle). Since all of these quantities are given except for the mass of the water, we can solve for m_w:

$$m_w = [(\mathcal{P}\,\Delta t/\Delta T) - m_K c_K]/c_w$$
$$= \{[(2 \text{ kW})(5.4 \times 60 \text{ s})/80 \text{ K}] -$$
$$(1.2 \text{ kg})(447 \text{ J/kg·K})\}/(4184 \text{ J/kg·K})$$
$$= 1.81 \text{ kg}.$$

Problem

33. A leaf absorbs sunlight with intensity 600 W/m². The leaf has a mass per unit area of 100 g/m², and its specific heat is 3800 J/kg·K. In the absence of any heat loss, at what rate would the leaf's temperature rise?

Solution

The derivative of Equation 19-5 with respect to time relates the rate of temperature rise to the rate of heat energy absorbed, $dQ/dt = mc(dT/dt)$. Using the values of dQ/dt and m given for unit areas of leaf, one finds $dT/dt = (600 \text{ W/m}^2)/(0.1 \text{ kg/m}^2)(3800 \text{ J/kg·K}) = 1.58$ K/s.

Problem

37. A thermometer of mass 83.0 g is used to measure the temperature of a 150-g water sample. The thermometer's specific heat is 0.190 cal/g·°C, and it reads 20.0°C before immersion in the water. The water temperature is initially 60.0°C. What does the thermometer read after it comes to equilibrium with the water?

Solution

If we assume that the thermometer and water are thermally insulated from their surroundings, Equation 19-6 (and its solution from Example 19-4) gives:

$$T = \frac{m_t c_t T_t + m_w c_w T_w}{m_t c_t + m_w c_w}$$

$$= \frac{(83.0)(0.190)(20.0°C) + (150)(1)(60.0°C)}{(83.0)(0.190) + (150)(1)}$$

$$= 56.2°C,$$

(we omitted common units in the numerator and denominator).

Sections 19-6 and 19-7 Heat Transfer and Thermal Energy Balance

Problem

41. Building heat loss in the United States is usually expressed in Btu/h. What is 1 Btu/h in SI?

Solution

The conversion to SI units is $(1 \text{ Btu/h})(1055 \text{ J/Btu}) \times (1 \text{ h}/3600 \text{ s}) = 0.293 \text{ W}$.

Problem

45. A biology lab's walk-in cooler measures 3.0 m × 2.0 m × 2.3 m and is insulated with 8.0-cm-thick styrofoam. If the surrounding building is at 20°C, at what average rate must the cooler's refrigeration unit remove heat in order to maintain 4.0°C in the cooler?

Solution

The total surface area (sides, top and bottom) of the cooler is $A = 2(3 \times 2 + 3 \times 2.3 + 2 \times 2.3) \text{ m}^2 = 35 \text{ m}^2$. A thickness of 8 cm of styrofoam of this area has a thermal resistance of $R = \Delta x/kA = (0.08 \text{ m})/(0.029 \text{ W/m·K})(35 \text{ m}^2) = 7.88 \times 10^{-2} \text{ K/W}$ (see Equation 19-9). Therefore, a heat-flow of magnitude $|\Delta T|/R = (20°C - 4°C)/(7.88 \times 10^{-2} \text{ K/W}) = 203 \text{ W}$ (see Equation 19-7) must be balanced by the refrigeration unit to maintain the desired steady state temperatures.

Problem

49. Repeat the preceding problem for a south-facing window where the average sunlight intensity is 180 W/m².

Solution

The difference in heat loss between R-factors of 19 and 2.1, for a window area of 40 ft² and given temperature difference, is $\Delta H = -A \, \Delta T(\mathfrak{R}_2^{-1} - \mathfrak{R}_1^{-1}) =$
$-(40 \text{ ft}^2)(15°F - 68°F)(2.1^{-1} - 19^{-1})(\text{ft}^2 \cdot \text{F}° \cdot \text{h/Btu})^{-1} =$
898 Btu/h. (A positive ΔH represents a greater heat-flow from inside the house to outside, or a loss of energy.) Over a winter month, $(898 \text{ Btu/h})(30 \text{ d})(24 \text{ h/d})(1 \text{ gal}/10^5 \text{ Btu}) = 6.47$ gal of oil would have to be consumed to compensate for this loss.

On the other hand, if a southern window location resulted in a net gain from solar power of $(180 \text{ W/m}^2)(1 \text{ Btu}/1055 \text{ J}) \times (3600 \text{ s/h})(40 \text{ ft}^2)(0.3048 \text{ m/ft})^2 = 2284 \text{ Btu/h}$, this would be equivalent to a savings of
$(2284 \text{ Btu/h})(30 \text{ d})(24 \text{ h/d})(1 \text{ gal}/10^5 \text{ Btu}) = 16.4$ gal of oil over a month. The resulting net savings is $16.4 - 6.47 = 9.97$ gal of oil for one winter month.

Problem

53. An electric stove burner has surface area 325 cm² and emissivity $e = 1.0$. The burner is at 900 K and the electric power input to the burner is 1500 W. If room temperature is 300 K, what fraction of the burner's heat loss is by radiation?

Solution

The net power radiated (emitted at T_1, absorbed at T_2) is
$\mathcal{P} = e\sigma A \, (T_1^4 - T_2^4)$
$= (1)(5.67 \times 10^{-8} \text{ W/m}^2 \cdot \text{K}^4)(3.25 \times 10^{-2} \text{ m}^2)(300 \text{ K})^4(3^4 - 1)$
$= 1194 \text{ W}$. This is 79.6% of the input power (1500 W).

Problem

57. Scientists worry that a nuclear war could inject enough dust into the upper atmosphere to reduce significantly the amount of solar energy reaching Earth's surface. If an 8% reduction in solar input occurred, what would happen to Earth's 287-K average temperature?

Solution

If we assume that the Earth's average temperature is proportional to the one-fourth power of the effective solar intensity ($T_{av} \sim S^{1/4}$), as explained in the text's application to the greenhouse effect and global warming, then reducing the intensity to $0.92S$ alters the average temperature according to $T'_{av} = T_{av}(0.92)^{1/4}$. This would result in a decrease in the present $T_{av} = 287$ K of $\Delta T_{av} = T_{av} - T'_{av} = [1 - (0.92)^{1/4}](287 \text{ K}) = 5.92 \text{ K}$.

Paired Problems

Problem

61. What is the power output of a microwave oven that can heat 430 g of water from 20°C to the boiling point in 5.0 minutes? Neglect the heat capacity of the container.

Solution

The average power supplied to the water is $\mathcal{P} = \Delta Q/\Delta t = mc\,\Delta T/\Delta t = (430\text{ g})(1\text{ cal/g·C°})(100°C - 20°C) \times (4.184\text{ J/cal})/(5 \times 60\text{ s}) = 480$ W. This is also the output of the microwave, if we neglect the power absorbed by the container and any leakage in the unit.

Problem

65. An enclosed rabbit hutch has a thermal resistance of 0.25 K/W. If you put a 50-W heat lamp in the hutch on a day when the outside temperature is $-15°C$, what will be the hutch temperature? Neglect the rabbit's metabolism.

Solution

The rate of heat loss of the hutch by conduction is $H = -\Delta T/R = -[(-15°C) - T]/(0.25\text{ K/W}) = (4\text{ W/K})(T + 15°C)$ (see Equations 19-7 and 9). If we neglect any heat loss by radiation and convection, and any heat generated by the rabbit, this is equal to the power supplied by the 50 W lamp at the equilibrium temperature of the hutch. Thus, $T = [50\text{ W}/(4\text{ W/K})] - 15°C = -2.5°C$. (Perhaps the rabbit would be more comfortable with a 100 W lamp.)

Supplementary Problems

Problem

69. My house currently burns 160 gallons of oil in a typical winter month when the outdoor temperature averages 15°F and the indoor temperature averages 66°F. Roof insulation consists of $\mathcal{R} = 19$ fiberglass, and the roof area is 770 ft². If I double the thickness of the roof insulation, by what percentage will my heating bills drop? A gallon of oil yields about 100,000 Btu of heat.

Solution

The rate of heat loss by conduction from the currently insulated house can be written as $H_0 = -\Delta T[(A/\mathcal{R})_{\text{roof}} + (A/\mathcal{R})_{\text{rest}}]$, where $\Delta T = 15°F - 66°F = -51$ F°, $A_{\text{roof}} = 770$ ft², $\mathcal{R}_{\text{roof}} = 19$ (ft²·F°·h/Btu), and A_{rest} and $\mathcal{R}_{\text{rest}}$ are the effective area and R-factor for the rest of the house (i.e., walls, windows, floor, etc.) If the R-factor for the roof is doubled, the rate of heat loss will be changed by $\Delta H = \Delta T A_{\text{roof}}/2\mathcal{R}_{\text{roof}}$, which can be calculated from the given data: $\Delta H = (-51\text{ F}°)(770\text{ ft}^2)/2(19\text{ ft}^2\cdot\text{F}°\cdot\text{h/Btu}) = -1.03\times10^3$ Btu/h. (A negative change represents a reduction in heat loss; or a drop in heating costs.) The original rate of

heat loss can be calculated from the given oil consumption: $H_0 = (160\text{ gal/mo.})(10^5\text{ Btu/gal})(1\text{ mo.}/30 \times 24\text{ h}) = 2.22\times10^4$ Btu/h. Thus, the extra insulation would result in a savings of $|\Delta H/H_0| = 1.03/22.2 = 4.65\%$.

Problem

73. At low temperatures the specific heat of a solid is approximately proportional to the cube of the temperature; for copper the specific heat is given by $c = 31(T/343\text{ K})^3$ J/g·K. When heat capacity is not constant, Equations 19-4 and 19-5 must be written in terms of the derivative dQ/dT and integrated to get the total heat involved in a temperature change. Find the heat required to bring a 40-g sample of copper from 10 K to 25 K.

Solution

If we write Equation 19-5 in the form $dQ = mc(T)\,dT$ and integrate, as suggested in the problem, we obtain

$$Q = m\int_{T_1}^{T_2} c(T)\,dT = m\int_{10\text{ K}}^{25\text{ K}} (31\text{ J/g·K})(T/343\text{ K})^3\,dT$$
$$= (40\text{ g})(31\text{ J/g·K})\frac{(25\text{ K})^4 - (10\text{ K})^4}{4(343\text{ K})^3} = 2.92\text{ J} = 0.699\text{ cal}.$$

Problem

77. A house is at 20°C on a winter day when the outdoor temperature is $-15°C$. Suddenly the furnace fails. Use the result of the previous problem to determine how long it will take the house temperature to reach the freezing point. The heat capacity of the house is 6.5 MJ/K, and its thermal resistance is 6.67 mK/W.

Solution

In the previous problem, the heat flow from the object is proportional to the temperature difference between it and the surroundings, $H = dQ/dt = -(T - T_0)/R$. This can be transformed into the desired equation by using the chain rule and the definition of heat capacity: $dQ/dt = (dQ/dT)(dT/dt) = C(dT/dt) = -(T - T_0)/R$. If we separate variables and integrate from initial values $t = 0$ and T_1, to arbitrary final values t and T, we obtain:

$$\int_{T_1}^{T} \frac{dT'}{T' - T_0} = \ln\left(\frac{T - T_0}{T_1 - T_0}\right) = -\int_0^t \frac{1}{RC}\,dt' = -\frac{t}{RC},$$

or $T - T_0 = (T_1 - T_0)\exp(-t/RC)$. The temperature at $t = 0$ was chosen to be T_1; as $t \to \infty$, $T - T_0 \to 0$, i.e., the body cools to the temperature of its surroundings.

This result can be applied to the house described in this problem: $(T - T_0)/(T_1 - T_0) = [0°C - (-15°C)]/[20°C - (-15°C)] = 3/7 = \exp(-t/RC)$, or $t = RC\ln(7/3) = (6.67\text{ mK/W})(6.5\text{ MJ/K})\ln(7/3) = 3.67\times10^4\text{ s} = 10.2$ h (enough time for the emergency service to arrive?).

● **CHAPTER 20** THE THERMAL BEHAVIOR OF MATTER

Section 20-1 Gases

Problem

1. Mars's atmosphere has a pressure only 0.0070 times that of Earth, and an average temperature of 218 K. What is the volume of 1 mole of the Martian atmosphere?

Solution

The molar volume of an ideal gas at STP for the surface of Mars can be calculated as in Example 20-1. However, expressing the ideal gas law for 1 mole of gas at the surfaces of Mars and Earth as a ratio, $P_M V_M/T_M = P_E V_E/T_E$, and using the previous numerical result, we find $V_M = (P_E/P_M)(T_M/T_E)V_E = (1/0.0070)(218/273)(22.4\times10^{-3} \text{ m}^3) = 2.56$ m³.

Problem

5. If 2.0 mol of an ideal gas are at an initial temperature of 250 K and pressure of 1.5 atm, (a) what is the gas volume? (b) The pressure is now increased to 4.0 atm, and the gas volume drops to half its initial value. What is the new temperature?

Solution

(a) From Equation 20-2:

$$V = \frac{nRT}{P} = \frac{(2 \text{ mol})(8.314 \text{ J/mol·K})(250 \text{ K})}{(1.5 \text{ atm})(1.013\times10^5 \text{ Pa/atm})}$$
$$= 2.74\times10^{-2} \text{ m}^3 = 27.4 \text{ L}.$$

(b) The ideal gas law in ratio form (for a fixed quantity of gas, $N_1 = N_2$) gives:

$$\frac{T_2}{T_1} = \frac{P_2 V_2}{P_1 V_1}, \text{ or } T_2 = \left(\frac{4.0 \text{ atm}}{1.5 \text{ atm}}\right)\left(\frac{0.5 V_1}{V_1}\right)(250 \text{ K}) = 333 \text{ K}.$$

Problem

9. A helium balloon occupies 8.0 L at 20°C and 1.0 atm pressure. The balloon rises to an altitude where air pressure is 0.65 atm and the temperature is −10°C. What is its volume when it reaches equilibrium at the new altitude?

Solution

Use the ideal gas law (Equation 20-1) in ratio form, to compare two different states: $P_1 V_1/P_2 V_2 = N_1 T_1/N_2 T_2$. Since the balloon contains the same number of molecules of gas (if none escape), $N_1 = N_2$, and $V_2 = (P_1/P_2)(T_2/T_1) V_1 = (1/0.65)(263/293)(8.0 \text{ L}) = 11.0$ L. (Note that absolute temperatures must be used, and that any consistent units for the ratio of the other quantities conveniently cancel.)

Problem

13. A 3000-ml flask is initially open while in a room containing air at 1.00 atm and 20°C. The flask is then closed, and immersed in a bath of boiling water. When the air in the flask has reached thermodynamic equilibrium, the flask is opened and air allowed to escape. The flask is then closed and cooled back to 20°C. (a) What is the maximum pressure reached in the flask? (b) How many moles escape when air is released from the flask? (c) What is the final pressure in the flask?

Solution

The initial conditions of the gas are $P_1 = 1$ atm, $V_1 = 3$ L, $T_1 = 293$ K, and $n_1 = P_1 V_1/RT_1 = (1 \text{ atm})(3 \text{ L})/(8.206\times10^{-2} \text{ L·atm/mol·K})(293 \text{ K}) = 0.125$ mol. (a) T_2 is the temperature of boiling water at 1 atm of pressure, or 373 K. Since the original quantity of gas was heated at constant volume, $P_2 = (T_2/T_1)P_1 = (373/293)(1 \text{ atm}) = 1.27$ atm, which is the maximum. (b) When the flask is opened at 373 K, the pressure decreases to 1 atm, so the quantity of gas remaining is $n_2 = P_1 V_1/RT_2 = n_1(T_1/T_2) = (0.125 \text{ mol})(293/373) = 0.0980$ mol. Therefore, the amount which escaped was $n_1 - n_2 = 0.0268$ mol. (c) The pressure of the remaining gas is $P_3 = n_2 RT_1/V_1 = (n_2/n_1)P_1 = (0.098/0.125)(1 \text{ atm}) = 0.786$ atm.

Problem

17. At what temperature would the thermal speed of nitrogen molecules in air be 1% of the speed of light? Why might it not be possible to achieve this temperature in N_2 gas?

Solution

With $v_{th} = 0.01 c = 3\times10^6$ m/s and $m = 28$ u for molecules of N_2, Equation 20-3 gives $T = mv_{th}^2/3k = (28 \times 1.66\times10^{-27} \text{ kg})(3\times10^6 \text{ m/s})^2/3(1.38\times10^{-23} \text{ J/K}) = 1.01\times10^{10}$ K. At this temperature, which is nearly 700 times the temperature at the center of the sun, N_2 molecules would dissociate instantly.

Problem

21. Because the correction terms ($n^2 a/V^2$ and $-nb$) in the van der Waals equation have opposite signs, there is a point at which the van der Waals and ideal gas equations predict the same temperature. For the gas of Example 20-3, at what pressure does that occur?

Solution

The van der Waals and ideal gas temperatures are the same when $(P + n^2 a/V^2)(V - nb) = PV$, or $P = na(V - nb)/bV^2$. If $n = 1$, $V = 2$ L and the values of a and b from Example 20-3 are substituted, one finds $P = 1.76$ MPa.

Section 20-2   Phase Changes

Problem

25. How much energy does it take to melt a 65-g ice cube?

Solution

The energy required for a solid-liquid phase transition at the normal melting point of water (0°C) is (Equation 20-6)
$Q = mL_f = (0.065 \text{ kg})(334 \text{ kJ/kg}) = 21.7 \text{ kJ}$, or 5.19 kcal. (See Table 20-1 for the heats of transformation.)

Problem

29. Find the energy needed to convert 28 kg of liquid oxygen at its boiling point into gas.

Solution

From Equation 20-6 and Table 20-1, $Q = mL_v =$ (28 kg)(213 kJ/kg) = 5.96 MJ.

Problem

33. What is the power of a microwave oven that takes 20 min to boil dry a 300-g cup of water initially at its boiling point?

Solution

The energy required to vaporize the water at 100°C in 20 min is $Q = mL_v$ (see Equation 20-6 and Table 20-1), so the average power supplied by the microwave oven was
$\mathcal{P} = Q/t = (0.3 \text{ kg})(2257 \text{ kJ/kg})/(20 \times 60 \text{ s}) = 564 \text{ W}$.

Problem

37. At its "thaw" setting a microwave oven delivers 210 W. How long will it take to thaw a frozen 1.8-kg roast, assuming the roast is essentially water and is initially at 0°C?

Solution

Using the same reasoning as in the solution to Problem 35, we find $t = mL_f/\mathcal{P} = (1.8 \text{ kg})(334 \text{ kJ/kg})/(210 \text{ W}) = 2.86 \times 10^3$ s = 47.7 min.

Problem

41. How much energy does it take to melt 10 kg of ice initially at $-10°C$? Consult Table 19-1.

Solution

Energy must be supplied to first raise the ice temperature to the melting point, and then change its phase. Thus,
$Q = mc_{ice} \Delta T + mL_f = (10 \text{ kg})[(2.05 \text{ kJ/kg·K})(10 \text{ K}) + 334 \text{ kJ/kg}] = 3.55$ MJ. (Combine Equations 19-5 and 20-6.)

Problem

45. What is the minimum amount of ice in Example 20-6 that will ensure a final temperature of 0°C?

Solution

In order for the final equilibrium temperature in Example 20-6 to be 0°C, the original 1 kg of water must lose at least $Q_2 = 62.8$ kJ of heat energy. (It could lose more, if some or all of it froze, but this would clearly require a greater amount

of original ice.) The minimum amount of original ice that could gain Q_2, without exceeding 0°C, would be just completely melted, so $Q_2 = m_{ice}(c_{ice} \Delta T + L_f) = m_{ice}[(2.05 \text{ kJ/kg·K})(0°C - (-10°C)) + 334 \text{ kJ/kg}]$. Therefore $m_{ice} = 62.8 \text{ kJ}/(354.5 \text{ kJ/kg}) = 177$ g. (Note: The maximum amount of original ice, which could produce a final temperature of 0°C, while freezing all the original water, is $m_w(c_w \Delta T_w + L_f)/c_{ice} \Delta T_{ice} = (62.8 + 334)\text{kJ}/(20.5 \text{ kJ/kg}) = 19.4$ kg. Between these limits, $m_{ice} = m_w c_w \Delta T_w/c_{ice} \Delta T_{ice} = (1 \text{ kg})(4.184/2.05)(15/10) = 3.06$ kg gives a final mixture with the original amounts of water and ice at 0°C. For 177 g $< m_{ice} <$ 3.06 kg, some of the original ice melts, and for 3.06 kg $< m_{ice} <$ 19.4 kg, some of the original water freezes.)

Problem

49. A bowl contains 16 kg of punch (essentially water) at a warm 25°C. What is the minimum amount of ice at 0°C that will cool the punch to 0°C?

Solution

Assume that the only heat transfer is between the punch and the ice. To cool to 0°C, $\Delta Q =$ (16 kg)(4.184 kJ/kg·K)(25°C − 0°C) = 1.67 MJ of heat must be extracted from the punch. A minimum mass $m = \Delta Q/L_f = 1.67 \text{ MJ}/(334 \text{ kJ/kg}) = 5.01$ kg of ice at 0°C could do this, but the punch would be diluted with 5.01 kg of meltwater. (To reduce the dilution, sufficient ice at a temperature below 0°C is needed.)

Section 20-3   Thermal Expansion

Problem

53. A Pyrex glass marble is 1.00000 cm in diameter at 20°C. What will be its diameter at 85°C?

Solution

The linear expansion coefficient for Pyrex glass is given in Table 20-2, so we can calculate the diameter of the marble from Equation 20-8. $\Delta L = \alpha L \Delta T = (3.2 \times 10^{-6} \text{ K}^{-1}) \times (1 \text{ cm})(85°C - 20°C) = 2.08 \times 10^{-4}$ cm, thus $L' = L + \Delta L = 1.00021$. (Note: We expressed the diameter at 85°C to the same accuracy as that given for 20°C.)

Problem

57. The tube in a mercury thermometer is 0.10 mm in diameter. What should be the volume of the thermometer bulb if a 1.0-mm rise is to correspond to a temperature change of 1.0°C? Neglect the expansion of the glass.

Solution

There should be a volume of mercury (V) in the reservoir bulb of the thermometer, such that the change in its volume $(\Delta V = \beta V \Delta T)$ over the full range of temperature equals the volume of the tube into which it expands ($\frac{1}{4}\pi d^2 L$). Here, we have neglected the expansion of the glass, as suggested since $3\alpha_{glass} \ll \beta_{mercury}$ in Table 20-2. Thus $\frac{1}{4}\pi d^2 L = \beta V \Delta T$, or $V = (\frac{1}{4}\pi d^2/\beta)(L/\Delta T)$. Since the gradation of the thermometer, $L/\Delta T = 1$ mm/C°, is given, we find $V = \frac{1}{4}\pi(0.1 \text{ mm})^2(1 \text{ mm/K})/(18\times10^{-5}\text{ K}^{-1}) = 43.6 \text{ mm}^3$.

Problem

61. Lake Erie's volume is 480 km³. If the lake temperature were uniform (it isn't!) and increased by 0.50°C, by how much would the volume change if the initial temperature were (a) 1°C and (b) 20°C?

Solution

The appropriate expansion coefficients for water, in the temperature ranges specified, are listed in Table 20-2. Thus, (a) $\Delta V = \beta V \Delta T = (-4.8\times10^{-5}\text{ K}^{-1})(480 \text{ km}^3)(0.5 \text{ K}) = -0.0115 \text{ km}^3$ (a drop equivalent to 3×10^9 gal), and (b) $\Delta V = (20\times10^{-5}\text{ K}^{-1}) \times (480 \text{ km}^3)(0.5 \text{ K}) = 0.048 \text{ km}^3$ (an increase over four times greater).

Paired Problems

Problem

65. What power is needed to melt 20 kg of ice in 6.0 min?

Solution

If the melting occurs at atmospheric pressure and the normal melting point, the heat of transformation from Table 20-1 requires $\mathscr{P} = Q/t = mL_f/t = $ (20 kg)(334 kJ/kg)/(6 × 60 s) = 18.6 kW.

Problem

69. Describe the composition and temperature of the equilibrium mixture after 1.0 kg of ice at −40°C is added to 1.0 kg of water at 5.0°C.

Solution

Assume that all the heat lost by the water is gained by the ice. The temperature of the water drops and that of the ice rises. If either reaches 0°C, a change of phase occurs, freezing or melting, depending on which reaches 0°C first. The water would lose $mc \Delta T = $ (1 kg)(4.184 kJ/kg·K)(5 K) = 20.9 kJ of heat cooling to 0°C, while the ice would gain (1 kg)(2.05 kJ/kg·K)(40 K) = 82.0 kJ of heat warming to 0°C. Evidently, the water reaches 0°C first, and can still lose 334 kJ of heat, more than enough to bring the ice to 0°C, if all of it were to freeze. In fact, only 82.0 − 20.9 = 61.1 kJ of heat is transferred during the change of phase, therefore only 61.1 kJ/(334 kJ/kg) = 0.183 kg of water freezes. The final mixture is at 0°C and contains 1.183 kg of ice and 1 − 0.183 = 0.818 kg of water.

Supplementary Problems

Problem

73. A solar-heated house (Fig. 20-25) stores energy in 5.0 tons of Glauber salt ($Na_2SO_4 \cdot 10H_2O$), a substance that melts at 90°F. The heat of fusion of Glauber salt is 104 Btu/lb, and the specific heats of the solid and liquid are, respectively, 0.46 Btu/lb·°F and 0.68 Btu/lb·°F. After a week of sunny weather, the storage medium is all liquid at 95°F. Then a cool, cloudy period sets in during which the house loses heat at an average rate of 20,000 Btu/h. (a) How long is it before the temperature of the storage medium drops below 60°F? (b) How much of this time is spent at 90°F?

Solution

(a) In cooling from 95°F to 60°F (including the solidification at 90°F), the medium exhausts heat $Q = $ $m[c_{liquid}(95°F - 90°F) + L_f + c_{solid}(90°F - 60°F)] = $ 1.21×10^6 Btu, where given values of m, the specific heats, and the heat of transformation were substituted. If all this heat were supplied to the house, which loses energy at the average rate of 2×10^4 Btu/h, it would take $(1.21\times10^6/2\times10^4)$ h = 60.6 h for this to occur. (b) The time spent during just the solidification at 90°F is $mL_f/H = $ (5 × 2000 lb)(104 Btu/lb) $\div (2\times10^4$ Btu/h) = 52.0 h.

Problem

77. Water's coefficient of volume expansion in the temperature range from 0°C to about 20°C is given approximately by $\beta = a + bT + cT^2$, where T is in Celsius and $a = -6.43\times10^{-5}$ °C^{-1}, $b = 1.70\times10^{-5}$ °C^{-2}, and $c = -2.02\times10^{-7}$ °C^{-3}. Show that water has its greatest density at approximately 4.0°C.

Solution

We do not actually need to differentiate the density or the volume ($\rho(T) = $ const. mass/$V(T)$), since Equation 20-7 shows that $dV/dT = \beta V = 0$ when $\beta(T) = 0$. Thus, the maximum density (or minimum volume) occurs for a temperature satisfying $a + bT + cT^2 = 0$. The quadratic formula gives $T = (-b \pm \sqrt{b^2 - 4ac})/2c$, or since both a and c are negative, $T = (b \mp \sqrt{b^2 - 4|a||c|})/2|c|$. Canceling a factor of 10^{-5} from the given coefficients, we find $T = (1.70 \mp \sqrt{(1.70)^2 - 4(6.43)(0.0202)})°C/0.0404 = 3.97°C$. (The other root, 80.2°C, can be discarded because it is outside the range of validity, $0 \leq T \leq 20°C$, of the original $\beta(T)$.) Thus, the maximum density of water occurs at a temperature close to 4°C. (That this represents a minimum volume can be verified by plotting $V(T)$, or from the second derivative, $d^2V/dT^2 = V(d\beta/dT) + \beta(dV/dT) = V(\beta^2 + d\beta/dT) = V(\beta^2 + b + 2cT) > 0$ for $T = 3.97°C$.)

Problem

81. Prove Equation 20-9 by considering a cube of side s and therefore volume $V = s^3$ that undergoes a small temperature change dT and corresponding length and volume changes ds and dV.

Solution

For a cubical volume $V = L^3$, the expansion coefficients are related by:

$$\beta = \frac{1}{V}\frac{dV}{dT} = \frac{1}{V}\frac{dV}{dL}\frac{dL}{dT} = \frac{1}{L^3}(3L^2)\frac{dL}{dT} = 3\frac{1}{L}\frac{dL}{dT} = 3\alpha.$$

(We used Equations 20-7 and 8 in differential form, with L in place of s, and the chain rule for differentiation.) Alternatively, use the binomial approximation for $dV = (s + ds)^3 - s^3 = 3s^2 ds$. Since $dV = \beta V\, dT$ and $ds = \alpha s\, dT$, one finds $3s^2(\alpha s\, dT) = \beta s^3 dT$, or $\beta = 3\alpha$.

●

● **CHAPTER 21**   HEAT, WORK, AND THE FIRST LAW OF THERMODYNAMICS

Section 21-1   The First Law of Thermodynamics

Problem

1. In a perfectly insulated container, 1.0 kg of water is stirred vigorously until its temperature rises by 7.0°C. How much work was done on the water?

Solution

Since the container is perfectly insulated thermally, no heat enters or leaves the water in it. Thus, $Q = 0$ in Equation 21-1. The change in the internal energy of the water is determined from its temperature rise, $\Delta U = mc\,\Delta T$ (see comments in Section 19-4 on internal energy), so $W = -\Delta U = -(1\text{ kg})(4.184\text{ kJ/kg·K})(7\text{ K}) = -29.3\text{ kJ}$. (The negative sign signifies that work was done on the water.)

Problem

5. An engine produces useful work at the rate of 45 kW and heats its environment at the rate of 95 kW. At what rate does the engine extract energy from its fuel?

Solution

If we assume that the engine operates in a cycle and choose it as "the system", then $dU/dt = 0$ and Equation 21-2 implies $dQ/dt = dW/dt$. Here, dW/dt is the rate that the engine performs work on its surroundings (45 kW in this problem), and dQ/dt is the net rate of heat flow into the engine from the surroundings. Since the system is just the engine, the net heat flow is the difference between the heat extracted from its fuel and the heat exhausted to the environment, i.e., $dQ/dt = (dQ/dt)_{in} - (dQ/dt)_{out} = (dQ/dt)_{in} - 95\text{ kW}$. Therefore, $(dQ/dt)_{in} = 45\text{ kW} + 95\text{ kW} = 140\text{ kW}$. (Note: If the system is considered to be the engine and its fuel, as in

Example 21-1, then dW/dt is still 45 kW, but $dQ/dt = -95$ kW (there is no heat input). The system's internal energy decreases because energy is extracted from the fuel, so $dU/dt = dQ/dt - dW/dt = -95\text{ kW} - 45\text{ kW} = -140\text{ W}$.)

Section 21-2   Thermodynamic Processes

Problem

9. Repeat the preceding problem for a process that follows the path ACB in Fig. 21-27.

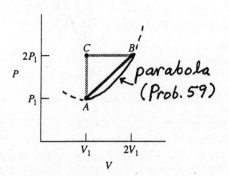

FIGURE 21-27   Problem 9.

Solution

AC is an isovolumic process, so $W_{AC} = 0$. CB is an isobaric process, so $W_{CB} = P_2(V_2 - V_1) = 2P_1(2V_1 - V_1) = 2P_1V_1$. Of course, $W_{ACB} = W_{AC} + W_{CB}$. (In the PV diagram, Fig. 21-27, the area under AC is zero, and that under CB, a rectangle, is $2P_1V_1$. Equation 21-3 could also be used.)

Problem

13. How much work does it take to compress 2.5 mol of an ideal gas to half its original volume while maintaining a constant 300 K temperature?

Solution

In an isothermal compression of a fixed quantity of ideal gas, work is done on the gas so W is negative in Equation 21-4. For the values given, $W = nRT \ln(V_2/V_1) =$ (25 mol)(8.314 J/mol·K)(300 K)ln($\frac{1}{2}$) $= -4.32$ kJ.

Problem

17. It takes 600 J to compress a gas isothermally to half its original volume. How much work would it take to compress it by a factor of 10 starting from its original volume?

Solution

For isothermal compressions starting from the same volume (and temperature), $W_{13}/W_{12} = \ln(V_3/V_1)/\ln(V_2/V_1)$ (see Equation 20-4). If $V_2 = V_1/2$, $V_3 = V_1/10$, and $W_{12} = -600$ J, then $W_{13} = (-600 \text{ J})(\ln 10)/(\ln 2) = -1.99$ kJ. (Note: $\ln x = -\ln (1/x)$.)

Problem

21. Repeat Problem 20 taking AB to be on an adiabat and using a specific heat ratio of $\gamma = 1.4$.

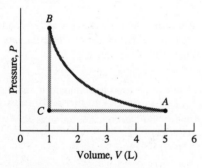

FIGURE 21-28 Problem 21.

Solution

(a) If AB represents an adiabatic process for an ideal gas, then the adiabatic law and the given values yield $P_B = P_A(V_A/V_B)^\gamma = (60 \text{ kPa})(5)^{1.4} = 571$ kPa. (b) The work done by the gas over the adiabat AB is $W_{AB} = (P_AV_A - P_BV_B)/(\gamma - 1) = [(60 \text{ kPa})(5 \text{ L}) - (571 \text{ kPa})(1 \text{ L})]/0.4 = -678$ J (see Equation 21-14). The process BC is isovolumic so $W_{BC} = 0$, and the process CA is isobaric so $W_{CA} = P_A(V_C - V_A) = (60 \text{ kPa})(5 \text{ L} - 1 \text{ L}) = 240$ J. The total work done by the gas is $W_{ABCA} = W_{AB} + W_{BC} + W_{CA} = -678 \text{ J} + 0 + 240 \text{ J} = -438$ J. The work done *on* the gas is the negative of this.

Problem

25. By how much must the volume of a gas with $\gamma = 1.4$ be changed in an adiabatic process if the kelvin temperature is to double?

Solution

$V/V_0 = (T_0/T)^{1/(\gamma-1)} = (0.5)^{1/0.4} = 0.177$ (Equation 21-13b).

Problem

29. A 2.0 mol sample of ideal gas with molar specific heat $C_V = \frac{5}{2}R$ is initially at 300 K and 100 kPa pressure. Determine the final temperature and the work done by the gas when 1.5 kJ of heat is added to the gas (a) isothermally, (b) at constant volume, and (c) isobarically.

Solution

(a) In an isothermal process, T is, of course, constant, so the final temperature is $T_2 = 300$ K. Since $\Delta U = 0$, $W = Q = 1.5$ kJ. (b) In an isovolumic process, $W = 0$ and $Q = nC_V\Delta T$. Therefore, $\Delta T = 1.5 \text{ kJ}/(2 \text{ mol})\frac{5}{2}R = 1.5 \text{ kJ}/(5 \times 8.314 \text{ J/K}) = 36.1$ K, and $T_2 = 300 \text{ K} + \Delta T = 336$ K. (c) In an isobaric process, $Q = nC_P \Delta T = n(C_V + R) \Delta T = n(\frac{7}{2} R) \Delta T$, so $\Delta T = 2Q/7nR = 2(1.5 \text{ kJ})/7(2 \times 8.314 \text{ J/K}) = 25.8$ K, and $T_2 = 326$ K. The work done is $W = P \Delta V = nR \Delta T = (R/C_P)Q = (\frac{2}{7})Q = 429$ J. (Refer to the relevant parts of Section 21-2 if necessary.)

Problem

33. The gas of Example 21-5 starts at state A in Fig. 21-20 and is heated at constant volume until its pressure has doubled. It's then compressed adiabatically until its volume is one-fourth its original value, then cooled at constant volume to 300 K, and finally allowed to expand isothermally to its original state. Find the net work done on the gas.

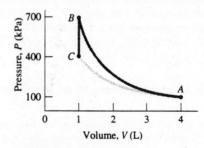

FIGURE 21-20 Problem 33.

Solution

The PV diagram for the cyclic process is shown. The work done *on* the gas, $-W_{ABCDA} = W_{ADCBA}$, can be calculated from the reversed cycle. AD is isothermal, so $W_{AD} = P_AV_A \ln(V_D/V_A) = (400 \text{ J})\ln(\frac{1}{4}) = 555$ J. DC and BA are isovolumic, so $W_{DC} = W_{BA} = 0$. CB is adiabatic, so $W_{CB} =$ $(P_C V_C - P_B V_B)/(\gamma - 1)$. Since $V_B = V_A$, $P_B = 2P_A$, and $P_C = P_B(V_B/V_C)^\gamma$, $W_{CB} = P_B V_B[(V_B/V_C)^{\gamma-1} - 1]/(\gamma - 1) =$ $2P_A V_A[(V_A/V_C)^{\gamma-1} - 1]/(\gamma - 1) =$ $2(400 \text{ J})(4^{0.4} - 1)/0.4 = 1482$ J. Finally, $W_{ADCBA} =$ -555 J $+ 0 + 1482$ J $+ 0 = 928$ J.

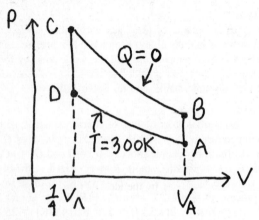

Problem 33 Solution.

Problem

37. A bicycle pump consists of a cylinder 30 cm long when the pump handle is all the way out. The pump contains air ($\gamma = 1.4$) at 20°C. If the pump outlet is blocked and the handle pushed until the internal length of the pump cylinder is 17 cm, by how much does the air temperature rise? Assume that no heat is lost.

Solution

If no heat is lost (or gained) by the gas, the compression is adiabatic and Equation 21-13b gives $TV^{\gamma-1} = T_0 V_0^{\gamma-1}$. Therefore, the temperature rise is $T - T_0 = \Delta T =$ $T_0[(V_0/V)^{\gamma-1} - 1]$. Since $V_0/V = (30 \text{ cm}/17 \text{ cm})$, $\Delta T =$ $[(30/17)^{0.4} - 1]$ (293 K) $= 74.7$ C°.

Section 21-3   Specific Heats of an Ideal Gas

Problem

41. What would be (a) the volume specific heat and (b) the specific heat ratio for a gas whose molecules had 9 degrees of freedom?

Solution

The equipartition theorem assigns $\frac{1}{2} kT$ averge internal energy for each degree of freedom of a molecule, so for N molecules, $U = N \times 9 \times \frac{1}{2} kT = (\frac{9}{2})nRT$. (Recall, $N = nN_A$ and $N_A k = R$, where N_A is Avogadro's number and n the number of moles.) Then (a) $C_V = (1/n)(dU/dT) = (\frac{9}{2})R$, and (b) $C_P/C_V = (C_V + R)/C_V = 1 + \frac{2}{9} = \frac{11}{9}$.

Problem

45. A gas mixture contains monatomic argon and diatomic oxygen. An adiabatic expansion that doubles its volume results in the pressure dropping to one-third of its original value. What fraction of the molecules are argon?

Solution

From the pressures and volumes in the described adiabatic expansion, $P_0 V_0^\gamma = (\frac{1}{3} P_0)(2V_0)^\gamma$, we can calculate that $\gamma =$ $\ln 3/\ln 2 = 1.58$. Then the result of Problem 43 gives $2.5 -$ $f_{Ar} = 1/0.58$, or $f_{Ar} = 79.0\%$.

Paired Problems

Problem

49. A 5.0 mol sample of ideal gas with $C_V = \frac{5}{2} R$ undergoes an expansion during which the gas does 5.1 kJ of work. If it absorbs 2.7 kJ of heat during the process, by how much does its temperature change? *Hint:* Remember that Equation 21-7 holds for *any* ideal gas process.

Solution

For any process connecting equilibrium states of an ideal gas, Equations 21-1 and 7 give $\Delta U = Q - W = nC_V \Delta T$, so $\Delta T = (2.7 \text{ kJ} - 5.1 \text{ kJ})/(5 \text{ mol})(\frac{5}{2} \times 8.314 \text{ J/mol·K}) =$ -23.1 K.

Problem

53. An ideal gas with $\gamma = 1.4$ is initially at 273 K and 100 kPa. The gas expands adiabatically until its temperature drops to 190 K. What is its final pressure?

Solution

The ideal gas law can be used to eliminate V and write the adiabatic law in terms of P and T: $PV^\gamma = P^{1-\gamma}(PV)^\gamma =$ $P^{1-\gamma}(nRT)^\gamma$, thus $P^{1-\gamma}T^\gamma$ is constant, or so is $PT^{\gamma/(1-\gamma)}$. For this problem, $P = P_0(T_0/T)^{\gamma/(1-\gamma)} =$ $(100 \text{ kPa})(273/190)^{-1.4/0.4} = 28.1$ kPa.

Supplementary Problems

Problem

57. An 8.5-kg rock at 0°C is dropped into a well-insulated vat containing a mixture of ice and water at 0°C. When equilibrium is reached there are 6.3 g less ice. From what height was the rock dropped?

Solution

The mechanical energy of the rock (originally gravitational potential energy) melted the ice (changed its internal energy) and no heat energy was transferred ($Q = 0$). Therefore, $-W = m_{rock}gh = \Delta U = m_{ice} L_f$, or $h = m_{ice} L_f/m_{rock}g =$ $(6.3 \text{ g})(334 \text{ J/g})/(8.5 \times 9.8 \text{ N}) = 25.3$ m. ($W < 0$ since the rock did work on the ice-water system.)

Problem

61. Show that the application of Equation 21-3 to an adiabatic process results in Equation 21-14.

Solution

The work done by an ideal gas undergoing an adiabatic process from state 1 to state 2 can be found by integration of the adiabatic law, $P = P_1 V_1^\gamma / V^\gamma$.

$$W_{12} = \int_{V_1}^{V_2} P \, dV = \int_{V_1}^{V_2} (P_1 V_1^\gamma) \frac{dV}{V^\gamma} = P_1 V_1^\gamma \left(\frac{V_2^{-\gamma+1} - V_1^{-\gamma+1}}{-\gamma + 1} \right)$$

$$= (P_1 V_1^\gamma V_1^{-\gamma+1} - P_2 V_2^\gamma V_2^{-\gamma+1})/(\gamma - 1),$$

which is Equation 21-14. (Note: $P_1 V_1^\gamma = P_2 V_2^\gamma$.)

Problem

65. Show that the work done by a van der Waals gas undergoing isothermal expansion from volume V_1 to V_2 is

$$W = nRT \ln\left(\frac{V_2 - nb}{V_1 - nb}\right) + an^2\left(\frac{1}{V_2} - \frac{1}{V_1}\right),$$

where a and b are the constants in Equation 20-5.

Solution

If we solve for P in the van der Waals equation of state, Equation 20-5, and substitute into Equation 21-3, we find

$$W = \int_{V_1}^{V_2} P \, dV = \int_{V_1}^{V_2} \left[\frac{nRT}{(V - nb)} - \frac{n^2 a}{V} \right] dV$$

$$= nRT \int_{V_1}^{V_2} \frac{dV}{V - nb} - n^2 a \int_{V_1}^{V_2} \frac{dV}{V^2}$$

$$= nRT \ln\left(\frac{V_2 - nb}{V_1 - nb}\right) + an^2\left(\frac{1}{V_2} - \frac{1}{V_1}\right),$$

where, for an isothermal process, T is a constant and can be taken out of the integral.

Problem

69. A cylinder of cross-sectional area A is closed by a massless piston. The cylinder contains n mol of ideal gas with specific heat ratio γ, and is initially in equilibrium with the surrounding air at temperature T_0 and pressure P_0. The piston is initially at height h_1 above the bottom of the cylinder. Sand is gradually sprinkled onto the piston until it has moved downward to a final height h_2. Find the total mass of the sand if the process is (a) isothermal and (b) adiabatic.

Solution

Initially, $P_1 = P_0$ (since the piston is massless) and $T_1 = T_0$. Finally, in equilibrium with its load of sand, M, the net force on the piston is zero, or $P_2 A = Mg + P_0 A$. Therefore, $M = (P_2 - P_0)A/g$. (a) In an isothermal process, $T_2 = T_1 = T_0$, and $P_2 V_2 = P_1 V_1$. Therefore, $P_2 - P_1 = P_1[(V_1/V_2) - 1] = P_0[(h_1/h_2) - 1]$, and $M = (P_0 A/g)[(h_1/h_2) - 1]$. (b) In an adiabatic process, $P_2 V_2^\gamma = P_1 V_1^\gamma$, or $P_2 - P_1 = P_1[(V_1/V_2)^\gamma - 1]$. Thus $M = (P_0 A/g)[(h_1/h_2)^\gamma - 1]$. (Note: the given data are related by the ideal gas law, $P_0 A h_1 = nRT_0$.)

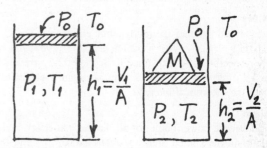

Problem 69 Solution.

● CHAPTER 22 THE SECOND LAW OF THERMODYNAMICS

Section 22-1 Reversibility and Irreversibility

Problem

1. The egg carton shown in Fig. 22-28 has places for one dozen eggs. (a) How many distinct ways are there to arrange six eggs in the carton? (b) Of these, what fraction correspond to all six eggs being in the left half of the carton? Treat the eggs as distinguishable, so an interchange of two eggs gives rise to a new state.

Solution

(a) There are twelve places for first egg, eleven for the second, etc., so the number of arrangments (states) of six distinguishable eggs into twelve places is $12 \times 11 \times 10 \times 9 \times 8 \times 7 = 12!/6! \approx 6.65 \times 10^5$. (b) If limited to the left side of the carton only, there are 6! arrangements, so the fraction is $6!/(12!/6!) = 1/924$.

FIGURE 22-28 Problem 1.

Sections 22-2 and 22-3 The Second Law and its Applications

Problem

5. What are the efficiencies of reversible heat engines operating between (a) the normal freezing and boiling points of water, (b) the 25°C temperature at the surface of a tropical ocean and deep water at 4°C, and (c) a 1000°C candle flame and room temperature?

Solution

The efficiency of a reversible engine, operating between two absolute temperatures, $T_h > T_c$, is given by Equation 22-3. (a) $e = 1 - T_c/T_h = 1 - 273/373 = 26.8\%$. (b) $e = (T_h - T_c)/T_h = \Delta T/T_h = 21/298$. (c) With room temperature at $T_c = 20°C$, $e = 980/1273 = 77.0\%$.

Problem

9. The maximum temperature in a nuclear power plant is 570 K. The plant rejects heat to a river where the temperature is 0°C in the winter and 25°C in the summer. What are the maximum possible efficiencies for the plant in these seasons? Why might the plant not achieve these efficiencies?

Solution

The winter and summer thermodynamic efficiencies are $1 - (T_c/T_h)$, or $1 - (273/570) = 52.1\%$ and $1 - (298/570) = 47.7\%$, respectively. As explained in Section 22-3, irreversible processes, transmission losses, etc., make actual efficiencies less than the theoretical maxima.

Problem

13. The electric power output of all the thermal electric power plants in the United States is about 2×10^{11} W, and these plants operate at an average efficiency around 33%. What is the rate at which all these plants use cooling water, assuming an average 5°C rise in cooling-water temperature? Compare with the 1.8×10^7 kg/s average flow at the mouth of the Mississippi River.

Solution

The total rate at which heat is exhausted by all power plants is

$$\frac{dQ_c}{dt} = \frac{d}{dt}(Q_h - W) = \frac{dW}{dt}\left(\frac{1}{e} - 1\right)$$

$$= (2 \times 10^{11} \text{ W})\left(\frac{1}{33\%} - 1\right) = 4 \times 10^{11} \text{ W}.$$

The mass rate of flow at which water could absorb this amount of energy, with only a 5°C temperature rise, is given by:

$$\frac{dQ_c}{dt} = \frac{dm}{dt} c_{\text{water}} \Delta T, \quad \text{or}$$

$$\frac{dm}{dt} = \frac{4 \times 10^{11} \text{ W}}{(4184 \text{ J/kg·K})(5°C)} = 1.91 \times 10^7 \text{ kg/s},$$

or about 1 Mississippi (a self-explanatory unit of river flow).

Problem

17. How much work does a refrigerator with a COP of 4.2 require to freeze 670 g of water already at its freezing point?

Solution

The amount of heat that must be extracted in order to freeze the water is $Q_c = mL_f = (0.67 \text{ kg})(334 \text{ kJ/kg}) = 224 \text{ kJ}$. The work consumed by the refrigerator while extracting this heat is $W = Q_c/\text{COP} = 224 \text{ kJ}/4.2 = 53.3 \text{ kJ}$. (See Equation 22-4.)

Problem

21. A heat pump consumes electrical energy at the rate P_e. Show that it delivers heat at the rate $(\text{COP} + 1)P_e$.

Solution

For each cycle of operation of the heat pump, $Q_h = Q_c + W$ and $\text{COP} = Q_c/W = (Q_h/W) - 1$. Therefore, $Q_h = (1 + \text{COP})W$, which if written in terms of rates, is the relation stated in the question.

Problem

25. A 0.20-mol sample of an ideal goes through the Carnot cycle of Fig. 22-30. Calculate (a) the heat Q_h absorbed, (b) the heat Q_c rejected, and (c) the work done. (d) Use these quantities to determine efficiency. (e) Find the maximum and minimum temperatures, and show explicitly that the efficiency as defined in Equation 22-1 is equal to the Carnot efficiency of Equation 22-3.

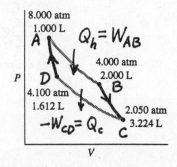

FIGURE 22-30 Problem 25 (Diagram is not to scale).

Solution

From the discussion of the efficiency of a Carnot engine in Section 22-2, (a) $Q_h = nRT_A \ln(V_B/V_A) = P_A V_A \ln(V_B/V_A) = (8)(1)(101.3 \text{ J})\ln 2 = 561.7 \text{ J}$, and (b) $Q_c = P_C V_C \times \ln(V_C/V_D) = (2.050)(3.224)(101.3 \text{ J})\ln(3.224/1.612) = 464.1 \text{ J}$ (we used 101.3 for the conversion factor of L·atm to J). (c) The work done in one cycle (from the first law) is $W = Q_h - Q_c = 97.66 \text{ J}$, resulting in an efficiency of (d) $e = W/Q_h = 0.1739$. (Note: For the Carnot cycle, $T_A = T_B$ and $T_C = T_D$, so for the adiabatic segments, $W_{BC} + W_{DA} = 0$. Thus, $W = W_{AB} + W_{BC} + W_{CD} + W_{DA} = Q_h - Q_c$, explicitly.) We find the maximum and minimum temperatures from the ideal gas law:

$$T_A = P_A V_A/nR = (8)(1)(101.3 \text{ J})/(0.2)(8.314 \text{ J/K})$$
$$= 487.4 \text{ K},$$

and

$$T_C = (2.050)(3.224)(101.3 \text{ J})/(0.2)(8.314 \text{ J/K}) = 402.6 \text{ K}.$$

These imply a Carnot efficiency of $e = 1 - T_C/T_A = 0.1739$, exactly as before. Equations 22-1 and 22-3 are identical because $Q_c/Q_h = T_C/T_A = 0.8261$, explicitly. (We did not round off until after completing all the calculations, and we labeled the states as in Fig. 22-6.)

Section 22-4 The Thermodynamic Temperature Scale

Problem

29. A Carnot engine operating between a vat of boiling sulfur and a bath of water at its triple point has an efficiency of 61.95%. What is the boiling point of sulfur?

Solution

From the efficiency of a reversible engine, $e = 1 - T_c/T_h$, and the triple point temperature 273.16 K, we find $T_h = T_c/(1 - e) = 273.16 \text{ K}/(1 - 61.95\%) = 718 \text{ K}$ (in agreement with Table 20-1, to three-figure accuracy).

Section 22-5 Entropy and the Quality of Energy

Problem

33. A 2.0-kg sample of water is heated to 35°C. If the entropy change is 740 J/K, what was the initial temperature?

Solution

Since the specific heat of water is approximately constant under normal conditions and no phase changes occur, Equation 22-9 can be solved for T_1 to yield $T_1 = T_2 e^{-\Delta S/mc} = (308 \text{ K}) \times \exp\{-(740 \text{ J/K})/(2 \text{ kg})(4.184 \text{ kJ/kg·K})\} = 282 \text{ K} = 8.93°\text{C}$.

Problem

37. The temperature of n moles of ideal gas is changed from T_1 to T_2 while the gas volume is held constant. Show that the corresponding entropy change is $\Delta S = nC_V \ln(T_2/T_1)$.

Solution

From the first law of thermodynamics ($dQ = dU + dW$) and the properties of an ideal gas ($dU = nC_V dT$ and $PV = nRT$), an infinitesimal entropy change is $dS = \frac{dQ}{T} = nC_V\frac{dT}{T} + \frac{P}{T}dV = nC_V\frac{dT}{T} + nR\frac{dV}{V}$. If we integrate from state 1 (T_1, V_1) to state 2 (T_2, V_2), we obtain $\Delta S = nC_V \ln(T_2/T_1) + nR \ln(V_2/V_1)$. For an isovolumic process (in which the gas does no work), ΔS is as given in the problem. (Of course, we could have started with $dQ = {}_nC_V dT$ at constant volume, but we wanted to display Δs.)

Problem

41. A 250-g sample of water at 80°C is mixed with 250 g of water at 10°C. Find the entropy changes for (a) the hot water, (b) the cool water, and (c) the system.

Solution

The equilibrium temperature for the mixture, assuming all the heat lost by the hot water is gained by the cold, is $T_{eq} = 45°\text{C}$. (a) $\Delta S_{\text{hot water}} = mc\ln(T_{eq}/T_{hot}) = (0.25 \text{ kg})(4.184 \text{ kJ/kg·K})\ln(318/353) = -109 \text{ J/K}$, (b) $\Delta S_{\text{cold water}} = (0.25 \text{ kg})(4.184 \text{ kJ/kg·K})\ln(318/283) = 122 \text{ J/K}$, and (c) $\Delta S_{tot} = -109 \text{ J/K} + 122 \text{ J/K} = 12.7 \text{ J/K}$.

Problem

45. Ideal gas occupying 1.0 cm³ is placed in a 1.0-m³ vacuum chamber, where it expands adiabatically. If 6.5 J of energy become unavailable to do work, what was the initial gas pressure?

Solution

From Equations 22-10, 11 and the ideal gas law, $nRT = E_{\text{unavailable}}/\ln(V_2/V_1) = 6.5 \text{ J}/\ln(10^6) = 0.470 \text{ J} = P_1 V_1$. Therefore, $P_1 = 0.470 \text{ J}/(10^{-6} \text{ m}^3) = 470 \text{ kPa}$.

Problem

49. Which would provide the greatest increase in efficiency of a Carnot engine, a 10 K increase in the maximum temperature or a 10 K decrease in the minimum temperature?

Solution

If we differentiate the efficiency with respect to T_c or T_h, respectively, we obtain: $de_c = -dT_c/T_h$, and $de_h = T_c\, dT_h/T_h^2$. If $dT_h = -dT_c$, then $de_h = (T_c/T_h)de_c < de_c$, since $T_c/T_h < 1$. The fact that a decrease in T_c produces a greater increase in efficiency than an increase of the same magnitude in T_h can also be demonstrated by direct substitution of $T_c - \Delta T$ or $T_h + \Delta T$ into Equation 22-3.

Problem

53. A reversible engine contains 0.20 mol of ideal monatomic gas, initially at 600 K and confined to 2.0 L. The gas undergoes the following cycle:

    - Isothermal expansion to 4.0 L.
    - Isovolumic cooling to 300 K.
    - Isothermal compression to 2.0 L.
    - Isovolumic heating to 600 K.

    (a) Calculate the net heat added during the cycle and the net work done. (b) Determine the engine's efficiency, defined as the ratio of the work done to only the heat *absorbed* during the cycle.

Solution

The P-V diagram for the cycle is as shown. For the isothermal expansion, $Q_{AB} = W_{AB} = nRT_A \ln(V_B/V_A) > 0$; for the isovolumic cooling, $W_{BC} = 0$, $Q_{BC} = \Delta U_{BC} = nC_V \Delta T_{BC} = \frac{3}{2}nR(T_C - T_B) < 0$; for the isothermal compression, $Q_{CD} = W_{CD} = nRT_C \ln(V_D/V_C) < 0$; and for the final isovolumic heating $W_{DA} = 0$, $Q_{DA} = \Delta U_{DA} = \frac{3}{2}nR(T_A - T_D) > 0$. For these processes, it is given that $V_B = 2V_A = V_C = 2V_D = 4$ L, $T_A = T_B = 600$ K, and $T_C = T_D = 300$ K.

(a) In a cyclic process ($\Delta U = 0$), the net heat added equals the net work done, $Q_{net} = W_{net} = nRT_A \ln(V_B/V_A) + nRT_C \ln(V_D/V_C) = nR(T_A - T_C)\ln(V_B/V_A) = (0.2 \text{ mol})(8.314 \text{ J/mol·K})(300 \text{ K})\ln 2 = 346$ J. (Note explicitly, that since $\Delta T_{BC} = -\Delta T_{DA}$, $W_{BC} + W_{DA} = 0 = \Delta U_{BC} + \Delta U_{DA} = Q_{BC} + Q_{DA}$, and therefore $W_{net} = W_{AB} + W_{CD} = Q_{AB} + Q_{CD} = Q_{net}$.) (b) The heat absorbed during the cycle (just the positive values of heat) is $Q_+ = Q_{AB} + Q_{DA} = nRT_A \ln(V_B/V_A) + \frac{3}{2} nR(T_A - T_B) = (0.2 \text{ mol})(3.314 \text{ J/mol·K})[(600 \text{ K})\ln 2 + 1.5(300 \text{ K})] = 1.44$ kJ. (Note: $Q_- = -Q_{BC} - Q_{CD} = 1.09$ kJ is the heat exhausted per cycle, and $Q_{net} = Q_+ - Q_-$.) The efficiency, as defined in this problem, is $W_{net}/Q_+ = 346 \text{ J}/1.44 \text{ kJ} = 24.0\%$. (Note: A Carnot engine operating between 600 K and 300 K has efficiency $1 - 300/600 = 50\%$. This is not a contradiction of Carnot's theorem, because the engine in this problem does *not* absorb and exhaust heat at constant temperatures.)

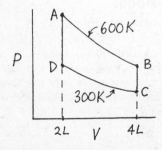

Problem 55 Solution.

Supplementary Problems

Problem

57. You're lying in a bathtub of water at 42°C. Suppose, in violation of the second law of thermodynamics, that the water spontaneously cooled to room temperature (20°C) and that the energy so released was transformed into your gravitational potential energy. Estimate the height to which you would rise above the bathtub.

Solution

A person taking a bath uses, on average, about 40 gal, or $(40 \text{ gal})(3.785 \times 10^{-3} \text{ m}^3/\text{gal})(10^3 \text{ kg/m}^3) = 151$ kg, of water (see Appendix C). (In contrast, 5 min under a low-flow shower head consumes about 15 gal of water.) The amount of thermal energy that would be released by an average bathtub full of water, which cooled from 42°C to 20°C, is $\Delta Q = mc_w \Delta T = (151 \text{ kg})(4184 \text{ J/kg·K})(22 \text{ K}) = 13.9$ MJ (see Equation 19-5 and Table 9-1). This amount is rather large compared to the increments of gravitational potential energy for ordinary objects near the Earth's surface, $\Delta U = mg \Delta y$ (see Equation 8-3). For a person of average mass 65 kg, it corresponds to a height difference of $\Delta y = (13.9 \text{ MJ})/(65 \times 9.81 \text{ N}) = 21.9$ km! (This energy is nearly half of the roughly 31.2 MJ required to launch 1 kg into a near Earth orbit (see Problem 9-45).)

Problem

61. Gasoline engines operate approximately on the **Otto cycle,** consisting of two adiabatic and two constant-volume segments. The Otto cycle for a particular engine is shown in Fig. 22-34. (a) If the gas in the engine has specific heat ratio γ, find the engine's efficiency, assuming all processes are reversible. (b) Find the maximum temperature, in terms of the minimum temperature $T_{\min}$. (c) How does the efficiency compare with that of a Carnot engine operating between the same two temperature extremes? *Note:* Fig. 22-34 neglects the intake of fuel-air and the exhaust of combustion products, which together involve essentially no net work.

FIGURE 22-34 Problem 61.

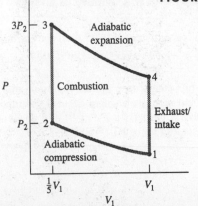

Solution

It is convenient to solve this problem beginning with part (b), because we will need to express the temperatures of all the numbered points in Fig. 22-34 in terms of T_1, the minimum. This is accomplished by use of the adiabatic and ideal gas laws, Equations 21-13a and b, and 20-2, respectively. First,

$$\frac{T_3}{T_4} = \left(\frac{V_4}{V_3}\right)^{\gamma-1} = 5^{\gamma-1} = \left(\frac{V_1}{V_2}\right)^{\gamma-1} = \frac{T_2}{T_1}.$$

Second,

$$\frac{P_3}{P_4} = \left(\frac{V_4}{V_3}\right)^{\gamma} = \left(\frac{V_1}{V_2}\right)^{\gamma} = \frac{P_2}{P_1}.$$

Last,

$$\frac{T_3}{T_2} = \frac{P_3}{P_2} = 3 = \frac{P_4}{P_1} = \frac{T_4}{T_1}.$$

Therefore, $T_2 = 5^{\gamma-1}T_1$, $T_4 = 3T_1$, and (the maximum temperature) $T_3 = 5^{\gamma-1}T_4 = 3 \times 5^{\gamma-1}T_1$. (a) Since no heat is transferred on the adiabatic segments, the heat input is $Q_h = Q_{23} = nC_V(T_3 - T_2) = nC_V 5^{\gamma-1}(3 - 1)\, T_1 = 2 \times 5^{\gamma-1}nC_V T_1$, and the heat exhaust is $Q_c = -Q_{41} =$

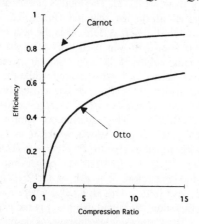

Problem 61 Solution.

$-nC_V(T_1 - T_4) = 2nC_V T_1$. Therefore, the efficiency is $e = 1 - Q_c/Q_h = 1 - 5^{1-\gamma}$. (c) A Carnot engine operating between T_3 and T_1 has efficiency $1 - T_1/T_3 = 1 - \frac{1}{3} \times 5^{1-\gamma}$, so $e_{\text{Otto}} < e_{\text{Carnot}}$.

Problem

65. A Carnot engine extracts heat from a block of mass m and specific heat c that is initially at temperature T_{h0} but which has no heat source to maintain that temperature. The engine rejects heat to a reservoir at a constant temperature T_c. The engine is operated so its mechanical power output is proportional to the temperature difference $T_h - T_c$:

$$P = P_0 \frac{T_h - T_c}{T_{h0} - T_c},$$

where T_h is the instantaneous temperature of the hot block and P_0 is the initial power output. (a) Find an expression for T_h as a function of time, and (b) determine how long it takes for the engine's power output to reach zero.

Solution

(a) In time dt, the engine extracts heat $dQ_h = -mc\, dT_h$ from the block, and does work $dW = \mathcal{P}\, dt$. Since it is a Carnot engine, $dW = e_{\max}\, dQ_h = [(T_h - T_c)/T_h](-mc\, dT_h) = \mathcal{P}\, dt$. The power is also assumed to be proportional to $T_h - T_c$, so the equation becomes $-mc\, dT_h/T_h = \mathcal{P}_0\, dt/(T_{h0} - T_c)$. Integrating from $t = 0$ and T_{h0} to t and T_h, we obtain

$$\int_{T_{h0}}^{T_h} \frac{dT'_h}{T'_h} = \ln\left(\frac{T_h}{T_{h0}}\right) = -\int_0^t \frac{\mathcal{P}_0\, dt}{mc(T_{h0} - T_c)} = -\frac{\mathcal{P}_0 t}{mc(T_{h0} - T_c)},$$

or $T_h = T_{h0} \exp\{-\mathcal{P}_0 t/mc(T_{h0} - T_c)\}$. (b) The power output is zero for $T_h = T_c$. This occurs at time $t_0 = (mc/\mathcal{P}_0)(T_{h0} - T_c)\ln(T_{h0}/T_c)$. (Note: The expression for $\mathcal{P}$ was originally assumed valid for $T_h \geq T_c$, or for times $t \leq t_0$. If we allow $T_h < T_c$, or $t > t_0$, then $dW = \mathcal{P}\, dt < 0$ becomes work input to an "engine" which acts like a refrigerator cooling the block.)

●

● PART 3 CUMULATIVE PROBLEMS

Problem

1. Figure 1 shows the thermodynamic cycle of a diesel engine. Note that this cycle differs from that of a gasoline engine (see Fig. 22-34) in that combustion takes place isobarically. As with the gasoline engine, the compression ratio r is the ratio of maximum to minimum volume; $r = V_1/V_2$. In addition, the so-called *cutoff ratio* is defined by $r_c = V_3/V_2$. Find an expression for the engine's efficiency, in terms of the ratios r and r_c and the specific heat ratio γ. Although your expression suggests that the diesel engine might be less efficient than the gasoline engine (see Problem 61 of Chap-

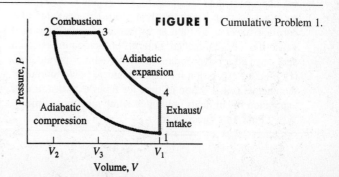

FIGURE 1 Cumulative Problem 1.

ter 22), the diesel's higher compression ratio more than compensates, giving it a higher efficiency.

Solution

From Table 21-1, the work done and heat absorbed during the four processes comprising the diesel cycle are:

$1 \rightarrow 2$ (adiabatic) $W_{12} = (P_1V_1 - P_2V_2)/(\gamma - 1)$, $Q_{12} = 0$,

$2 \rightarrow 3$ (isobaric) $W_{23} = P_2(V_3 - V_2)$,

$$Q_{23} = nC_P(T_3 - T_2) \equiv Q_h,$$

$3 \rightarrow 4$ (adiabatic) $W_{34} = (P_3V_3 - P_4V_4)/(\gamma - 1)$, $Q_{34} = 0$,

$4 \rightarrow 1$ (isovolumic) $W_{41} = 0$, $Q_{41} = nC_V(T_1 - T_4) \equiv -Q_c$.

For the whole cycle, the work done is

$$W = \frac{P_1V_1 - P_2V_2 + P_3V_3 - P_4V_4}{(\gamma - 1)} + P_3V_3 - P_2V_2$$

$$= \frac{\gamma(P_3V_3 - P_2V_2) - P_4V_4 + P_1V_1}{(\gamma - 1)},$$

while the heat added can be written as $Q_h = nC_P(T_3 - T_2) = \gamma(P_3V_3 - P_2V_2)/(\gamma - 1)$, where we used the ideal gas law, $nT = PV/R$, and $C_P/R = C_P/(C_P - C_V) = \gamma/(\gamma - 1)$. Therefore,

$$W = Q_h - \frac{(P_4V_4 - P_1V_1)}{(\gamma - 1)} = Q_h - \frac{(P_4V_4 - P_1V_1)Q_h}{\gamma(P_3V_3 - P_2V_2)}$$

and the efficiency is $e = W/Q_h =$
$1 - (P_4V_4 - P_1V_1)/\gamma(P_3V_3 - P_2V_2)$. The adiabatic law can now be used to express every product in terms of P_2V_2 and the compression and cutoff ratios: $P_2 = P_3$, $V_1 = V_4$, $V_1/V_2 = r$, $V_3/V_2 = r_c$, $P_1V_1^\gamma = P_2V_2^\gamma$, $P_3V_3^\gamma = P_4V_4^\gamma$, so $P_1V_1 = P_2V_2r^{1-\gamma}$, $P_3V_3 = P_2V_2r_c$, and $P_4V_4 = P_2V_2r^{1-\gamma}r_c^\gamma$. Finally, $e = 1 - r^{1-\gamma}(r_c^\gamma - 1)/\gamma(r_c - 1)$.

Problem

5. The ideal Carnot engine shown in Fig. 4 operates between a heat reservoir and a block of ice with mass M. An external energy source maintains the reservoir at a constant temperature T_h. At time $t = 0$ the ice is at its melting point T_0, but it is insulated from everything except the engine, so it is free to change state and temperature. The engine is operated in such a way that it extracts heat from the reservoir at a constant rate P_h. (a) Find an expression for the time t_1 at which the ice is all melted, in terms of the quantities given and any other appropriate thermodynamic parameters. (b) Find an expression for the mechanical power output of the engine as a function of time for times $t > t_1$. (c) Your expression in (b) holds only up to some maximum time t_2. Why? Find an expression for t_2.

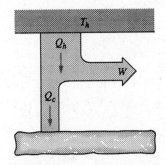

FIGURE 4 Cumulative Problem 5.

Solution

(a) The ice acts as a constant temperature cold reservoir at $T_c = T_0 = 273$ K, while it is melting for times $0 \leq t \leq t_1$. The Carnot engine exhausts heat at a constant rate, over this time interval, so the simplified form of Equation 22-2 gives $\mathcal{P}_c = (T_c/T_h)\mathcal{P}_h$, (where the first and second laws of thermodynamics have been written in terms of rate of heat flow or power, as in Example 22-2). The total heat exhausted is just sufficient to melt all the ice, therefore $\mathcal{P}_ct_1 = (T_0/T_h)\mathcal{P}_ht_1 = ML_f$, or $t_1 = ML_fT_h/\mathcal{P}_hT_0$. (Here, L_f is the heat of fusion of water and Equation 20-6 was used.) (b) After the ice is melted, the temperature of the cold reservoir increases with time. So does the rate of heat exhausted, while the mechanical power of the engine, $\mathcal{P} = \mathcal{P}_h - \mathcal{P}_c$, decreases. From Equation 19-5 in terms of rates, $\mathcal{P}_cdt = (T_c/T_h)\mathcal{P}_hdt = McdT_c$, or $dT_c/T_c = (\mathcal{P}_h/McT_h)\,dt$. This equation holds for a time interval $t_1 \leq t \leq t_2$ (see part (c) below), where $T_c = T_0$ at $t = t_1$, and can be integrated to give $\ln(T_c/T_0) = (\mathcal{P}_h/McT_h)(t - t_1)$, or $T_c = T_0\exp\{\mathcal{P}_h(t - t_1)/McT_h\}$. The power output of the Carnot engine is therefore $\mathcal{P} = e\mathcal{P}_h = (1 - T_c/T_h)\mathcal{P}_h = [1 - (T_0/T_h)\exp\{\mathcal{P}_h(t - t_1)/McT_h\}]\mathcal{P}_h$, where e is the Carnot efficiency (Equation 22-3) and $\exp\{\ldots\}$ is the exponential function. (c) The Carnot engine operates as described above only as long as $T_c < T_h$. When $T_c = T_h$ at time t_2, the power output has dropped to zero. Then, from part (b), $\ln(T_h/T_0) = (\mathcal{P}_h/McT_h)(t_2 - t_1)$, or $t_2 = t_1 + (McT_h/\mathcal{P}_h)\ln(T_h/T_0)$.

PART 4 ELECTROMAGNETISM

● CHAPTER 23 ELECTRIC CHARGE, FORCE, AND FIELD

Section 23-2 Electric Charge
Problem
1. Suppose the electron and proton charges differed by one part in one billion. Estimate the net charge you would carry.

Solution
Nearly all of the mass of an atom is in its nucleus, and about one half of the nuclear mass of the light elements in living matter (H, O, N, and C) is protons. Thus, the number of protons in a 65 kg average-sized person is approximately $\frac{1}{2}(65 \text{ kg})/(1.67\times10^{-27} \text{ kg}) \approx 2\times10^{28}$, which is also the number of electrons, since an average person is electrically neutral. If there were a charge imbalance of $|q_{proton} - q_{electron}| = 10^{-9}e$, a person's net charge would be about $\pm2\times10^{28} \times 10^{-9} \times 1.6\times10^{-19}$ C $= \pm3$ 2C, or several coulombs (huge by ordinary standards).

Section 23-3 Coulomb's Law
Problem
5. If the charge imbalance of Problem 1 existed, what would be the approximate force between you and another person 10 m away? Treat the people as point charges, and compare the answer with your weight.

Solution
The magnitude of the Coulomb force between two point charges of 3.2 C (see solution to Problem 1), at a distance of 10 m, is $kq^2/r^2 = (9\times10^9 \text{ N·m}^2/\text{C}^2)(3.2 \text{ C}/10 \text{ m})^2 = 9.22\times10^8$ N. This is approximately 1.45 million times the weight of an average-sized 65 kg person.

Problem
9. Two charges, one twice as large as the other, are located 15 cm apart and experience a repulsive force of 95 N. What is the magnitude of the larger charge?

Solution
The product of the charges is $q_1 q_2 = r^2 F_{Coulomb}/k = (0.15 \text{ m})^2(95 \text{ N})/(9\times10^9 \text{ N·m}^2/\text{C}^2) = 2.38\times10^{-10}$ C^2. If one charge is twice the other, $q_1 = 2q_2$, then $\frac{1}{2}q_1^2 = 2.38\times10^{-10}$ C and $q_1 = \pm21.8$ μC.

Problem
13. A 9.5-μC charge is at $x = 16$ cm, $y = 5.0$ cm, and a -3.2-μC charge is at $x = 4.4$ cm, $y = 11$ cm. Find the force on the negative charge.

Solution
Denote the positions of the charges by $\mathbf{r}_1 = (16\hat{\imath} + 5\hat{\jmath})$ cm for $q_1 = 9.5$ μC, and $\mathbf{r}_2 = (4.4\hat{\imath} + 11\hat{\jmath})$ cm for $q_2 = -3.2$ μC. The vector from q_1 to q_1 is $\mathbf{r} = \mathbf{r}_2 - \mathbf{r}_1$, and a unit vector in this direction is $\hat{\mathbf{r}} = (\mathbf{r}_2 - \mathbf{r}_1)/|\mathbf{r}_2 - \mathbf{r}_1|$. The vector form of Coulomb's law for the electric force of q_1 on q_2 is $\mathbf{F}_{12} = kq_1 q_2(\mathbf{r}_2 - \mathbf{r}_1)/|\mathbf{r}_2 - \mathbf{r}_1|^3$. (This gives the Coulomb force between two point charges, as a function of their positions, and is a convenient form to memorize because of its direct applicability.) Substituting the given values for this problem, we find:

$$\mathbf{F}_{12} = \left(\frac{9\times10^9 \text{ N·m}^2}{\text{C}^2}\right)(9.5 \text{ }\mu\text{C})(-3.2 \text{ }\mu\text{C})$$

$$\times \frac{(4.4\hat{\imath} + 11\hat{\jmath} - 16\hat{\imath} - 5\hat{\jmath}) \text{ cm}}{[(4.4 - 16)^2 + (11 - 5)^2]^{3/2} \text{ cm}^3}$$

$$= (14.2\hat{\imath} - 7.37\hat{\jmath}) \text{ N},$$

with magnitude 16.0 N and direction $\theta = -27.3°$ to the x axis (negative angle measured CW).

Problem
17. A 60-μC charge is at the origin, and a second charge is on the positive x axis at $x = 75$ cm. If a third charge placed at $x = 50$ cm experiences no net force, what is the second charge?

Solution
In order for the net force to be zero at a position between the first two charges, they must both have the same sign, i.e., $q_1 = 60$ μC at $x_1 = 0$ and $q_2 > 0$ at $x_2 = 75$ cm. (Then the separate forces of the first two charges on the third are in opposite directions.) Therefore, for the third charge q_3 at $x_3 = 50$ cm, $F_{3x} = kq_3[q_1(x_3 - x_1)^{-2} - q_2(x_2 - x_3)^{-2}] = kq_3[60 \text{ }\mu\text{C}(50 \text{ cm})^{-2} - q_2(25 \text{ cm})^{-2}] = 0$ implies $q_2 = 60 \text{ }\mu\text{C}(25/50)^2 = 15 \text{ }\mu\text{C}$.

Problem

21. Four identical charges q form a square of side a. Find the magnitude of the electric force on any of the charges.

Solution

By symmetry, the magnitude of the force on any charge is the same. Let's find this for the charge at the lower left corner, which we take as the origin, as shown. Then $\mathbf{r}_1 = 0$, $\mathbf{r}_2 = a\hat{\mathbf{j}}$, $\mathbf{r}_3 = a(\hat{\mathbf{i}} + \hat{\mathbf{j}})$, $\mathbf{r}_4 = a\hat{\mathbf{i}}$, and

$$\mathbf{F}_1 = kq^2\left[\frac{\mathbf{r}_1 - \mathbf{r}_2}{|\mathbf{r}_1 - \mathbf{r}_2|^3} + \frac{\mathbf{r}_1 - \mathbf{r}_3}{|\mathbf{r}_1 - \mathbf{r}_3|^3} + \frac{\mathbf{r}_1 - \mathbf{r}_4}{|\mathbf{r}_1 - \mathbf{r}_4|^3}\right]$$

$$= kq^2\left[\frac{-a\hat{\mathbf{j}}}{a^3} - \frac{a(\hat{\mathbf{i}} + \hat{\mathbf{j}})}{2\sqrt{2}a^3} - \frac{a\hat{\mathbf{i}}}{a^3}\right]$$

$$= -\frac{kq^2}{a^2}(\hat{\mathbf{i}} + \hat{\mathbf{j}})\left(1 + \frac{1}{2\sqrt{2}}\right),$$

(Use the vector form of Coulomb's law in the solution to Problem 13, and the superposition principle). Since $|\hat{\mathbf{i}} + \hat{\mathbf{j}}| = \sqrt{2}$, $|\mathbf{F}_1| = (kq^2/a^2)\sqrt{2}(1 + 1/2\sqrt{2}) = (kq^2/a^2)(\sqrt{2} + \frac{1}{2}) = 1.91kq^2/a^2$.

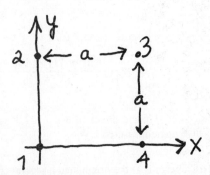

Problem 21 Solution.

Section 23-4 The Electric Field

Problem

25. An electron placed in an electric field experiences a 6.1×10^{-10} N electric force. What is the field strength?

Solution

From Equation 23-2a, $E = F/e = 6.1\times10^{-10}$ N$/1.6\times10^{-19}$ C $= 3.81\times10^9$ N/C. (The field strength is the magnitude of the field.)

Problem

29. The electron in a hydrogen atom is 0.0529 nm from the proton. What is the proton's electric field strength at this distance?

Solution

The proton in a hydrogen atom behaves like a point charge, for an electron one Bohr radius away (see solution to Problem 7), so Equation 23-3 gives $E = ke/a_0^2 = (9\times10^9$ N·m^2/C$^2)(1.6\times10^{-19}$ C$)/(5.29\times10^{-11}$ m$)^2 = 5.15\times10^{11}$ N/C.

Section 23-5 Electric Fields of Charge Distributions

Problem

33. A proton is at the origin and an ion is at $x = 5.0$ nm. If the electric field is zero at $x = -5$ nm, what is the charge on the ion?

Solution

The proton, charge e, is at $\mathbf{r}_p = 0$, and the ion, charge q, is at $\mathbf{r}_I = 5\hat{\mathbf{i}}$ nm. The field at point $\mathbf{r} = -5\hat{\mathbf{i}}$ nm is given by Equation 23-4, with spacial factors written as in the solutions to Problems 13 or 31:

$$\mathbf{E}(\mathbf{r}) = \sum_i kq_i\frac{(\mathbf{r} - \mathbf{r}_i)}{|\mathbf{r} - \mathbf{r}_i|^3}$$

$$= ke\frac{(-5\hat{\mathbf{i}} \text{ nm})}{(5 \text{ nm})^3} + kq\frac{(-5\hat{\mathbf{i}} \text{ nm} - 5\hat{\mathbf{i}} \text{ nm})}{(10 \text{ nm})^3}.$$

Therefore, $\mathbf{E} = 0$ implies $2q/(10)^3 = -e/(5)^3$, or $q = -4e$. (Note how we used the general expression for the electric field, at position $\mathbf{r}$, due to a distribution of static point charges at positions $\mathbf{r}_i$.)

Problem

37. A dipole lies on the y axis, and consists of an electron at $y = 0.60$ nm and a proton at $y = -0.60$ nm. Find the electric field (a) midway between the two charges, (b) at the point $x = 2.0$ nm, $y = 0$, and (c) at the point $x = -20$ nm, $y = 0$.

Solution

We can use the result of Example 23-6, with y replaced by x, and x by $-y$ (or equivalently, $\hat{\mathbf{j}}$ by $\hat{\mathbf{i}}$, and $\hat{\mathbf{i}}$ by $-\hat{\mathbf{j}}$). Then $\mathbf{E}(x) = 2kqa\,\hat{\mathbf{j}}(a^2 + x^2)^{-3/2}$, where $q = e = 1.6\times10^{-19}$ C and $a = 0.6$ nm. (Look at Fig. 23-15 rotated 90° CW.) The constant $2kq = 2(9\times10^9$ N·m^2/C$^2)(1.6\times10^{-19}$ C$) = (2.88$ GN/C)(nm)^2. (a) At $x = 0$, $\mathbf{E}(0) = 2kq\hat{\mathbf{j}}/a^2 = (2.88$ GN/C)$\hat{\mathbf{j}}/(0.6)^2 = (8.00$ GN/C)$\hat{\mathbf{j}}$. (b) For $x = 2$ nm, $\mathbf{E} = (2.88$ GN/C)$\hat{\mathbf{j}}(0.6)(0.6^2 + 2^2)^{-3/2} = (190$ MN/C)$\hat{\mathbf{j}}$. (c) At $x = 20$ nm, $\mathbf{E} = (2.88$ GN/C)$(0.6)(0.6^2 + 20^2)^{-3/2} = (216$ kN/C)$\hat{\mathbf{j}}$.

Problem

41. A point dipole lies at the origin, with its dipole moment vector $\mathbf{p}$ making an angle θ with the x axis, as shown in Fig. 23-39. By resolving $\mathbf{p}$ into components and applying

Equations 23-5a and 23-5b to the x and y components, respectively, show that the electric field at an arbitrary point P on the y axis is given by

$$\mathbf{E} = \frac{kp}{y^3}(-\hat{\mathbf{i}}\cos\theta + 2\hat{\mathbf{j}}\sin\theta).$$

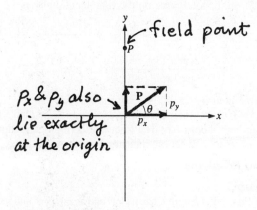

FIGURE 23-39 Problem 41.

Solution

The field point (P in Fig. 23-39) is on the perpendicular bisector of the x component of the dipole moment, $p_x = p\cos\theta$, so Equation 23-5a gives $E_x\hat{\mathbf{i}} = -kp_x\hat{\mathbf{i}}/y^3$. The field point is along the dipole axis for the y component, $p_y = p\sin\theta$, so Equation 23-5b gives $E_y\hat{\mathbf{j}} = 2kp_y\hat{\mathbf{j}}/y^3$. (Note the change of x into y in Equation 23-5b, since here, the field point and the dipole component are along the y axis.) The total field is $\mathbf{E} = E_x\hat{\mathbf{i}} + E_y\hat{\mathbf{j}} = k(-p_x\hat{\mathbf{i}} + 2p_y\hat{\mathbf{j}})y^{-3} = kp(-\hat{\mathbf{i}}\cos\theta + 2\hat{\mathbf{j}}\sin\theta)y^{-3}.$

Problem

45. The rods shown in Fig. 23-41 are both 15 cm long and both carry 1.2 μC of charge. Find the magnitude and direction of the electric field at point P.

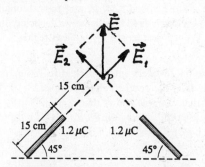

FIGURE 23-41 Problem 45 Solution.

Solution

The magnitude of the electric field from each rod is the same (given in Example 23-7, with $a = \ell = 15$ cm), and the directions are parallel to each rod, as shown on Fig. 23-41. By symmetry, $E_{1x} = -E_{2x}$ and $E_{1y} = E_{2y}$, so the net field of both rods is just $\mathbf{E} = \mathbf{E}_1 + \mathbf{E}_2 = 2E_{1y}\hat{\mathbf{j}} = 2(kQ/a(a+\ell))\cos 45°\,\hat{\mathbf{j}} = \hat{\mathbf{j}}(9\times10^9\ \text{N·m}^2/\text{C}^2)(1.2\ \mu\text{C})\sqrt{2}/(0.15\times 0.30\ \text{m}^2) = (339\ \text{kN/C})\hat{\mathbf{j}}.$

Problem

49. A uniformly charged ring is 1.0 cm in radius. The electric field on the axis 2.0 cm from the center of the ring has magnitude 2.2 MN/C and points toward the ring center. Find the charge on the ring.

Solution

From Example 23-8, the electric field on the axis of a uniformly charged ring is $kQx(x^2 + a^2)^{-3/2}$, where x is positive away from the center of the ring. For the given ring in this problem, $-2.2\ \text{MN/C} = kQ(2\ \text{cm})(4\ \text{cm}^2 + 1\ \text{cm}^2)^{-3/2}$, or $Q = (-2.2\ \text{MN/C})(5.59\ \text{cm}^2)/(9\times10^9\ \text{N·m}^2/\text{C}) = -0.137\ \mu\text{C}.$

Problem

53. The electric field 22 cm from a long wire carrying a uniform line charge density is 1.9 kN/C. What will be the field strength 38 cm from the wire?

Solution

For a very long wire ($\ell \gg 38$ cm), Example 23-9 shows that the magnitude of the radial electric field falls off like $\frac{1}{r}$. Therefore, $E(38\ \text{cm})/E(22\ \text{cm}) = 22\ \text{cm}/38\ \text{cm}$; or $E(38\ \text{cm}) = (22/38)1.9\ \text{kN/C} = 1.10\ \text{kN/C}.$

Section 23-6 Matter in Electric Fields

Problem

57. In this famous 1909 experiment that demonstrated quantization of electric charge, R. A. Millikan suspended small oil drops in an electric field. With a field strength of 20 MN/C, what mass drop can be suspended when the drop carries a net charge of 10 elementary charges?

Solution

In equilibrium under the gravitational and electrostatic forces, $mg = qE$, or $m = (10\times1.6\times10^{-19}\ \text{C}) \times (2\times10^7\ \text{N/C})/(9.8\ \text{m/s}^2) = 3.27\times10^{-12}\ \text{kg}$. (Because this is so small, the size of such a drop may be better appreciated in terms of its radius, $R = (3m/4\pi\rho_{\text{oil}})^{1/3}$. Millikan used oil of density 0.9199 g/cm³, so $R = 9.46\ \mu\text{m}$ for this drop.)

Problem

61. A uniform electric field **E** is set up between two metal plates of length ℓ and spacing d, as shown in Fig. 23-46. An electron enters the region midway between the plates moving horizontally with speed v, as shown. Find an expression for the minimum speed the electron needs to get through the region without hitting either plate. Neglect gravity.

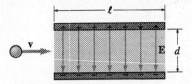

FIGURE 23-46 Problem 61.

Solution

The maximum vertical deflection, during the transit time $t = \ell/v$, can be $d/2$, for the electron to pass through. Thus, $y = \frac{1}{2}at^2 = \frac{1}{2}(eE/m)(\ell/v)^2 < \frac{1}{2}d$, or $v > \ell\sqrt{eE/md}$.

Problem

65. What is the line charge density on a long wire if a 6.8-μg particle carrying 2.1 nC describes a circular orbit about the wire with speed 280 m/s?

Solution

The solution to Problem 63 reveals that $\lambda = -mv^2/2kq = -(6.8\times10^{-9} \text{ kg})(280 \text{ m/s})^2/2(9\times10^9 \text{ N·m}^2/\text{C}^2)$ $(2.1\times10^{-9} \text{ C}) = -14.1 \ \mu\text{C/m}$. (In this case, the force on a positively charged orbiting particle is attractive for a wire with negative linear charge density.)

Problem

69. Two identical dipoles, each of charge q and separation a, are a distance x apart as shown in Fig. 23-48. By considering forces between pairs of charges in the different dipoles, calculate the net force between the dipoles. (a) Show that, in the limit $a \ll x$, the force has magnitude $6kp^2/x^4$, where $p = qa$ is the dipole moment. (b) Is the force attractive or repulsive?

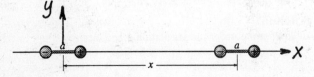

FIGURE 23-48 Problem 69.

Solution

All the forces are along the same line, so take the origin at the center of the left-hand dipole and the positive x axis in the direction of the right-hand dipole in Fig. 23-48. The right-

hand dipole has charges $+q$ at $x + a/2$, $-q$ at $x - a/2$, each of which experiences a force from both charges of the left-hand dipole, which are $+q$ at $a/2$ and $-q$ at $-a/2$. (There are forces between four pairs of changes.) The Coulomb force on a charge in the right-hand dipole, due to one in the left-hand one, is $kq_rq_\ell(x_r - x_\ell)\hat{\mathbf{i}}/|x_r - x_\ell|^3$ (see solution to Problem 13), so the total force on the right-hand dipole is

$$F_x = kg^2\hat{\mathbf{i}}\left[\frac{1}{x^2} - \frac{1}{(x+a)^2} - \frac{1}{(x-a)^2} + \frac{1}{x^2}\right]$$
$$= -\frac{2kq^2a^2(3x^2 - a^2)}{x^2(x^2 - a^2)^2}\hat{\mathbf{i}}.$$

(a) In the limit $a \ll x$, $F_x \to -2kq^2a^2(3x^2)\hat{\mathbf{i}}/x^6 = -6kq^2a^2\hat{\mathbf{i}}/x^4 = -6kp^2\hat{\mathbf{i}}/x^4$, where $p = qa$ is the dipole moment of both dipoles. (b) The force on the right-hand dipole is in the negative x direction, indicating an attractive force.

Paired Problems

Problem

73. A thin rod of length ℓ has its left end at $x = -\ell$ and its right end at the origin. It carries a line charge density given by $\lambda = \lambda_0\dfrac{x^2}{\ell^2}$, where λ_0 is a constant. Find the electric field at the origin.

Solution

The electric field at the origin, due to an element of charge $dq = \lambda \, dx$, located at x, where $-\ell \le x \le 0$, is $d\mathbf{E} = k\lambda\hat{\mathbf{i}} \, dx/x^2 = k(\lambda_0/\ell^2)\hat{\mathbf{i}} \, dx$. The total field is the integral of this over the rod,

$$\mathbf{E}(0) = \int_{-\ell}^{0} (k\lambda_0/\ell^2)\hat{\mathbf{i}} \, dx = (k\lambda_0/\ell^2)\hat{\mathbf{i}}[0 - (-\ell)] = (k\lambda_0/\ell)\hat{\mathbf{i}}.$$

Problem

77. Ink-jet printers work by deflecting moving ink droplets with an electric field so they hit the right place on the paper. Droplets in a particular printer have mass 1.1×10^{-10} kg, charge 2.1 pC, speed 12 m/s, and pass through a uniform 97-kN/C electric field in order to be deflected through a $10°$ angle. What is the length of the field region?

Solution

Suppose the ink droplets enter the field region perpendicular to the field, as in the geometry of Example 23-10. Then the analysis of that example shows that $v_y/v_x = \tan\theta = qE_y\Delta x/mv_x^2$, so $\Delta x = mv_x^2\tan\theta/qE_y = (0.11 \ \mu\text{g})(12 \text{ m/s})^2 \times \tan 10°/(2.1 \text{ pC})(97 \text{ kN/C}) = 1.37$ cm.

Supplementary Problems

Problem

81. A 3.8-g particle with a 4.0-μC charge experiences a downward force of 0.24 N in a uniform electric field. Find the electric field, assuming that the gravitational force is *not* negligible.

Solution

If gravity and Coulomb forces both act, then $\mathbf{F}_{net} = -mg\hat{\mathbf{j}} + q\mathbf{E} = -(0.24 \text{ N})\hat{\mathbf{j}}$, where $\hat{\mathbf{j}}$ is upward. Thus, $\mathbf{E} = (mg - 0.24 \text{ N})\hat{\mathbf{j}}/q = [(3.8\times10^{-3} \text{ kg})(9.8 \text{ m/s}^2) - 0.24 \text{ N}]\hat{\mathbf{j}}/(4.0 \ \mu\text{C}) = -50.7\hat{\mathbf{j}}$ kN/C.

Problem

85. The dipole moment of a water molecule is 6.2×10^{-30} C·m. A water molecule is located 1.5 nm from a proton, with its dipole moment vector aligned as shown in Fig. 23-52. (a) Use Equation 23-5b to find the force the molecule exerts on the proton. (b) Now find the net force on the dipole in the proton's nonuniform electric field by considering that the dipole consists of two opposite charges q separated by a distance d, such that $qd = 6.2\times10^{-30}$ C·m. Take the limit as d becomes very small, and show that the force has the same magnitude as that of part (a), as required by Newton's third law.

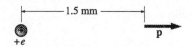

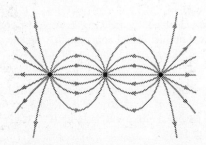

FIGURE 23-52 Problem 85.

Solution

(a) Newton's third law holds for Coulomb forces, so the force on the dipole due to the field of the proton is the negative of the force on the proton due to the field of the dipole. The latter is given by Equation 23-5b, for the orientation of **p** shown in Fig. 23-52, $\mathbf{F}_{proton} = e\mathbf{E}_{dipole} = 2kep\hat{\mathbf{i}}/|x|^3$. (Equation 23-5b was written for a dipole at the origin, with x axis in the direction of **p**, so the proton is at $x = -1.5$ nm in Fig. 23-52.) Numerically, $\mathbf{F}_{proton} = 2(9\times10^9 \text{ N·m}^2/\text{C}^2) \times (1.6\times10^{-19} \text{ C})(6.2\times10^{-30} \text{ C·m})\hat{\mathbf{i}}/(1.5\times10^{-9}\text{m})^3 = 5.29$ pN$\hat{\mathbf{i}}$. (b) The attractive force on the negative charge of the dipole, at distance $|x| - d/2$ from the proton, is greater than the repulsive force on the positive charge of the dipole, at distance $|x| + d/2$, so the net force on the pair is attractive (to the left in Fig. 23-52, or opposite to **p**). Thus, using Equation 23-3 for the field of the proton, we find $\mathbf{F}_{dipole} = -keq\hat{\mathbf{i}}[(|x| - d/2)^{-2} - (|x| + d/2)^{-2}] = -2keq|x|d\hat{\mathbf{i}}(x^2 - d^2/4)^{-2}$. In the limit $d \to 0$, with $qd \to p$, $\mathbf{F}_{dipole} \to -2kep\hat{\mathbf{i}}/|x|^3 = -\mathbf{F}_{proton} = -5.29$ pN$\hat{\mathbf{i}}$, as expected from Newton's third law. (Note: The equation for the force on a point dipole in a nonuniform external field was not developed in the text because it involves a vector differential operator (i.e., $\mathbf{p}\cdot\nabla$, pronounced "pee dot del") which is equivalent to the limiting procedure used above.)

● CHAPTER 24 GAUSS'S LAW

Section 24-1 Electric Field Lines

Problem

1. What is the net charge shown in Fig. 24-38? The magnitude of the middle charge is 3 μC.

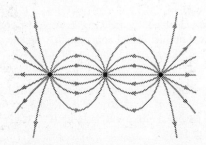

FIGURE 24-38 Problem 1 Solution.

Solution

The number of lines of force emanating from (or terminating on) the positive (or negative) charges is the same (14 in Fig. 24-38), so the middle charge is $-3 \ \mu$C and the outer ones are $+3 \ \mu$C. The net charge shown is therefore $3 + 3 - 3 = 3 \ \mu$C. This is reflected by the fact that 14 lines emerge from the boundary of the figure.

Section 24-2 Electric Flux

Problem

5. A flat surface with area 2.0 m² is in a uniform electric field of 850 N/C. What is the electric flux through the surface when it is (a) at right angles to the field, (b) at 45° to the field, and (c) parallel to the field?

Solution

(a) When the surface is perpendicular to the field, its normal is either parallel or anti-parallel to $\mathbf{E}$. Then Equation 24-1 gives $\Phi = \mathbf{E}{\cdot}\mathbf{A} = EA\cos(0° \text{ or } 180°) = \pm(850 \text{ N/C})(2 \text{ m}^2) = \pm1.70 \text{ kN·m}^2/\text{C}$. (b) $\Phi = \mathbf{E}{\cdot}\mathbf{A} = EA\cos(45° \text{ or } 135°) = \pm(1.70 \text{ kN·m}^2/\text{C})(0.866) = \pm1.20 \text{ kN·m}^2/\text{C}$. (c) $\Phi = EA\cos 90° = 0$.

Problem

9. What is the flux through the hemispherical open surface of radius R shown in Fig. 24-40? The uniform field has magnitude E. *Hint:* Don't do a messy integral! Think about the flux through the open end of the hemisphere.

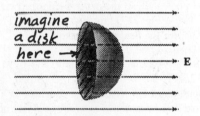

FIGURE 24-40 Problem 9 Solution.

Solution

All of the lines of force going through the hemisphere also go through an equitorial disk covering its edge in Fig. 24-40. Therefore, the flux through the disk (normal in the direction of $\mathbf{E}$) equals the flux through the hemisphere. Since $\mathbf{E}$ is uniform, the flux through the disk is just $\pi R^2 E$. (Note: Gauss's Law gives the same result, since the flux through the closed surface, consisting of the hemisphere plus the disk, is zero. See Section 24-3.)

Section 24-3   Gauss's Law

Problem

13. A 2.6-μC charge is at the center of a cube 7.5 cm on each side. What is the electric flux through one face of the cube? *Hint:* Think about symmetry, and don't do an integral.

Solution

The symmetry of the situation guarantees that the flux through one face is $\frac{1}{6}$ the flux through the whole cubical surface, so $\Phi_{\text{face}} = \frac{1}{6}\oint_{\text{cube}} \mathbf{E}{\cdot}d\mathbf{A} = q_{\text{enclosed}}/6\varepsilon_0 = (2.6 \ \mu\text{C})/6(8.85\times10^{-12} \text{ C}^2/\text{N·m}^2) = 49.0 \text{ kN·m}^2/\text{C}$.

Section 24-4   Using Gauss's Law

Problem

17. The electric field at the surface of a uniformly charged sphere of radius 5.0 cm is 90 kN/C. What would be the field strength 10 cm from the surface?

Solution

The electric field due to a uniformly charged sphere is like the field of a point charge for points outside the sphere, i.e., $E(r) \simeq 1/r^2$ for $r \geq R$. Thus, at 10 cm from the surface, $r = 15$ cm and $E(15 \text{ cm}) = (\frac{5}{15})^2 E(5 \text{ cm}) = (90 \text{ kN/C})/9 = 10 \text{ kN/C}$.

Problem

21. A 10-nC point charge is located at the center of a thin spherical shell of radius 8.0 cm carrying -20 nC distributed uniformly over its surface. What are the magnitude and direction of the electric field (a) 2.0 cm, (b) 6.0 cm, and (c) 15 cm from the point charge?

Solution

The total electric field, the superposition of the fields due to the point charge and the spherical shell, is spherically symmetric about the center. Inside the shell ($r < R = 8$ cm), its field is zero, so the total field is just due to the 10 μC point charge. Outside ($r > R$), the shell's field is like that of a point charge of -20 μC at the same central location as the 10 μC charge. (This situation is described in Example 24-3.) (a) and (b) For $r = 2$ cm or 6 cm $< R$, $E = kq_{\text{pt}}/r^2 = (9\times10^9 \text{ N·m}^2/\text{C}^2)\times (10 \text{ nC})/(2 \text{ cm or } 6 \text{ cm})^2 = 225 \text{ kN/C}$ or 25.0 kN/C, respectively, directed radially outward. (c) For $r = 15$ cm $> R$, $E = k(q_{\text{pt}} + q_{\text{shell}})/r^2 = (9\times10^9 \text{ N·m}^2/\text{C}^2)(10 \text{ nC} - 20 \text{ nC})/(15 \text{ cm})^2 = -4.00 \text{ kN/C}$, directed radially inward.

Problem

25. A spherical shell 30 cm in diameter carries a total charge 85 μC distributed uniformly over its surface. A 1.0-μC point charge is located at the center of the shell. What is the electric field strength (a) 5.0 cm from the center and (b) 45 cm from the center? (c) How would your answers change if the charge on the shell were doubled?

Solution

(a) The field due to the shell is zero inside, so at $r = 5$ cm, the field is due to the point charge only. Thus, $\mathbf{E} = kq\hat{\mathbf{r}}/r^2 = (9\times10^9 \text{ N·m}^2/\text{C}^2)(1 \ \mu\text{C})\hat{\mathbf{r}}/(0.05 \text{ m})^2 = (3.60\times10^6 \text{ N/C})\hat{\mathbf{r}}$. (b) Outside the shell, its field is like that of a point charge, so at $r = 45$ cm, $\mathbf{E} = k(q + Q)\hat{\mathbf{r}}/r^2 = (9\times10^9 \text{ N·m}^2/\text{C}^2) \times (86 \ \mu\text{C})\hat{\mathbf{r}}/(0.45 \text{ m})^2 = (3.82\times10^6 \text{ N/C})\hat{\mathbf{r}}$. (c) If the charge on the shell were doubled, the field inside would be unaffected, while the field outside would approximately double, $E = k(1.0 \ \mu\text{C} + 2\times85 \ \mu\text{C})/(45 \text{ cm})^2 = 7.60 \text{ MN/C}$.

Problem

29. A long solid rod 4.5 cm in radius carries a uniform volume charge density. If the electric field strength at the surface of the rod (not near either end) is 16 kN/C, what is the volume charge density?

Solution

If the rod is long enough to approximate its field using line symmetry, we can equate the flux through a length ℓ of its surface (Equation 24-8) to the charge enclosed. The latter is the charge density (a constant) times the volume of a length ℓ of rod. Thus, $2\pi R\ell E = q_{\text{enclosed}}/\varepsilon_0 = \rho\pi R^2\ell/\varepsilon_0$, or $\rho = 2\varepsilon_0 E/R = 2(8.85\times10^{-12}\ \text{C}^2/\text{N}\cdot\text{m}^2)(16\ \text{kN/C})/(4.5\ \text{cm}) = 6.29\ \mu\text{C/m}^3$. (This is the magnitude of ρ, since the direction of the field at the surface, radially inward or outward, was not specified.)

Problem

33. A long, thin wire carries a uniform line charge density $\lambda = -6.8\ \mu\text{C/m}$. It is surrounded by a thick concentric cylindrical shell of inner radius 2.5 cm and outer radius 3.5 cm. What uniform volume charge density in the shell will result in zero electric field outside the shell?

Solution

In order to have $\mathbf{E} = 0$ outside the shell, it is only necessary for the charge per unit length of shell to cancel that of the wire, i.e., $\lambda_{\text{shell}} = +6.8\ \mu\text{C/m}$ (see Gauss's law and Equation 24-8, with $q_{\text{enclosed}} = 0$). A uniform charge density which guarantees this is $\rho = \lambda_{\text{shell}}\ell/V$, where V is the volume of length ℓ of shell. Thus, $\rho = \lambda_{\text{shell}}\ell/\pi(r_2^2 - r_1^2)\ell = (6.8\ \mu\text{C/m})/\pi(3.5^2 - 2.5^2)\times10^{-4}\ \text{m}^2 = 3.61\times10^{-3}\ \text{C/m}^3$.

Problem

37. Figure 24-46 shows sections of three infinite flat sheets of charge, each carrying surface charge density with the same magnitude σ. Find the magnitude and direction of the electric field in each of the four regions shown.

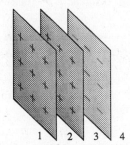

FIGURE 24-46 Problem 37.

Solution

The field from each sheet has magnitude $\sigma/2\varepsilon_0$ and points away from the positive sheets and toward the negative sheet. Take the x axis perpendicular to the sheets, to the right in

Fig. 24-46. Superposition gives the field in each of the four regions, as shown.

First sheet: $\quad = \quad \overset{-\sigma\hat{\imath}/2\varepsilon_0}{\longleftarrow}\ (+)\ \overset{\sigma\hat{\imath}/2\varepsilon_0}{\longrightarrow}\ (+)\ \overset{\sigma\hat{\imath}/2\varepsilon_0}{\longrightarrow}\ (-)\ \overset{\sigma\hat{\imath}/2\varepsilon_0}{\longrightarrow}$
$\qquad\qquad\qquad\qquad (+)\qquad\qquad (+)\qquad\qquad (-)$

Second sheet: $\quad \overset{-\sigma\hat{\imath}/2\varepsilon_0}{\longleftarrow}\ (+)$
$\qquad\quad \overset{-\sigma\hat{\imath}/2\varepsilon_0}{\longleftarrow}\ (+)\quad (+)\ \overset{\sigma\hat{\imath}/2\varepsilon_0}{\longrightarrow}\ (-)\ \overset{\sigma\hat{\imath}/2\varepsilon_0}{\longrightarrow}$
$\qquad\qquad\qquad\qquad\qquad\qquad (+)\qquad\qquad (-)$

Third sheet: $\quad \overset{\sigma\hat{\imath}/2\varepsilon_0}{\longrightarrow}\ (+)\ \overset{\sigma\hat{\imath}/2\varepsilon_0}{\longrightarrow}\ (+)\ \overset{\sigma\hat{\imath}/2\varepsilon_0}{\longrightarrow}\ (-)\ \overset{-\sigma\hat{\imath}/2\varepsilon_0}{\longleftarrow}$
$\qquad\qquad\qquad\qquad (+)\qquad\qquad (+)\qquad\qquad (-)$

Sum: $\quad \overset{-\sigma\hat{\imath}/2\varepsilon_0}{\longleftarrow}\ (+)\ \overset{\sigma\hat{\imath}/2\varepsilon_0}{\longrightarrow}\ (+)\ \overset{3\sigma\hat{\imath}/2\varepsilon_0}{\longrightarrow}\ (-)\ \overset{\sigma\hat{\imath}/2\varepsilon_0}{\longrightarrow}$
$\qquad\qquad\qquad (+)\qquad\qquad (+)\qquad\qquad (-)$

Section 24-5 Fields of Arbitrary Charge Distributions

Problem

41. The electric field strength on the axis of a uniformly charged disk is given by $E = 2\pi k\sigma(1 - x/\sqrt{x^2 + a^2})$, with σ the surface charge density, a the disk radius, and x the distance from the disk center. If $a = 20$ cm, (a) for what range of x values does treating the disk as an infinite sheet give an approximation to the field that is good to within 10%? (b) For what range of x values is the point-charge approximation good to 10%?

Solution

(Note: The expression given, for the field strength on the axis of a uniformly charged disk, holds only for positive values of x.) (a) For small x, using the field strength of an infinite sheet, $E_{\text{sheet}} = \sigma/2\varepsilon_0 = 2\pi k\sigma$, produces a fractional error less than 10% if $|E_{\text{sheet}} - E|/E < 0.1$. Since $E_{\text{sheet}} > E$, this implies that $E_{\text{sheet}}/E < 1.1$ or $2\pi k\sigma/2\pi k\sigma(1 - x/\sqrt{x^2 + a^2}) < 1.1$. The steps in the solution of this inequality are: $1.1x < 0.1\sqrt{x^2 + a^2}$, $1.21x^2 < 0.01(x^2 + a^2)$, $x < a\sqrt{0.01/1.20} = 9.13\times10^{-2}\ a$. For $a = 20$ cm, $x < 1.83$ cm. (b) For large x, the point charge field, $E_{\text{pt}} = kq/x^2 = k\pi\sigma a^2/x^2$, is good to 10% for $|E_{\text{pt}} - E|/E < 0.1$. The solution of this inequality is simplified by defining an angle ϕ, such that $\cos\phi = x/\sqrt{x^2 + a^2}$ and $\tan\phi = a/x$. In terms of ϕ, one finds $E = 2\pi k\sigma(1 - \cos\phi)$, $E_{\text{pt}} = k\pi\sigma\tan^2\phi$, and $E_{\text{pt}}/E = \tan^2\phi/2(1 - \cos\phi)$. Furthermore, $\tan^2\phi = \sin^2\phi/\cos^2\phi = (1 - \cos\phi)(1 + \cos\phi)/\cos^2\phi$, so $E_{\text{pt}}/E = (1 + \cos\phi)/2\cos^2\phi$. The range $0 \le x < \infty$ corresponds to $0 < \phi \le \pi/2$, so $E_{\text{pt}}/E > 1$ and the inequality becomes $E_{\text{pt}}/E = (1 + \cos\phi)/2\cos^2\phi < 1.1$, or $2.2\cos^2\phi - \cos\phi - 1 > 0$. The quadratic formula for the positive root gives $\cos\phi > (1 + \sqrt{1 + 8.8})/4.4 = 0.939$, or $\phi < 20.2°$. This implies $x = a/\tan\phi > a/\tan 20.2° = 2.72\ a$. For $a = 20$ cm, $x > 54.5$ cm.

Section 24-6 Gauss's Law and Conductors

Problem

45. A net charge of 5.0 μC is applied on one side of a solid metal sphere 2.0 cm in diameter. After electrostatic equilibrium is reached, what are (a) the volume charge density inside the sphere and (b) the surface charge density on the sphere? Assume there are no other charges or conductors nearby. (c) Which of your answers depends on this assumption, and why?

Solution

(a) The electric field within a conducting medium, in electrostatic equilibrium, is zero. Therefore, Gauss's law implies that the net charge contained in any closed surface, lying within the metal, is zero. (b) If the volume charge density is zero within the metal, all of the net charge must reside on the surface of the sphere. If the sphere is electrically isolated, the charge will be uniformly distributed (i.e., spherically symmetric), so $\sigma = Q/4\pi R^2 = (5\ \mu C)/4\pi(1\ \text{cm})^2 = 3.98\times10^{-3}\ \text{C/m}^2$. (c) Spherical symmetry for σ depends on the proximity of other charges and conductors.

Problem

49. An irregular conductor containing an irregular, empty cavity carries a net charge Q. (a) Show that the electric field inside the cavity must be zero. (b) If you put a point charge inside the cavity, what value must it have in order to make the surface charge density on the outer surface of the conductor everywhere zero?

Solution

(a) When there is no charge inside the cavity, the flux through any closed surface within the cavity (S_1) is zero, hence so is the field. (b) If the surface charge density on the outer surface (and also the electric field there) is to vanish, then the net charge inside a gaussian surface containing the conductor (S_2) is zero. Thus, the point charge in the cavity must equal $-Q$. (Note: The argument in part (a) depends on the conservative nature of the electrostatic field (see Section 25-1), for then positive flux on one part of S_1 canceling negative flux on another part is ruled out.)

Problem 49 Solution.

Problem

53. A conducting sphere 2.0 cm in radius is concentric with a spherical conducting shell with inner radius 8.0 cm and outer radius 10 cm. The small sphere carries 50 nC charge and the shell has no net charge. Find the electric field strength (a) 1.0 cm, (b) 5.0 cm, (c) 9.0 cm, and (d) 15 cm from the center.

Solution

If we assume the two-conductor system is isolated and in electrostatic equilibrium, then the field has spherical symmetry. Gauss's law requires that the field inside the conducting material be zero (for $0 \le r < 2$ cm and 8 cm $< r < 10$ cm in this problem), and that, since the shell is neutral, the field elsewhere is like that from a point charge of 50 μC located at the center of symmetry ($r = 0$). Thus, (a) $E(1\ \text{cm}) = 0$, (b) $E(5\ \text{cm}) = kq/r^2 = (9\times10^9\ \text{N·m}^2/\text{C}^2)(50\ \mu C)/(5\ \text{cm})^2 = 180\ \text{kN/C}$, (c) $E(9\ \text{cm}) = 0$, and (d) $E(15\ \text{cm}) = kq/r^2 = (\frac{1}{9})E(5\ \text{cm}) = 20\ \text{kN/C}$.

Paired Problems

Problem

57. A point charge q is at the center of a spherical shell of radius R carrying charge $2q$ spread uniformly over its surface. Write expressions for the electric field strength at (a) $\frac{1}{2}R$ and (b) $2R$.

Solution

As explained in Example 24-3, (a) for $r = \frac{1}{2}R < R$, $q_{\text{enclosed}} = q$ and $E = kq/(\frac{1}{2}R)^2 = 4kq/R^2$, and (b) for $r = 2R > R$, $q_{\text{enclosed}} = q + 2q$ and $E = 3kq/(2R)^2 = 3\,kq/4R^2$.

Problem

61. An early (and incorrect) model for the atom pictured its positive charge as spread uniformly throughout the spherical atomic volume. For a hydrogen atom of radius 0.0529 nm, what would be the electric field due to such a distribution of positive charge (a) 0.020 nm from the center and (b) 0.20 nm from the center?

Solution

(a) Inside a uniformly charged spherical volume, $E = kQr/R^3 = (9\ \text{GN·m}^2/\text{C}^2)(1.6\times10^{-19}\ \text{C})(0.02\ \text{nm})/(0.0529\ \text{nm})^3 = 195\ \text{GN/C}$ (see Equation 24-7). (b) Outside, the field is like that of a point charge, $E = kQ/r^2 = (9\ \text{GN·m}^2/\text{C}^2)(1.6\times10^{-19}\ \text{C})/(0.2\ \text{nm})^2 = 36.0\ \text{GN/C}$ (see Equation 24-6).

Supplementary Problems

Problem

65. Repeat Problem 10 for the case $\mathbf{E} = E_0\left(\dfrac{y}{a}\right)^2 \hat{\mathbf{k}}$.

Solution

Since the electric field depends only on y, break up the square in Fig. 24-41 into strips of area $d\mathbf{A} = \pm a\,dy\,\hat{\mathbf{k}}$, of length a parallel to the x axis and width dy, the normal to which could be $\pm\hat{\mathbf{k}}$. The electric flux through the square is

$$\Phi = \int_{\text{square}} \mathbf{E}\cdot d\mathbf{A} = \pm\int_0^a E_0\left(\frac{y}{a}\right)^2 a\,dy$$
$$= \pm\left(\frac{E_0}{a}\right)\int_0^a y^2\,dy = \frac{1}{3}E_0 a^2.$$

Problem

69. Repeat Problem 36 for the case when the charge density in the slab is given by $\rho = \rho_0|x/d|$, where ρ_0 is a constant.

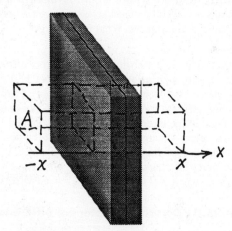

FIGURE 24·45 Problem 69 Solution.

Solution

Gauss's law, plane symmetry, and Equation 24-9 can be used to find the electric field strength, but we must integrate to get the charge enclosed by the gaussian surface. We use charge elements that are thin parallel sheets of the same area as the face of the gaussian surface and of thickness dx, as shown. (a) Inside the slab ($|x| < d/2$), $2EA = \varepsilon_0^{-1}\int_{-x}^x \rho A\,dx = (\rho_0 A/\varepsilon_0 d)(\int_{-x}^0 (-x)\,dx + \int_0^x x\,dx) = \rho_0 A x^2/\varepsilon_0 d$, hence $E = \rho_0 x^2/2\varepsilon_0 d$. (b) Outside, $2EA = \varepsilon_0^{-1}\int_{-d/2}^{d/2} \rho A\,dx = (\rho_0 A/\varepsilon_0 d)(\int_{-d/2}^0 (-x)\,dx + \int_0^{d/2} x\,dx) = (\rho_0 A/\varepsilon_0 d)(d/2)^2$, hence $E = \rho_0 d/8\varepsilon_0$. (This is equivalent, of course, to the field strength outside an infinite sheet with $\sigma = \rho_0 d/4$.)

Problem

73. A thick spherical shell of inner radius a and outer radius b carries a charge density given by $\rho = \dfrac{ce^{-r/a}}{r^2}$, where a and c are constants. Find expressions for the electric field strength for (a) $r < a$, (b) $a < r < b$, and (c) $r > b$.

Solution

Spherical symmetry, Equation 24-5 and Gauss's law give a field strength of $E(r) = q_{\text{enclosed}}/4\pi\varepsilon_0 r^2$, where $q_{\text{enclosed}} = \int_0^r \rho\,dV$ is the charge within a concentric spherical surface of radius r, and $dV = 4\pi r^2\,dr$ is the volume element for a thin shell with this surface. (a) For $r < a$, $q_{\text{enclosed}} = 0$ hence $E(r) = 0$. (b) For $a \leq r \leq b$, $q_{\text{enclosed}} = 4\pi c\int_a^r e^{-r/a}\,dr = 4\pi ac(e^{-1} - e^{r/a})$ hence $E(r) = ac(e^{-1} - e^{-r/a})/\varepsilon_0 r^2$. (c) For $r > b$, $q_{\text{enclosed}} = 4\pi c\int_a^b e^{-r/a}\,dr$ hence $E(r) = ac(e^{-1} - e^{-b/a})/\varepsilon_0 r^2$.

Problem

77. Two flat, parallel, closely spaced metal plates of area 0.080 m² carry total charges of -2.1 μC and $+3.8$ μC. Find the surface charge densities on the inner and outer faces of each plate.

Solution

If the thickness and separation of the plates is small compared to their lateral dimensions ($\sqrt{0.08\ \text{m}^2} \approx 28$ cm), we can assume that the electric field near the plates (edge effects neglected) is uniform and normal to the plates. (Of course, far away, the field goes like $1/r^2$.) The general case, where the plates carry arbitrary charges q_1 and q_2, can be viewed as the superposition of three simpler cases: a neutral plate, a charged plate, and two oppositely charged plates (see last paragraph in Section 24-6). In each of the three regions,

Problem 77 Solution.

left of, between, and right of the plates (the thickness, assumed negligible, does not add to any region), the electric field is the sum of three contributions, as shown. The final diagram, together with Equation 24-11 (which gives $\sigma = \varepsilon_0 E$ at each surface), shows that the charge densities on the outer surfaces are equal, $\sigma^{out} = \frac{1}{2}(q_1 + q_2)/A$, and that the charge densities on the inner surfaces are equal and opposite, $\sigma^{im} = \pm \frac{1}{2}(q_2 - q_1)/A$. Numerically $\sigma^{out} = \frac{1}{2}(-2.1 + 3.8)\mu C \div 0.08$ m^2 = $10.6\,\mu C/m^2$, and $\sigma^{in} = \pm \frac{1}{2}(3.8 - (-2.1))\mu C \div 0.08$ m^2 = $\pm 36.9\,\mu C/m^2$. (Note: the direction of the fields is shown for positive total charge, with the more positive plate on the right.) ●

● CHAPTER 25 ELECTRIC POTENTIAL

Section 25-2 Potential Difference

Problem

1. How much work does it take to move a 50-μC charge against a 12-V potential difference?

Solution

The potential difference and the work per unit charge, done by an external agent, are equal in magnitude, so $W = q\,\Delta V = (50\,\mu C)(12\,V) = 600\,\mu J$. (Note: Since only magnitudes are needed in this problem, we omitted the subscripts $A \rightarrow B$.)

Problem

5. Find the magnitude of the potential difference between two points located 1.4 m apart in a uniform 650 N/C electric field, if a line between the points is parallel to the field.

Solution

For ℓ in the direction of a uniform electric field, Equation 25-2b gives $|\Delta V| = E\ell = (650\,N/C)(1.4\,m) = 910\,V$. (See note in solution to Problem 1. Since $dV = -\mathbf{E}\cdot\mathbf{d\ell}$, the potential always decreases in the direction of the electric field.)

Problem

9. A proton, an alpha particle (a bare helium nucleus), and a singly ionized helium atom are accelerated through a potential difference of 100 V. Find the energy each gains.

Solution

The energy gained is $q\,\Delta V$ (see Example 25-1). The proton and singly-ionized helium atom have charge e, so they gain 100 eV $= (1.6\times10^{-19}\,C)(100\,V) = 1.6\times10^{-17}$ J, while the α-particle has charge $2e$ and gains twice this energy.

Problem

13. A 12-V car battery stores 2.8 MJ of energy. How much charge can move between the battery terminals before it is totally discharged? Assume the potential difference remains at 12 V, an assumption that is not realistic.

Solution

A charge q, moving through a potential difference ΔV, is equivalent to electrostatic potential energy $\Delta U = q\,\Delta V$, stored in the battery. Thus, $q = 2.8$ MJ/12 V $= 2.33\times10^5$ C.

Problem

17. A 5.0-g object carries a net charge of 3.8 μC. It acquires a speed v when accelerated from rest through a potential difference V. A 2.0-g object acquires twice the speed under the same circumstances. What is its charge?

Solution

The speed acquired by a charge q, starting from rest at point A and moving through a potential difference of V, is given by $\frac{1}{2}mv_B^2 = q(V_A - V_B) = qV$, or $v_B = \sqrt{2V(q/m)}$. (This is the work-energy theorem for the electric force. A positive charge is accelerated in the direction of decreasing potential.) If the second object acquires twice the speed of the first object, moving through the same potential difference, it must have four times the charge to mass ratio, q/m. Thus, $q_2 = 4(q_1/m_1)m_2 = 4(3.8\,\mu C)(2\,g/5\,g) = 6.08\,\mu C$.

Section 25-3 Calculating Potential Difference

Problem

21. Points A and B lie 20 cm apart on a line extending radially from a point charge Q, and the potentials at these points are $V_A = 280$ V, $V_B = 130$ V. Find Q and the distance r between A and the charge.

Solution

Since $V_A = kQ/r_A$ and $V_B = kQ/r_B$, division yields $r_B = (V_A/V_B)r_A = (280/130)r_A = 2.15r_A$. But $r_B - r_A = 20$ cm, so $r_A = (20\,cm)(2.15 - 1)^{-1} = 17.3$ cm. Then $Q = V_A r_A/k = (280\,V)(17.3\,cm)/(9\times10^9\,N\cdot m^2/C^2) = 5.39$ nC.

Problem

25. A thin spherical shell of charge has radius R and total charge Q distributed uniformly over its surface. What is the potential at its center?

Solution

The potential at the surface of the shell is kQ/R (as in Example 25-3). The electric field inside a uniformly charged shell is zero, so the potential anywhere inside is a constant, equal, therefore, to its value at the surface.

Problem

29. The potential difference between the surface of a 3.0-cm-diameter power line and a point 1.0 m distant is 3.9 kV. What is the line charge density on the power line?

Solution

If we approximate the potential from the line by that from an infinitely long charged wire, Equation 25-5 can be used to find λ: $\lambda = 2\pi\varepsilon_0\,\Delta V_{A\to B}/\ln(r_A/r_B) = (3.9\text{ kV}) \times [2(9\times10^9\text{ N}\cdot\text{m}^2/\text{C}^2)\ln(100/1.5)]^{-1} = 51.6\text{ nC/m}$. (Note: $\Delta V_{A\to B} = V_B - V_A$ so B is at the surface of the wire and A is 100 cm distant.)

Problem

33. Find the potential 10 cm from a dipole of moment $p = 2.9$ nC·m (a) on the dipole axis, (b) at 45° to the axis, and (c) on the perpendicular bisector. The dipole separation is much less than 10 cm.

Solution

Equation 25-7 gives the potential from a point dipole as a function of distance and angle from the dipole axis. For the dipole moment and distance given, $V(r, \theta) = kp\cos\theta/r^2 = (9\text{ GN}\cdot\text{m}^2/\text{C}^2)(2.9\text{ nC-m})\cos\theta/(10\text{ cm})^2 = (2.61\text{ kV})\cos\theta$. For the three given angles,
(a) $V = (2.61\text{ kV})\cos 0° = 2.61$ kV;
(b) $V = (2.61\text{ kV})\cos 45° = 1.85$ kV; and
(c) $V = (2.61\text{ kV})\cos 90° = 0$.

Problem

37. A thin ring of radius R carries a charge $3Q$ distributed uniformly over three-fourths of its circumference, and $-Q$ over the rest. What is the potential at the center of the ring?

Solution

The result in Example 25-6 did not depend on the ring being uniformly charged. For a point on the axis of the ring, the geometrical factors are the same, and $\int_{\text{ring}} dq = Q_{\text{tot}}$ for any arbitrary charge distribution, so $V = kQ_{\text{tot}}(x^2 + a^2)^{-1/2}$ still holds. Thus, at the center $(x = 0)$ of a ring of total charge $Q_{\text{tot}} = 3Q - Q = 2Q$, and radius $a = R$, the potential is $V = 2kQ/R$.

Problem

41. (a) Find the potential as a function of position in the electric field $\mathbf{E} = E_0(\hat{\mathbf{i}} + \hat{\mathbf{j}})$, where $E_0 = 150$ V/m. Take the zero of potential at the origin. (b) Find the potential difference from the point $x = 2.0$ m, $y = 1.0$ m to the point $x = 3.5$ m, $y = -1.5$ m.

Solution

(a) Equation 25-2b gives the potential for a uniform field. Take the zero of potential at the origin (point A in Equation 25-2b) and let $\ell = \mathbf{r} = x\hat{\mathbf{i}} + y\hat{\mathbf{i}} + z\hat{\mathbf{k}}$ be the vector from the origin to the field point (point B in Equation 25-2b). Then $\Delta V_{A\to B} = V_B - V_A = V(\mathbf{r}) - 0 = V(x, y) = -E_0(\hat{\mathbf{i}} + \hat{\mathbf{j}}) \cdot \mathbf{r} = -E_0(x + y)$. (The potential is independent of z, so we wrote $V(\mathbf{r}) = V(x, y)$.) (b) $V(3.5\text{ m}, -1.5\text{ m}) - V(2.0\text{ m}, 1.0\text{ m}) = -(150\text{ V/m})(3.5\text{ m} - 1.5\text{ m} - 2.0\text{ m} - 1.0\text{ m}) = 150$ V.

Section 25-4   Potential Difference and the Electric Field

Problem

45. The potential in a certain region is given by $V = axy$, where a is a constant. (a) Determine the electric field in the region. (b) Sketch some equipotentials and field lines.

Solution

(a) The x and y components of the electric field can be found from Equation 25-10: $E_x = -\partial V/\partial x = -\partial/\partial x(axy) = -ay$, and $E_y = -\partial V/\partial y = -ax$. Thus $\mathbf{E} = -a(y\hat{\mathbf{i}} + x\hat{\mathbf{j}})$. (The field has no z component.) (b) See chart below.

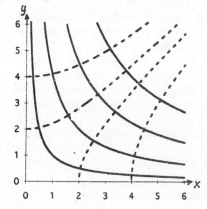

Problem 45 Solution.

Problem

49. The electric potential in a region of space is given by $V = 2xy - 3zx + 5y^2$, with V in volts and the coordinates in meters. If point P is at $x = 1$ m, $y = 1$ m, $z = 1$ m, find (a) the potential at P and (b) the x, y, and z components of the electric field at P.

Solution

(a) Direct substitution gives $V(P) = 2(1)(1) - 3(1)(1) + 5(1)^2 = 4$ V. (b) Use of Equation 25-10 gives $E_x = -\partial V/\partial x = -2y + 3z$, $E_y = -\partial V/\partial y = -2x - 10y$, and $E_z = -\partial V/\partial z = 3x$. At $P(x = y = z = 1)$, $E_x = 1$ V/m, $E_y = -12$ V/m, and $E_z = 3$ V/m.

Problem

53. The electric potential in a region is given by $V = -V_0(r/R)$, where V_0 and R are constants, r is the radial distance from the origin, and where the zero of potential is taken at $r = 0$. Find the magnitude and direction of the electric field in this region.

Solution

Since $V = -(V_0/R)r$ depends only on r, the field is spherically symmetric and its direction is radial. From Equation 25-10 (which applies unmodified for the radial coordinate), $E_r = -dV/dr = V_0/R$, and $\mathbf{E} = (V_0/R)\hat{\mathbf{r}}$ ($\hat{\mathbf{r}}$ is a unit vector radially outward).

Section 25-5 Potentials of Charged Conductors

Problem

57. Two metal spheres each 1.0 cm in radius are far apart. One sphere carries 38 nC of charge, the other -10 nC. (a) What is the potential on each? (b) If the spheres are connected by a thin wire, what will be the potential on each once equilibrium is reached? (c) How much charge must move between the spheres in order to achieve equilibrium?

Solution

(a) Since the spheres are far apart (approximately isolated), we can use Equation 25-11 to find their potentials: $V_1 = kQ_1/R_1 = (9 \text{ GN·m}^2/\text{C}^2)(38 \text{ nC})/(1 \text{ cm}) = 34.2 \text{ kV}$ and $V_2 = kQ_2/R_2 = -9 \text{ kV}$. (b) When connected by a thin wire, the spheres reach electrostatic equilibrium with the same potential, so $V = kQ_1'/R_1 = kQ_2'/R_2$. Since the radii are equal, so must be the charges, $Q_1' = Q_2'$. The total charge is 38 nC $-$ 10 nC $= 28$ nC $= Q_1' + Q_2' = 2Q_1'$ (if we assume that the wire is so thin that it has a negligible charge), so $Q_1' = Q_2' = 14$ nC. Then $V' = k(14 \text{ nC})/(1 \text{ cm}) = 12.6 \text{ kV}$. (c) In this process, the first sphere loses $38 - 14 = 24$ nC to the second.

Paired Problems

Problem

61. Three 50-pC charges sit at the vertices of an equilateral triangle 1.5 mm on a side. How much work would it take to bring a proton from very far away to the midpoint of one of the triangle's sides?

Solution

Two of the charges are at distances of $\frac{1}{2}(1.5 \text{ mm}) = 0.75$ mm, and the third is at $\sqrt{3}$ (0.75 mm) from the midpoint of one side. Therefore, the potential at this point (Equation 25-6) is $V(P) = k(50 \text{ pC})(1 + 1 + 1/\sqrt{3}) \div (0.75 \text{ mm}) = 1.55$ kV. The work it would take to move a proton of charge e from infinity (where the potential is zero) to this point is $eV = 1.55$ keV $= (1.6 \times 10^{-19} \text{ C})(1.55 \text{ kV}) = 2.47 \times 10^{-16}$ J.

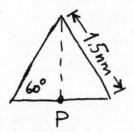

Problem 61 Solution.

Problem

65. A 2.0-cm-radius metal sphere carries 75 nC and is surrounded by a concentric spherical conducting shell of radius 10 cm carrying -75 nC. (a) Find the potential difference between the shell and the sphere. (b) How would your answer change if the shell charge were changed to $+150$ nC?

Solution

(a) The electric field outside the sphere (radius $R_1 = 2$ cm), but inside the shell (inner radius $R_2 = 10$ cm), is only due to the charge q_1, on sphere (recall Gauss's law), and equals kq_1/r^2 radially outward. The potential difference is $V_1 - V_2 = -\int_{R_2}^{R_1} kq_1 \, dr/r^2 = kq_1(R_1^{-1} - R_2^{-1}) = (9 \text{ GN·m}^2/\text{C}^2)(75 \text{ nC})[(2 \text{ cm})^{-1} - (10 \text{ cm})^{-1}] = 27.0 \text{ kV}$. (We used Equation 25-2a, with A at R_2 and B at R_1. Note that $\mathbf{E} \cdot d\mathbf{r} = kq_1 \, dr/r^2$ regardless of the choice of points A and B.) (b) Adding more charge to the conducting shell does not affect the field inside, nor the potential difference between points inside, like $V_1 - V_2$ in part (a). The field outside the shell and the potential relative to infinity would change.

Problem

69. The potential as a function of position in a certain region is given by $V(x) = 3x - 2x^2 - x^3$, with x in meters and V in volts. Find (a) all points on the x axis where $V = 0$, (b) an expression for the electric field, and (c) all points on the x axis where $\mathbf{E} = 0$.

Solution

(a) The expression for the potential can be factored into $V(x) = x(x + 3)(1 - x)$, so $V(x) = 0$ at $x = 0$, 1 m, and -3 m. (b) V is independent of y and z, hence $\mathbf{E} = E_x\hat{\mathbf{i}}$ has only an x component, and $E_x = -dV/dx = 3x^2 + 4x - 3$. (c) $E_x = 0$ for $x = (-2 \pm \sqrt{4 + 9})/3 = 0.535$ m and -1.87 m.

Supplementary Problems

Problem

73. The potential on the axis of a uniformly charged disk at 5.0 cm from the disk center is 150 V; the potential 10 cm from disk center is 110 V. Find the disk radius and its total charge.

Solution

Combining the given data with the potential in Exercise 25-7, we find $150 \text{ V} = 2kQ/a^2(\sqrt{(5 \text{ cm})^2 + a^2} - 5 \text{ cm})$ and $110 \text{ V} = 2kQ/a^2(\sqrt{(10 \text{ cm})^2 + a^2} - 10 \text{ cm})$. The charge can be eliminated by division,

$$\left(\frac{150}{110}\right) = \frac{\sqrt{1 + (a/5 \text{ cm})^2} - 1}{\sqrt{4 + (a/5 \text{ cm})^2} - 2}.$$

Several lines of algebra to remove the square roots finally yields $a = (5 \text{ cm})\sqrt{105 \times 209/52} = 14.2 \text{ cm}$. We can now solve for Q from either of the first two equations, $Q = (Va^2/2k)[\sqrt{x^2 + a^2} - x]^{-1} = 1.67 \text{ nC}$.

Problem

77. A thin rod of length ℓ lies on the x axis with its center at the origin. It carries a line charge density given by $\lambda = \lambda_0(x/\ell)^2$, where λ_0 is a constant. (a) Find an expression for the potential on the x axis for $x > \ell/2$. (b) Integrate the charge density to find the total charge on the rod. (c) Show that your answer for (a) reduces to the potential of a point charge whose charge is the answer to (b), for $x \gg \ell$.

Solution

(a) For points P on the x axis at $x > \ell/2$, the potential from a charge element $dq = \lambda \, dx'$ at x' along the rod $(-\ell/2 \leq x' \leq \ell/2)$ is $dV = k \, dq/|x - x'| = k\lambda \, dx'/(x - x')$. (We used x' for the variable position of dq, and took the potential relative to zero at infinity.) The potential due to the entire rod, for $\lambda = \lambda_0(x'/\ell)^2$, is

$$V(x) = \int_{-\ell/2}^{\ell/2} dV = \frac{k\lambda_0}{\ell^2} \int_{-\ell/2}^{\ell/2} \frac{x'^2 \, dx'}{(x - x')}$$

$$= \frac{k\lambda_0}{\ell^2} \left| -x^2 \ln(x - x') - xx' - \frac{x'^2}{2} \right|_{-\ell/2}^{\ell/2}$$

$$= \frac{k\lambda_0}{\ell^2} \left[x^2 \ln\left(\frac{x + \ell/2}{x - \ell/2}\right) - x\ell \right].$$

(Use partial fractions, $x'^2/(x - x') = -x - x' + x^2/(x - x')$, or standard tables to evaluate the integral.)

(b) The total charge on the rod is $Q = \int dq = \int_{-\ell/2}^{\ell/2} (\lambda_0/\ell^2)x^2 \, dx = (\lambda_0/\ell^2)(\frac{2}{3})(\ell/2)^3 = \lambda_0\ell/12$. (c) For $x \gg \ell$, we can expand the logarithms, $\ln(1 \pm \ell/2x) = \pm(\ell/2x) - \frac{1}{2}(\ell/2x)^2 \pm \frac{1}{3}(\ell/2x)^3 - \dots$. Therefore:

$$V(x) = (k\lambda_0/\ell^2)[x^2 \ln(1 + \ell/2x) - x^2 \ln(1 - \ell/2x) - x\ell]$$

$$= \left(\frac{k\lambda_0}{\ell^2}\right)\left[x^2\left(\frac{\ell}{2x} - \frac{1}{2}\left(\frac{\ell}{2x}\right)^2 + \frac{1}{3}\left(\frac{\ell}{2x}\right)^3 - \dots\right) \right.$$

$$\left. -x^2\left(-\frac{\ell}{2x} - \frac{1}{2}\left(\frac{\ell}{2x}\right)^2 - \frac{1}{3}\left(\frac{\ell}{2x}\right)^3 - \dots\right) - x\ell \right]$$

$$= (k\lambda_0/\ell^2)\left[x^2\left(\frac{\ell}{x} + \frac{2}{3}\left(\frac{\ell}{2x}\right)^3 + \dots\right) - x\ell \right]$$

$$= k\lambda_0\ell/12x = kQ/x,$$

as expected.

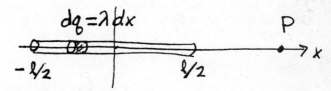

Problem 77 Solution.

Problem

81. An open-ended cylinder of radius a and length $2a$ carries charge q spread uniformly over its surface. Find the potential on the cylinder axis at its center. *Hint:* Treat the cylinder as a stack of charged rings, and integrate.

Solution

The cylinder can be considered to be composed of rings of radius a, width dx, and charge $dq = (q/2a) \, dx$. The potential at the center of the cylinder (which we take as the origin, with x axis along the cylinder axis) due to a ring at $x(-a \leq x \leq a)$ is $dV = k \, dq/\sqrt{x^2 + a^2}$ (see Example 25-6). The whole potential at the center follows from integration:

$$V = \int_{-a}^{a} \left(\frac{kq}{2a}\right)\frac{dx}{\sqrt{x^2 + a^2}} = \frac{kq}{2a} \ln\left(\frac{a + \sqrt{2}a}{-a + \sqrt{2}a}\right)$$

$$= \frac{kq}{a} \ln(1 + \sqrt{2}) = 0.881\frac{kq}{a}.$$

● **CHAPTER 26** ELECTROSTATIC ENERGY AND CAPACITORS

26-1 Energy of a Charge Distribution

Problem

1. Three point charges, each of $+q$, are moved from infinity to the vertices of an equilateral triangle of side ℓ. How much work is required?

Solution

The sentence preceding Example 26-1 allows us to rewrite Equation 26-1 (for the electrostatic energy of a distribution of point charges) as $W = \sum_{\text{pairs}} kq_i q_j / r_{ij}$. For three equal charges (three different pairs) at the corners of an equilateral triangle ($r_{ij} = \ell$ for each pair) $W = 3kq^2/\ell$.

Problem

5. Suppose two of the charges in Problem 1 are held in place, while the third is allowed to move freely. If this third charge has mass m, what will be its speed when it's far from the other two charges?

Solution

With one charge removed to infinity, the potential energy is reduced to that of just one pair of charges, $W_f = kq^2/\ell$. The initial potential energy was $W_i = 3kq^2/\ell$ (see Problem 1), so the kinetic energy of the charge at infinity (from the conservation of energy) is $K = W_i - W_f = 2kq^2/\ell$. Thus, $v = \sqrt{2K/m} = q/\sqrt{\pi\varepsilon_0 m\ell}$.

Section 26-2 Two Isolated Conductors

Problem

9. Two square conducting plates 25 cm on a side and 5.0 mm apart carry charges $\pm 1.1\ \mu C$. Find (a) the electric field between the plates, (b) the potential difference between the plates, and (c) the stored energy.

Solution

(a) The electric field between two closely spaced, oppositely charged, parallel conducting plates is approximately uniform (directed from the positive to the negative plate), with strength $E = \sigma/\varepsilon_0 = q/\varepsilon_0 A = (1.1\ \mu C)/(8.85\ \text{pF/m})(0.25\ \text{m})^2 = 1.99\ \text{MV/m}$. (See the last paragraph of Section 24-6.) (b) Since E is uniform, $V = Ed = (1.99\ \text{MV/m})(5\ \text{mm}) = 9.94\ \text{kV}$. (See Section 26-2.) (c) The energy stored is $U = \frac{1}{2} q^2 d/\varepsilon_0 A = \frac{1}{2} qV = \frac{1}{2}(1.1\ \mu C)(9.94\ \text{kV}) = 5.47\ \text{mJ}$. (See Equation 26-2, and note that $U = \frac{1}{2}\varepsilon_0 E^2 A d$.)

Problem

13. A conducting sphere of radius a is surrounded by a concentric spherical shell of radius b. Both are initially uncharged. How much work does it take to transfer charge from one to the other until they carry charges $\pm Q$?

Solution

When a charge q (assumed positive) is on the inner sphere, the potential difference between the spheres is $V = kq(a^{-1} - b^{-1})$. (See the solution to Problem 25-65(a).) To transfer an additional charge dq from the outer sphere requires work $dW - V\,dq$, so the total work required to transfer charge Q (leaving the spheres oppositely charged) is $W = \int_0^Q V\,dq = \int_0^Q kq\,dq(a^{-1} - b^{-1}) = \frac{1}{2}kQ^2(a^{-1} - b^{-1})$. (Incidentally, this shows that the capacitance of this spherical capacitor is $1/k(a^{-1} - b^{-1}) = ab/k(b - a)$; see Equation 26-8.)

Section 26-3 Energy and the Electric Field

Problem

17. A car battery stores about 4 MJ of energy. If all this energy were used to create a uniform electric field of 30 kV/m, what volume would it occupy?

Solution

In a uniform field, Equation 26-4 can be written as $U = \frac{1}{2}\varepsilon_0 E^2 \times$ (Volume of field region). Therefore, the volume is $2(4\ \text{MJ})/(8.85\ \text{pF/m})(30\ \text{kV/m})^2 = 1.00\times10^9\ \text{m}^3 = 1\ \text{km}^3$.

Problem

21. The electric field strength as a function of position x in a certain region is given by $E = E_0(x/x_0)$, where $E_0 = 24\ \text{kV/m}$ and $x_0 = 6.0\ \text{m}$. Find the total energy stored in a cube 1.0 m on a side, located between $x = 0$ and $x = 1.0$ m. (The field strength is independent of y and z.)

Solution

Since there is no y or z dependence, the volume element of the cube can be written as $dV = \ell^2\,dx$, where $\ell = 1$ m is the cube's edge. Then $U = \int u\,dV = \int_0^\ell \frac{1}{2}\varepsilon_0 (E_0/x_0)^2 x^2 \ell^2\,dx = \frac{1}{2}\varepsilon_0 (E_0/x_0)^2 \ell^5/3$. Numerically, $U = \frac{1}{6}(8.85\ \text{pF m})(24\ \text{kV/m})^2(1\ \text{m})^5/(6\ \text{m})^2 = 23.6\ \mu J$.

Problem

25. Two 4.0-mm-diameter water drops each carry 15 nC. They are initially separated by a great distance. Find the change in the electrostatic potential energy if they are brought together to form a single spherical drop. Assume all charge resides on the drops' surfaces.

Solution

The initial electrostatic energy of two isolated spherical drops, with charge Q on their surfaces and radii R, is $U_i = 2(\frac{1}{2}kQ^2/R)$ (see Problem 23 and Example 26-3). Together, a drop of charge $2Q$, radius $2^{1/3}R$, and energy $U_f = \frac{1}{2}k(2Q)^2/(2^{1/3}R) = 2^{2/3}\,kQ^2/R$, is created. The work required is the difference in energy, $W = U_f - U_i = (2^{2/3} - 1)kQ^2/R = (0.587)(9\times10^9 \text{ m/F})(1.5\times10^{-8} \text{ C})^2/(2\times10^{-3} \text{ m}) = 5.95\times10^{-4}$ J.

Section 26-4 Capacitors

Problem

29. The "memory" capacitor in a VCR has a capacitance of 4.0 F and is charged to 3.5 V. What is the charge on its plates?

Solution

The definition of capacitance (Equation 26-5) gives the magnitude of the charge on either plate, $Q = CV = 4.0$ F$\times3.5$ V $= 14.0$ C. (This is a very large capacitor.)

Problem

33. Find the capacitance of a parallel-plate capacitor consisting of circular plates 20 cm in radius separated by 1.5 mm.

Solution

For a (closely spaced) parallel plate capacitor, with circular plates, Example 26-4 shows that $C = \varepsilon_0 \pi r^2/d = (8.85 \text{ pF/m})\pi(20 \text{ cm})^2/(1.5 \text{ mm}) = 741$ pF.

Problem

37. Figure 26-25 shows a capacitor consisting of two electrically connected plates with a third plate between them, spaced so its surfaces are a distance d from the other plates. The plates have area A. Neglecting edge effects, show that the capacitance is $2\varepsilon_0 A/d$.

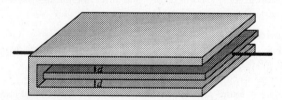

FIGURE 26-25 Problem 37.

Solution

When the third (middle) plate is positively charged, the electric field (not near an edge) is approximately uniform and away from the plate, with magnitude $E = \sigma/\varepsilon_0$. Since half of the total charge Q is on either side (by symmetry), $\sigma = Q/2A$. The potential difference between the third plate

and the outer two plates (which are both at the same potential and carry charges of $-Q/2$ on their inner surfaces) is $V = Ed = \sigma d/\varepsilon_0 = Qd/2\varepsilon_0 A$. Therefore the capacitance is $C = Q/V = 2\varepsilon_0 A/d$. (The arrangement is like two capacitors in parallel.)

Section 26-5 Energy Storage in Capacitors

Problem

41. Which can store more energy, a 1-μF capacitor rated at 250 V or a 470 pF capacitor rated at 3 kV?

Solution

The first capacitor stores $U_C = \frac{1}{2}(1 \ \mu\text{F})(250 \text{ V})^2 = 31.3$ m of energy, while the second only $\frac{1}{2}(470 \text{ pF})(3 \text{ kV})^2 = 2.12$ mJ, about 14.8 times less. (See Equation 26-8b.)

Problem

45. A camera flashtube requires 5.0 J of energy per flash. The flash duration is 1.0 ms. (a) What is the power used by the flashtube *while it is actually flashing?* (b) If the flashtube operates at 200 V, what size capacitor is needed to supply the flash energy? (c) If the flashtube is fired once every 10 s, what is its *average* power consumption?

Solution

(a) $\mathcal{P}_{\text{flash}} = W/t = 5 \text{ J}/1 \text{ ms} = 5$ kW. (b) $U = \frac{1}{2}CV^2$, so $C = 2U/V^2 = 2(5 \text{ J})/(200 \text{ V})^2 = 250 \ \mu\text{F}$. (c) $\mathcal{P}_{\text{av}} = 5 \text{ J}/10 \text{ s} = 0.5$ W, only 10^{-4} times $\mathcal{P}_{\text{flash}}$.

Problem

49. The cylindrical capacitor of Example 26-5 is charged to a voltage V. Obtain an expression for the energy density as a function of radial position in the capacitor, and integrate to show explicitly that the stored energy is $\frac{1}{2}CV^2$.

Solution

The electric field in the capacitor is approximately $\mathbf{E} = \lambda\hat{\mathbf{r}}/2\pi\varepsilon_0 r$, where $\hat{\mathbf{r}}$ is the radial unit vector in cylindrical coordinates (see Example 25-4). (The assumption of line symmetry neglects fringing fields at the ends of the capacitor.) The energy density is $u = \frac{1}{2}\varepsilon_0 E^2 = \lambda^2/8\pi^2\varepsilon_0 r^2$. We can take the volume element to be a cylindrical shell of radius r, thickness dr, and length L, so $dV = 2\pi rL\, dr$. Then the stored energy is

$$U = \int u\, dV = \int_a^b \frac{\lambda^2 2\pi rL\, dr}{8\pi^2\varepsilon_0 r^2} = \frac{\lambda^2 L}{4\pi\varepsilon_0}\int_a^b \frac{dr}{r}$$

$$= \frac{\lambda^2 L}{4\pi\varepsilon_0}\ln\left(\frac{b}{a}\right).$$

Reference to Example 26-5 shows that this is precisely $\frac{1}{2}CV^2$.

Section 26-6 Connecting Capacitors Together

Problem

53. You're given three capacitors: 1.0 μF, 2.0 μF, and 3.0 μF. Find (a) the maximum, (b) the minimum, and (c) two intermediate values of capacitance you could achieve with various combinations of all three capacitors.

Solution

The capacitors can be connected (a) all in parallel: $1 + 2 + 3 = 6\ \mu$F; (b) all in series: $1/1 + 1/2 + 1/3 = 11/6$, or $6/11 = 0.545\ \mu$F; (c) one in parallel with the other two in series:

$$1 + \frac{2\times 3}{2 + 3} = \frac{11}{5} = 2.20\ \mu\text{F},$$

$$2 + \frac{1\times 3}{1 + 3} = \frac{11}{4} = 2.75\ \mu\text{F},$$

$$3 + \frac{1\times 2}{1 + 2} = \frac{11}{3} = 3.67\ \mu\text{F},$$

or one in series with the other two in parallel:

$$\frac{1(2 + 3)}{1 + 2 + 3} = \frac{5}{6} = 0.933\ \mu\text{F},$$

$$\frac{2(1+ 3)}{2 + 1 + 3} = \frac{4}{3} = 1.33\ \mu\text{F},$$

$$\frac{3(1 + 2)}{3 + 1 + 2} = \frac{3}{2} = 1.50\ \mu\text{F}.$$

Problem

57. What is the equivalent capacitance in Fig. 26-29?

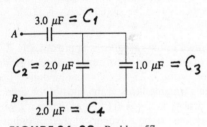

FIGURE 26-29 Problem 57.

Solution

Number the capacitors as shown. Relative to points A and B, C_1, C_4, and the combination of C_2 and C_3 are in series, so the capacitance is given by $C_{AB}^{-1} = C_1^{-1} + C_4^{-1} + C_{23}^{-1}$. C_{23} is a parallel combination, hence $C_{23} = C_2 + C_3$, therefore $C_{AB}^{-1} = (3\ \mu\text{F})^{-1} + (2\ \mu\text{F})^{-1} + (2\ \mu\text{F} + 1\ \mu\text{F})^{-1}$, or $C_{AB} = \frac{6}{7}\ \mu\text{F} = 0.857\ \mu\text{F}$.

Problem

61. A variable "trimmer" capacitor used to make fine adjustments has a capacitance range from 10 to 30 pF. The trimmer is in parallel with a capacitor of about 0.001 μF. Over what percentage range can the capacitance of the combination be varied?

Solution

"Capacitors in parallel add" (Equation 26-9a), so the combination covers a range from 1010 to 1030 pF, or about $\pm 10/1020 \approx \pm 1\%$ from the central value.

Section 26-7 Capacitors and Dielectrics

Problem

65. A 470-pF capacitor consists of two circular plates 15 cm in radius, separated by a sheet of polystyrene. (a) What is the thickness of the sheet? (b) What is the working voltage?

Solution

(a) With reference to Equations 26-6, 26-11, and Table 26-1, one finds that $C = \kappa C_0 = \kappa\varepsilon_0 A/d$, or $d = \kappa\varepsilon_0 A/C = (2.6)(8.85\ \text{pF/m})\pi(0.15\ \text{m})^2/470\ \text{pF} = 3.46$ mm. (Since this is much less than the radius of the plates, the parallel plate approximation (plane symmetry) is a good one.) (b) The dielectric breakdown field for polystyrene is $E_{max} = 25$ kV/mm, so the maximum voltage for this capacitor is $V_{max} = E_{max}d = (25\ \text{kV/mm})(3.46\ \text{mm}) = 86.5$ kV. (Note: in practice, the working voltage would be less than this by a comfortable safety margin.)

Problem

69. The capacitor of the preceding problem is connected to its 900-V charging battery and left connected as the plexiglass sheet is inserted, so the potential difference remains at 900 V. What are (a) the charge on the plates and (b) the stored energy both before and after the plexiglass is inserted?

Solution

(a) The capacitance before and after the insertion of the plexiglass insulation is $C_0 = \varepsilon_0 A/d = (8.85\ \text{pF/m})(76\ \text{cm}^2) \div (1.2\ \text{mm}) = 56.1$ pF, and $C = \kappa C_0 = (3.4)(56.1\ \text{pF}) = 191$ pF, as found previously. Therefore, since the voltage stays at 900 V in this case (due to the battery), $Q_0 = C_0(900\ \text{V}) = 50.4$ nC, and $Q = C(900\ \text{V}) = \kappa Q_0 = 172$ nC, before and after insertion, respectively. (b) The stored energy is $U_0 = \frac{1}{2}C_0(900\ \text{V})^2 = 22.7\ \mu$J before, and $U = \frac{1}{2}C(900\ \text{V})^2 = \kappa U_0 = 77.2\ \mu$J after. (The difference between this situation and the one in the previous problem is that the battery does additional work moving more charge to the capacitor plates, while maintaining the constant voltage. Equation 26-11 applies to an isolated capacitor only.)

Paired Problems

Problem

73. A 20-μF air-insulated parallel-plate capacitor is charged to 300 V. The capacitor is then disconnected from the charging battery, and its plate separation is doubled. Find the stored energy (a) before and (b) after the plate separation increases. Where does the extra energy come from?

Solution

(a) Initially, the stored evergy is $U_0 = \frac{1}{2} C_0 V_0^2 = \frac{1}{2}(20\ \mu F)(300\ V)^2 = 0.9$ J. (b) Disconnected from the battery, the charge stays constant, but the capacitance is halved when the separation is doubled ($C = \varepsilon_0 A/2d = C_0/2$). Therefore, the stored energy is doubled, since $U = Q^2/2C = Q^2/2(C_0/2) = 2U_0 = 1.8$ J. Work must be done, against the attractive force between the oppositely charged plates, to increase their separate.

Supplementary Problems

Problem

77. A typical lightning flash transfers 30 C across a potential difference of 30 MV. Assuming such flashes occur every 5 s in the thunderstorm of Example 26-2, roughly how long could the storm continue if its electrical energy were not replenished?

Solution

The energy in the thunderstorm of Example 26-2 was about 1.4×10^{11} J, while the energy in a lightning flash is $qV = (30\ C)(30\ MV) = 9 \times 10^8$ J. Thus, there is energy for about $1.4 \times 10^{11}/9 \times 10^8 = 156$ flashes, which at a rate of one flash in 5 s, would last for 156×5 s = 13 min.

Problem

81. An air-insulated parallel-plate capacitor of capacitance C_0 is charged to voltage V_0 and then disconnected from the charging battery. A slab of material with dielectric constant κ, whose thickness is essentially equal to the capacitor spacing, is then inserted halfway into the capacitor (Fig. 26-33). Determine (a) the new capacitance, (b) the stored energy, and (c) the force on the slab in terms of C_0, V_0, κ, and the capacitor plate length L.

Solution

(a) In so far as fringing fields can be neglected, the electric field between the plates is uniform, $E = V/d$ (but when the dielectric is inserted, $V \neq V_0$ and E depend on x). In fact, on the left side, where the slab has penetrated, $E = (1/\kappa)(\sigma_\ell/\varepsilon_0)$, and on the right, $E = \sigma_r/\varepsilon_0$, where σ_ℓ and σ_r are the charge densities on the left and right sides. Thus, $\sigma_\ell = \kappa\varepsilon_0 E$ and $\sigma_r = \varepsilon_0 E$, and the charge can be written (in terms of geometrical variables superposed on Fig. 26-33) as

$q = \sigma_\ell wx + \sigma_r w(L - x) = \epsilon_0 Ew(\kappa x + L - x) = \varepsilon_0(V/d)w(\kappa x + L - x)$. From Equation 26-5, $C = q/V = C_0(\kappa x + L - x)/L$, where $C_0 = \varepsilon_0 A/d$ and $A = Lw$. Although the question specifies $x = \frac{1}{2}L$, for which value the capacitance is $\frac{1}{2}C_0(\kappa + 1)$, we give C as a function of x, because we will need to differentiate with respect to x in part (c). (b) When the battery is disconnected, the capacitor is isolated and the charge on it is a constant, $q = q_0$. The stored energy is (Equation 26-8a) $U = q^2/2C = q_0^2 L/2C_0(\kappa x + L - x) = U_0 L/(\kappa x + L - x)$, where $U_0 = \frac{1}{2}q_0^2/C_0 = \frac{1}{2}C_0 V_0^2$. For $x = \frac{1}{2}L$, the energy is $C_0 V_0^2/(\kappa + 1)$. (c) The force on a part of an isolated system is related to the potential energy of the system by Equation 8-9. The force on the slab is therefore

$$F_x = -\frac{dU}{dx} = -\frac{d}{dx}\left(\frac{U_0 L}{\kappa x + L - x}\right) = \frac{U_0 L(\kappa - 1)}{(\kappa x + L - x)^2},$$

in the direction of increasing x (so as to pull the slab into the capacitor). For $x = \frac{1}{2}L$, the magnitude of the force is $2C_0 V_0^2(\kappa - 1)/L(\kappa + 1)^2$. It turns out that if we rewrite the force, for any value of x, in terms of the voltage for that x, using $q_0 = C_0 V_0 = CV = C_0 V(\kappa x + L - x)/L$, the expression can be used in the succeeding problem. Thus,

$$F_x = \frac{C_0 V_0^2 L(\kappa - 1)}{2(\kappa x + L - x)^2} = \frac{C_0}{2}\left(\frac{V}{L}\right)^2 L(\kappa - 1) = \frac{C_0 V^2(\kappa - 1)}{2L}.$$

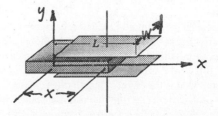

FIGURE 26-33 Problem 81 Solution.

Problem

85. Equation 26-2 gives the potential energy of a pair of oppositely charged plates. (a) Differentiate this expression with respect to the plate spacing to find the magnitude of the attractive force between the plates. (b) Compare with the answer you would get by multiplying one plate's charge by the electric field between the plates. Why do your answers differ? Which is right?

Solution

(a) Equation 26-2 gives the potential energy of two isolated oppositely charged plates, $U(x) = Q^2 x/2\varepsilon_0 A$, where x is their separation. Equation 8-9, $F_x = -dU/dx$, implies an attractive force of $F_x = -Q^2/2\varepsilon_0 A$ acting between the plates. (b) Multiplying the charge by the total electric field between the plates gives one $Q(\sigma/\varepsilon_0) = Q^2/\varepsilon_0 A$, an expression equal to twice the magnitude of the force. The total field includes the field of both plates, whereas the force on one plate depends on only the field of the other plate. (In general, the force per unit area on the surface charge distribution on a conductor is $\frac{1}{2}\sigma E = \sigma^2/2\varepsilon_0$ for the same reason.)

Problem

89. A TV antenna cable consists of two 0.50-mm-diameter wires spaced 12 mm apart. Estimate the capacitance per unit length of this cable, neglecting dielectric effects of the insulation.

Solution

The capacitance per unit length for a bifilar cable, in air, when the diameter of the wires is small compared to their separation, is calculated in the solution to Problem 25-75:

$$\frac{C}{L} = \frac{\lambda}{\Delta V} = \frac{\pi\varepsilon_0}{\ln((b/a) - 1)} = \frac{\pi(8.85 \text{ pF/m})}{\ln((12/.25) - 1)} = 7.22 \text{ pF/m}.$$

●

● CHAPTER 27 ELECTRIC CURRENT

Section 27-1 Electric Current

Problem

1. A wire carries 1.5 A. How many electrons pass through the wire in each second?

Solution

The current is the amount of charge passing a given point in the wire, per unit time, so in one second, $\Delta q = I \Delta t = (1.5 \text{ A})(1 \text{ s}) = 1.5 \text{ C}$. The number of electrons in this amount of charge is $1.5 \text{ C}/1.6\times10^{-19} \text{ C} = 9.38\times10^{18}$.

Problem

5. Electrons in the Stanford Linear Accelerator are accelerated to nearly the speed of light. These high-energy electrons are produced in pulses containing 5×10^{11} electrons each, lasting 1.6 μs. (a) Assuming an electron speed essentially that of light, what is the physical length of each pulse? (b) What is the peak current (i.e., the rate of charge flow while a pulse is going by)? (c) If the accelerator produces 180 pulses per second, what is the average current?

Solution

(a) At a velocity of the speed of light, a pulse width of 1.6 μs corresponds to a physical length of $(1.6 \ \mu s)(3\times10^8 \text{ m/s}) = 480$ m. (If the pulse lasts for a time Δt, the electrons have moved a distance $c \Delta t$.) (b) The amount of charge passing a given point during the time of one pulse is the peak current, $I_{peak} = \Delta q/\Delta t = (5\times10^{11})(1.6\times10^{-19} \text{ C})/(1.6 \ \mu s) = 50$ mA. (This is the charge in one pulse divided by the time for it to pass by.) (c) The average current is the charge in 180 pulses divided by one second, or $I_{av} = (180)(5\times10^{11})(1.6\times10^{-19} \text{ C})/1 \text{ s} = 14.4 \ \mu A$. (Note: In one second, there is a pulse passing a given point for a fraction $180\times1.6 \ \mu s/1$ s of the time, and no current during the rest of that second, so $I_{av} = (180\times1.6\times10^{-6})I_{peak}$.)

Problem

9. What is the drift speed in a silver wire carrying a current density of 150 A/mm^2? Each silver atom contributes 1.3 free electrons.

Solution

Calculating the density of conduction electrons as in Example 27-1, and using Equation 27-3a for the drift speed, we find $n = (1.3)(10.5\times10^3 \text{ kg/m}^3)/(107.87 \text{ u/ion}) \times (1.66\times10^{-27} \text{ kg/u}) = 7.62\times10^{28} \text{ m}^{-3}$, and $v_d = J/ne = (150 \text{ A/mm}^2)/(7.62\times10^{28} \text{ m}^{-3})(1.6\times10^{-19} \text{ C}) = 1.23$ cm/s.

Problem

13. A plasma used in fusion research contains 5.0×10^{18} electrons and an equal number of protons per cubic meter. Under the influence of an electric field the electrons drift in one direction at 40 m/s, while the protons drift in the opposite direction at 6.5 m/s. (a) What is the current density? (b) What fraction of the current is carried by the electrons?

Solution

The proton current density is $J_p = nev_{d,p} = (5.0\times10^{18} \text{ m}^{-3}) \times (1.6\times10^{-19} \text{ C})(6.5 \text{ m/s}) = 5.20 \text{ A/m}^2$ (positive in the direction of $v_{d,p}$), and the electron current density is $J_e = n(-e)v_{d,e} = (5.0\times10^{18} \text{ m}^{-3})(-1.6\times10^{-19} \text{ C})(-40 \text{ m/s}) = 32.0 \text{ A/m}^2$. (a) The total current density is $J_p + J_e = 37.2 \text{ A/m}^2$, and (b) the fraction carried by electrons is $32.0/37.2 = 86.0\%$.

Problem

17. What electric field is necessary to drive a 7.5-A current through a silver wire 0.95 mm in diameter?

Solution

From Ohm's law (which applies to silver) and the definition of current density (which we assume is uniform in the wire) one finds $E = \rho J = \rho I/\frac{1}{4}\pi d^2 = (1.59\times10^{-8} \ \Omega\cdot\text{m}) \times (7.5 \text{ A})/\frac{1}{4}\pi(0.95 \text{ mm})^2 = 0.168$ V/m.

Problem

21. Use Table 27-1 to determine the conductivity of (a) copper and (b) sea water.

Solution

Equations 27-4a and b show that the conductivity and the resistivity are reciprocals of one another. Thus, (a) $\rho^{-1} = \sigma = (1.68 \times 10^{-8} \ \Omega \cdot m)^{-1} = 5.95 \times 10^{7} (\Omega \cdot m)^{-1}$ for copper, and (b) $\sigma = (0.22 \ \Omega \cdot m)^{-1} = 4.55 (\Omega \cdot m)^{-1}$ for typical seawater. (The salinity of open-ocean water varies between 33 and 37%, but can vary from 1 to 80% in shallow coastal waters.)

Section 27-3 Resistance and Ohm's Law

Problem

25. What is the resistance of a heating coil that draws 4.8 A when the voltage across it is 120 V?

Solution

The macroscopic form of Ohm's Law is probably applicable to the heating coil, which is typically a coil of wire. Equation 27-7 gives $R = V/I = 120 \ \text{V}/4.8 \ \text{A} = 25 \ \Omega$.

Problem

29. What current flows when a 45-V potential difference is imposed across a 1.8-kΩ resistor?

Solution

If the resistor obeys Ohm's law, $I = V/R = 45 \ \text{V}/1.8 \ \text{k}\Omega = 25 \ \text{mA}$.

Problem

33. A cylindrical iron rod measures 88 cm long and 0.25 cm in diameter. (a) Find its resistance. If a 1.5-V potential difference is applied between the ends of the rod, find (b) the current, (c) the current density, and (d) the electric field in the rod.

Solution

(a) Equation 27-6 gives the resistance of a uniform object of Ohmic material, $R = \rho \ell / A = (9.71 \times 10^{-8} \ \Omega \cdot m) \times (88 \ \text{cm})/\frac{1}{4}\pi(0.25 \ \text{cm})^2 = 17.4 \ \text{m}\Omega$ (see Table 27-1 for the resistivity of iron). (b) Equation 27-7 (Ohm's law) gives $I = V/R = 1.5 \ \text{V}/17.4 \ \text{m}\Omega = 86.2 \ \text{A}$. (c) Equation 27-3a gives $J = I/\frac{1}{4}\pi d^2 = 86.2 \ \text{A}/\frac{1}{4}\pi(0.25 \ \text{cm})^2 = 17.6 \ \text{MA/m}^2$. (d) Equation 27-4b gives $E = \rho J = (9.71 \times 10^{-8} \ \Omega \cdot m) \times (17.6 \ \text{MA/m}^2) = 1.70 \ \text{V/m}$. (The quantities were calculated in the order querried; alternatively, in reverse order, $E = V/\ell$, $J = E/\rho$, $I = JA$, and $R = V/I$.)

Problem

37. Engineers call for a power line with a resistance per unit length of 50 mΩ/km. What wire diameter is required if the line is made of (a) copper or (b) aluminum? (c) If the costs

of copper and aluminum wire are \$1.53/kg and \$1.34/kg, which material is more economical? The densities of copper and aluminum are 8.9 g/cm^3 and 2.7 g/cm^3, respectively.

Solution

From Equation 27-6, $R/\ell = \rho/(\frac{1}{4}\pi d^2)$, so $d = 2\sqrt{\rho/\pi(R/\ell)}$. With resistivities from Table 27-1, (a) $d_{Cu} = 2\sqrt{1.68 \times 10^{-8} \ \Omega \cdot m/(50\pi \ m\ell/km)} = 2.07 \ \text{cm}$, and (b) $d_{Al} = \sqrt{2.65/1.68} \ d_{Cu} = 2.60 \ \text{cm}$ (see Problem 35). (c) With these diameters, the cost of one meter of wire is $\frac{1}{4}\pi d_{Cu}^2$ (1 m)(8.9 g/cm^3)(\$1.53/kg) = \$4.58 for copper, and \$1.92 for aluminum.

Section 27-4 Electric Power

Problem

41. A car's starter motor draws 125 A with 11 V across its terminals. What is its power consumption?

Solution

Equation 27-8 gives the power supplied to the motor, $\mathcal{P} = VI = (11 \ \text{V})(125 \ \text{A}) = 1.38 \ \text{kW}$.

Problem

45. What is the resistance of a standard 120-V, 60-W light bulb?

Solution

The bulb's resistance, from Equation 27-9b, is $R = V^2/\mathcal{P} = (120 \ \text{V})^2/60 \ \text{W} = 240 \ \Omega$, at its operating temperature. (This equation, with average values of voltage and power, can be used for an ac-resistor.)

Problem

49. How much total energy could the battery of Problem 3 supply?

Solution

If the battery is a typical 12 V automotive model, it can supply a total energy of (12 V)(80 A)(1 h) = (0.960 kW·h) × (3600 s/h) = 3.46 MJ. (Note that the total energy can be thought of as either the power times 1 h of operating time, $\mathcal{P} \ \Delta t = VI \ \Delta t$, or the total charge times the potential difference, $\Delta qV = (I \ \Delta t)V$ as in Problem 3.)

Problem

53. A 2000-horsepower electric railroad locomotive gets its power from an overhead wire with 0.20 Ω/km. The potential difference between wire and track is 10 kV. Current returns through the track, whose resistance is negligible. (a) How much current does the locomotive draw? (b) How far from the power plant can the train go before 1% of the energy is lost in the wire?

Solution

(a) If we neglect possible energy losses in the locomotive's engine, $\mathcal{P}_{in} = \mathcal{P}_{out}$, or $VI = 2000$ hp. Thus, $I =$ (2000 hp)(746 W/hp)/(10^4 V) $= 149$ A. (b) The power loss in getting current to and from the locomotive is $\mathcal{P}_{loss} = I^2 R$, where I is the current from part (a) and R is the resistance of the feed/return circuit. The high voltage wire has resistance (0.2 Ω/km)ℓ, where its length, ℓ, is also the distance from the power station, and the resistance of the return rail is negligible (see Problem 28). Then $\mathcal{P}_{loss} = 1\%\ \mathcal{P}_{in}$ implies $I^2 R = 0.01 VI$, or $\ell = (0.01)(10^4$ V)/(149 A)(0.2 Ω/km) $=$ 3.35 km.

Problem

57. What is the resistance of a column of mercury 0.75 m long and 1.0 mm in diameter?

Solution

At ordinary temperatures, metallic mercury obeys Ohm's law, so Equation 27-6 and Table 27-1 give a resistance of $R = \rho \ell / A = (9.84 \times 10^{-7}\ \Omega \cdot \text{m})(0.75\ \text{m})/\frac{1}{4}\pi(10^{-3}\ \text{m})^2 =$ 0.940 Ω.

Problem

61. A 240-V electric motor is 90% efficient, meaning that 90% of the energy supplied to it ends up as mechanical work. If the motor lifts a 200-N weight at 3.1 m/s, how much current does it draw?

Solution

The electrical power input to the motor is $\mathcal{P}_{in} = VI$, while the mechanical power output (lifting the weight) is $\mathcal{P}_{out} = Fv = 90\%\ \mathcal{P}_{in}$. Thus, $I = Fv/0.9\ V =$ (200 N)(3.1 m/s)/(0.9×240 V) $= 2.87$ A.

Problem

65. The electric car of Fig. 27-24 converts 70% of its electrical energy supply into mechanical energy available at the wheels. The car weighs 640 kg and has a 96-V battery. How much current does the motor draw when the car is climbing a 10° slope at 45 km/h? Neglect friction and air resistance.

Solution

If friction and air resistance are neglected, the motor only supplies power to work against gravity, $\mathcal{P}_{out} = F_\parallel v =$ ($mg \sin 10°$)v, which equals 70% of the input electrical power from the battery, $\mathcal{P}_{in} = VI$. Therefore, $I = mg \sin 10° v/0.7\ V$ $= (640 \times 9.8$ N)(45 m/3.6 s)$\sin 10°$/(0.7×96 V) $= 203$ A.

Problem

69. Figure 27-25 shows a resistor made from a truncated cone of material with uniform resistivity ρ. Consider the cone to be made of thin slices of thickness dx, like the one shown; Equation 27-6 shows that the resistance of each slab is $dR = \rho\ dx/A$. By integrating over all such slices, shows that the resistance between the two flat faces is $R = \rho \ell / \pi ab$. (This method assumes the equipotentials are planes, which is only approximately true.)

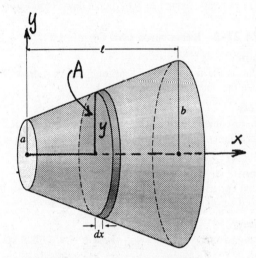

FIGURE 27-25 Problem 69 Solution.

Solution

As suggested, consider that the total resistance R equals $\int_{cone} dR$, where $dR = \rho\ dx/A$ is the resistance of a thin disk as shown on Fig. 27-25. The area of such a disk is $A = \pi y^2$, where $y = a + (b - a)x/\ell$ is the radius and $0 \leq x \leq \ell$ as shown. Then

$$R = \int_0^\ell \frac{\rho\ dx}{\pi[a + (b - a)x/\ell]^2}$$
$$= \frac{\rho}{\pi}\left| -\left(\frac{\ell}{b - a}\right)\frac{1}{[a + (b - a)x/\ell]}\right|_0^\ell$$
$$= \frac{\rho}{\pi}\left(\frac{\ell}{b - a}\right)\left(\frac{1}{a} - \frac{1}{b}\right) = \frac{\rho \ell}{\pi ab}.$$

(This result depends on the condition that the flat faces and parallel circular cross-sections of the cone are equipotential surfaces.)

● **CHAPTER 28** ELECTRIC CIRCUITS

Section 28-1 Circuits and Symbols

Problem

1. Sketch a circuit diagram for a circuit that includes a resistor R_1 connected to the positive terminal of a battery, a pair of parallel resistors R_2 and R_3 connected to the lower-voltage end of R_1, then returned to the battery's negative terminal, and a capacitor across R_2.

Solution

A literal reading of the circuit specifications results in connections like those in sketch (a). Because the connecting wires are assumed to have no resistance (a real wire is represented by a separate resistor), a topologically equivalent circuit diagram is shown in sketch (b).

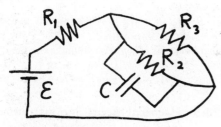

Problem 1 Solution (a).

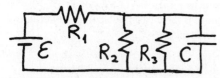

Problem 1 Solution (b).

Problem

5. A 1.5-V battery stores 4.5 kJ of energy. How long can it light a flashlight bulb that draws 0.60 A?

Solution

The average power, supplied by the battery to the bulb, multiplied by the time equals the energy capacity of the battery. For an ideal battery, $\mathscr{P} = \mathscr{E}I$, therefore $\mathscr{E}It = 4.5$ kJ, or $t = 4.5$ kJ$/(1.5$ V$)(0.60$ A$) = 5{\times}10^3$ s $= 1.39$ h.

Problem

9. What resistance should be placed in parallel with a 56-kΩ resistor to make an equivalent resistance of 45 kΩ?

Solution

The solution for R_2 in Equation 28-3a is $R_2 = R_1 R_{parallel} \div (R_1 - R_{parallel}) = (56$ kΩ$)(45)/(56 - 45) = 229$ kΩ.

Problem

13. What is the internal resistance of the battery in the preceding problem?

Solution

The solution of the preceding problem (or the reasoning of Example 28-2) gives $R_{int} = 0.02\,\Omega$ (i.e. $(12$ V $- 6$ V$)/300$ A).

Problem

17. A partially discharged car battery can be modeled as a 9-V emf in series with an internal resistance of 0.08 Ω. Jumper cables are used to connect this battery to a fully charged battery, modeled as a 12-V emf in series with a 0.02-Ω internal resistance. How much current flows through the discharged battery?

Solution

Terminals of like polarity are connected with jumpers of negligible resistance. Kirchhoff's voltage law gives $\mathscr{E}_1 - \mathscr{E}_2 - IR_1 - IR_2 = 0$, or $I = (\mathscr{E}_1 - \mathscr{E}_2)/(R_1 + R_2) = (12 - 9)$ V$/(0.02 + 0.08)\,\Omega = 30$ A.

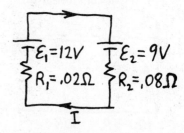

Problem 17 Solution.

Problem

21. How many 100-W, 120-V light bulbs can be connected in parallel before they below a 20-A circuit breaker?

Solution

The circuit breaker is activated if $I = 120$ V$/R_{min} > 20$ A, or if $R_{min} < 6\,\Omega$. The resistance of each light bulb is $R = V^2/\mathscr{P} = (120$ V$)^2/100$ W $= 144\,\Omega$, and n bulbs in parallel have resistance $R_{\parallel} = R/n$. Therefore $R_{\parallel} \geq R_{min}$ implies $n \leq 144/6 = 24$, so more than 24 bulbs would blow the circuit.

Problem

25. In the circuit of Fig. 28-48, R_1 is a variable resistor, and the other two resistors have equal resistances R. (a) Find an expression for the voltage across R_1, and (b) sketch a graph of this quantity as a function of R_1 as R_1 varies from 0 to $10R$. (c) What is the limiting value as $R_1 \rightarrow \infty$?

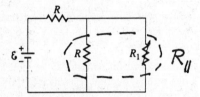

FIGURE 28-48 Problem 25.

Solution

(a) The resistors in parallel have an equivalent resistance of $R_\parallel = RR_1/(R + R_1)$. The other R, and $R_\parallel$, is a voltage divider in series with $\mathcal{E}$, so Equation 28-2 gives $V_\parallel = \mathcal{E}R_\parallel/(R + R_\parallel) = \mathcal{E}R_1/(R + 2R_1)$. (b) and (c) If $R_1 = 0$ (the second resistor shorted out), $V_\parallel = 0$, while if $R_1 = \infty$ (open circuit), $V_\parallel = \frac{1}{2}\mathcal{E}$ (the value when R_1 is removed). If $R_1 = 10R$, $V_\parallel = (10/21)\mathcal{E}$ (as in Problem 24).

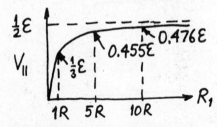

FIGURE 28-48 Problem 25 Solution.

Section 28-4 Kirchhoff's Laws and Multiloop Circuits

Problem

29. In the circuit of Fig. 28-50 it makes no difference whether the switch is open or closed. What is $\mathcal{E}_3$ in terms of the other quantities shown?

Solution

If the switch is irrelevant, then there is no current through its branch of the circuit. Thus, points A and B must be at the same potential, and the same current flows through R_1 and R_2. Kirchhoff's voltage law applied to the outer loop, and to the left-hand loop, gives $\mathcal{E}_1 - IR_1 - IR_2 + \mathcal{E}_2 = 0$, and $\mathcal{E}_1 - IR_1 + \mathcal{E}_3 = 0$, respectively. Therefore,

$$\mathcal{E}_3 = IR_1 - \mathcal{E}_1 = \left(\frac{\mathcal{E}_1 + \mathcal{E}_2}{R_1 + R_2}\right)R_1 - \mathcal{E}_1 = \frac{\mathcal{E}_2R_1 - \mathcal{E}_1R_2}{R_1 + R_2}.$$

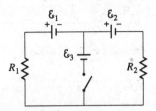

FIGURE 28-50 Problem 29 Solution.

Problem

33. Find all three currents in the circuit of Fig. 28-16 with the values given, but with battery $\mathcal{E}_2$ reversed.

Solution

The general solution for this circuit given in Problem 75, with the particular values of emf's and resistors in this problem, yields currents of $I_1 = [(4 + 1)6 - 1(-9)] \text{ A}/14 = 2.79$ A, $I_2 = [1 \times 6 - (2 + 1)(-9)] \text{ A}/14 = 2.36$ A, and $I_3 = [4 \times 6 + 2(-9)] \text{ A}/14 = 0.429$ A. Or, one could retrace the reasoning of Example 28-6, with $\mathcal{E}_2 = -9$ V replacing the original value in loop 2. Then, everything is the same until the equation $-9 + 2(6 - 3I_3) - I_3 = 0$, or $I_3 = (\frac{3}{7})$ A, $I_2 = \frac{1}{2}(6 - 3 \times \frac{3}{7})$ A $= (33/14)$ A, and $I_1 = I_2 + I_3 = (39/14)$ A.

Section 28-5 Electrical Measuring Instruments

Problem

37. A voltmeter with 200-kΩ resistance is used to measure the voltage across the 10-kΩ resistor in Fig. 28-54. By what percentage is the measurement in error because of the finite meter resistance?

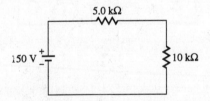

FIGURE 28-54 Problem 37.

Solution

The voltage across the 10 kΩ resistor in Fig. 28-54 is $(150 \text{ V})(10)/(10 + 5) = 100$ V (the circuit is just a voltage divider as described by Equations 28-2a and b), as would be measured by an ideal voltmeter with infinite resistance. With the real voltmeter connected in parallel across the 10 kΩ resistor, its effective resistance is changed to $R_\parallel = (10 \text{ k}\Omega) \times (200 \text{ k}\Omega)/(210 \text{ k}\Omega) = 9.52$ kΩ, and the voltage reading is only $(150 \text{ V})(9.52)/(9.52 + 5) = 98.4$ V, or about 1.64% lower.

Problem

41. You have an ammeter with 10-Ω resistance whose fullscale reading is 1.0 mA. What resistance should you put in series with it to make a voltmeter that reads 25 V full scale?

Solution

A potential difference of $(1.0 \text{ mA})(10 \ \Omega) = 10 \text{ mV}$ across the meter will cause a full-scale deflection. The series resistor R which will accomplish the desired conversion must divide a voltage of 25 V, such that 10 mV appears across the meter. Equation 28-2b gives $10 \text{ mV} = (25 \text{ V})(10 \ \Omega)/(10 \ \Omega + R)$, or $R = (10 \ \Omega)[(25 \text{ V}/10 \text{ mV}) - 1] = 25 \text{ k}\Omega - 10 \ \Omega = 24.99 \text{ k}\Omega$. A very precise resistor is required.

Section 28-6 Circuits with Capacitors

Problem

45. Show that the quantity RC has the units of time (seconds).

Solution

The SI units for the time constant, RC, are $(\Omega)(\text{F}) = (\text{V/A})(\text{C/V}) = (\text{s/C})(\text{C}) = \text{s}$, as stated.

Problem

49. Figure 28-57 shows the voltage across a capacitor that is charging through a 4700-Ω resistor in the circuit of Fig. 28-27. Use the graph to determine (a) the battery voltage, (b) the time constant, and (c) the capacitance.

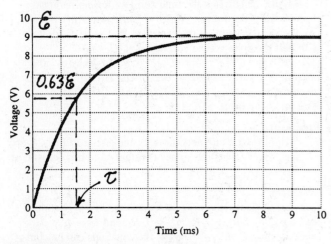

FIGURE 28-57 Problem 49 Solution.

Solution

(a) For the circuit considered, the voltage across the capacitor asymptotically approaches the battery voltage after a long time (compared to the time constant). In Fig. 28-57, this is about 9 V. (b) The time constant is the time it takes the capacitor voltage to reach $1 - e^{-1} = 63.2\%$ of its asymptotic

value, or 5.69 V in this case. From the graph, $\tau \approx 1.5$ ms. (c) The time constant is RC, so $C = 1.5 \text{ ms}/4700 \ \Omega = 0.319 \ \mu\text{F}$.

Problem

53. A capacitor is charged until it holds 5.0 J of energy. It is then connected across a 10-kΩ resistor. In 8.6 ms, the resistor dissipates 2.0 J. What is the capacitance?

Solution

Equation 28-8 gives a voltage $V = V_0 e^{-t/RC}$ for a capacitor discharging through a resistor. If 2 J is dissipated in time t, the energy stored in the capacitor drops from $U_0 = 5$ J to $U = 3$ J (assuming there are no losses due to radiation, etc.). Since $U = \frac{1}{2}CV^2$, $U_0/U = (V_0/V)^2 = e^{2t/RC}$ and we may solve for C: $C = 2t/R\ln(U_0/U) = 2(8.6 \text{ ms})/(10 \text{ k}\Omega)\ln(5/3) = 3.37 \ \mu\text{F}$.

Problem

57. In the circuit for Fig. 28-60 the switch is initially open and the capacitor is uncharged. Find expressions for the current I supplied by the battery (a) just after the switch is closed and (b) a long time after the switch is closed.

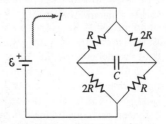

FIGURE 28-60 Problem 57.

Solution

(a) Just after the switch is closed, the uncharged capacitor acts instantaneously like a short circuit and the resistors act like two parallel pairs in series. The effective resistance of the combination is $2 \times (R)(2R)/(R + 2R) = 4R/3$, and the current supplied by the battery is $I(0) = 3\mathcal{E}/4R$. (b) A long time after the switch is closed, the capacitor is fully charged and acts like an open circuit. Then the resistors act like two series pairs in parallel, with an effective resistance of $(\frac{1}{2}) \times (R + 2R) = 3R/2$. The battery current is $I(\infty) = 2\mathcal{E}/3R$.

Problem

61. A battery's voltage is measured with a voltmeter whose resistance is 1000 Ω; the result is 4.36 V. When the measurement is repeated with a 1500-Ω meter the result is 4.41 V. What are (a) the battery voltage and (b) its internal resistance?

Solution

The internal resistance of the battery (R_i) and the resistance of the voltmeter (R_m) are in series with the battery's emf, so the current is $I = \mathcal{E}/(R_i + R_m)$. The potential drop across the meter (its reading) is $V_m = IR_m = \mathcal{E}R_m/(R_i + R_m)$. From the given data, 4.36 V $= \mathcal{E}(1\ \mathrm{k\Omega})/(R_i + 1\ \mathrm{k\Omega})$ and 4.41 V $= \mathcal{E}(1.5\ \mathrm{k\Omega})/(R_i + 1.5\ \mathrm{k\Omega})$, which can be solved simultaneously for $\mathcal{E}$ and R_i. One obtains $R_i + 1\ \mathrm{k\Omega} = \mathcal{E}(1\ \mathrm{k\Omega}/4.36\ \mathrm{V})$ and $R_i + 1.5\ \mathrm{k\Omega} = \mathcal{E}(1.5\ \mathrm{k\Omega}/4.41\ \mathrm{V})$, or

$$\mathcal{E} = (1.5\ \mathrm{k\Omega} - 1\ \mathrm{k\Omega})\left(\frac{1.5\ \mathrm{k\Omega}}{4.41\ \mathrm{V}} - \frac{1\ \mathrm{k\Omega}}{4.36\ \mathrm{V}}\right)^{-1} = 4.51\ \mathrm{V}$$

and $R_i = (4.51\ \mathrm{V})(1\ \mathrm{k\Omega}/4.36\ \mathrm{V}) - 1\ \mathrm{k\Omega} = 35.2\ \Omega$.

Problem

65. In Fig. 28-63 what are the meter readings when (a) an ideal voltmeter or (b) an ideal ammeter is connected between points A and B?

Solution

(a) An ideal voltmeter has $R_m = \infty$ (AB open circuited), so V_{AB} is just the voltage across the 5.6 kΩ resistor. This is part of a voltage divider (in series with the 4.7 kΩ resistor), so Equation 28-2 gives $V_{AB} = (24\ \mathrm{V})(5.6)/(5.6 + 4.7) = 13.0\ \mathrm{V}$. (b) An ideal ammeter has $R_m = 0$ (AB short circuited), so I_{AB} is just the current through the 3.3 kΩ resistor. This is part of parallel combination, $R_\| = (3.3\ \mathrm{k\Omega})(5.6)/(3.3 + 5.6) = 2.08\ \mathrm{k\Omega}$, which, in series with the 4.7 kΩ resistor, draws a total current of $I_{tot} = 24\ \mathrm{V}/(2.08 + 4.7)\ \mathrm{k\Omega} = 3.54\ \mathrm{mA}$. Now, two resistors in parallel form a current divider, each one taking a fraction of the total current, given by $I_1 = (R_\|/R_1)I_{tot}$, and $I_2 = (R_\|/R_2)I_{tot}$, respectively. (This follows directly from Ohm's law: $V_\| = I_{tot}R_\| = I_1 R_1 = I_2 R_2$.) Therefore, in the case of the ideal ammeter, $I_{AB} = (R_\|/3.3\ \mathrm{k\Omega})I_{tot} = (2.08/3.3)(3.54\ \mathrm{mA}) = 2.23\ \mathrm{mA}$.

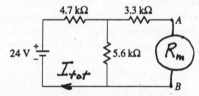

FIGURE 28-63 Problem 65 Solution.

Supplementary Problems

Problem

69. Suppose the currents into and out of a circuit node differed by 1 μA. If the node consists of a small metal sphere with diameter 1 mm, how long would it take for the electric field around the node to reach the breakdown field in air (3 MV/m)?

Solution

The charge on the node (whether positive or negative) accumulates at a rate of 1 μA = 1 μC per second, so $|q(t)| = (1\ \mu\mathrm{A})t$ (where we assume that $q(0) = 0$). If the node is treated approximately as an isolated sphere, the electric field strength at its surface, $k|q|/r^2 = k(1\ \mu\mathrm{A})t/r^2$, equals the breakdown field for air, when
$t = (3\ \mathrm{MV/m})(0.5\ \mathrm{mm})^2/(9\times10^9\ \mathrm{m/F})(1\ \mu\mathrm{A}) = 83.3\ \mu\mathrm{s}$.

Problem

73. A parallel-plate capacitor is insulated with a material of dielectric constant κ and resistivity ρ. Since the resistivity is finite, the capacitor "leaks" charge and can be modeled as an ideal capacitor in parallel with a resistor. (a) Show that the time constant of the capacitor is independent of its dimensions (provided the spacing is small enough that the usual parallel-plate approximation applies) and is given by $\varepsilon_0\kappa\rho$. (b) If the insulating material is polystyrene ($\kappa = 2.6$, $\rho = 10^{16}\ \Omega\cdot\mathrm{m}$), how long will it take for the stored energy in the capacitor to decrease by a factor of 2?

Solution

(a) In the parallel plate approximation, the electric field between the plates is constant, $E = V/d$, the leakage current density is uniform, $J = E/\rho$, and the capacitance is $C = \kappa\varepsilon_0 A/d$. These relations imply that the resistance of the dielectric slab between the plates is $R = V/I = V/JA = V/(EA/\rho) = \rho d/A$, and the time constant for discharging through the dielectric is $\tau = RC = (\rho d/A)(\kappa\varepsilon_0 A/d) = \kappa\varepsilon_0\rho$. (b) The stored energy in the capacitor, $U_C = \frac{1}{2}CV_C^2$, decays with half the voltage time constant, i.e., $U_C = \frac{1}{2}C \times (V_0 e^{-t/RC})^2 = \frac{1}{2}CV_0^2 e^{-2t/RC} = U_0 e^{-t/(RC/2)}$. To decay by 50% takes time $t = (RC/2)\ln 2 = (\kappa\varepsilon_0\rho/2)\ln 2$. With the values given for polystyrene, $t = (2.6)(8.85\ \mathrm{pF/m})(10^{16}\ \Omega\cdot\mathrm{m}) \times \frac{1}{2}\ln 2 = 7.97\times10^4\ \mathrm{s} = 22.2\ \mathrm{h}$.

Problem

77. Write the loop and node laws for the circuit of Fig. 28-66, and show that the time constant for this circuit is $R_1 R_2 C/(R_1 + R_2)$.

Solution

Consider the loops and node added to Fig. 28-66. Kirchhoff's laws are $\mathcal{E} = I_1 R_1 + I_2 R_2$, $V_C = I_2 R_2$, and $I_C = I_1 - I_2$. Since $V_C = q/C$ and $I_C = dq/dt$, the equations can be combined to yield

$$\mathcal{E} - I_1 R_1 - I_2 R_2 = \mathcal{E} - (I_C + I_2)R_1 - I_2 R_2$$

$$= \mathcal{E} - I_C R_1 - \left(\frac{V_C}{R_2}\right)(R_1 + R_2)$$

$$= \mathcal{E} - I_C R_1 - \frac{q}{CR_2/(R_1 + R_2)} = 0.$$

This is exactly in the same form as the first equation, solved in the text, in the section "The RC Circuit: Charging" (with $I \rightarrow I_C$, $R \rightarrow R_1$ and $C \rightarrow CR_2/(R_1 + R_2)$), so the time constant for the circuit is $\tau = CR_1R_2/(R_1 + R_2)$ (the ratio of the coefficients of I_C and q).

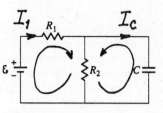

FIGURE 28-66   Problem 77 Solution.   ●

● **CHAPTER 29**   THE MAGNETIC FIELD

Section 29-2   Electric Charge and the Magnetic Field

Problem

1. (a) What is the minimum magnetic field needed to exert a 5.4×10^{-15}-N force on an electron moving at 2.1×10^7 m/s? (b) What magnetic field strength would be required if the field were at 45° to the electron's velocity?

Solution

(a) From Equation 29-1b, $B = F/ev \sin\theta$, which is a minimum when $\sin\theta = 1$ (the magnetic field perpendicular to the velocity). Thus, $B_{min} =$ $(5.4 \times 10^{-15}$ N$)/(1.6 \times 10^{-19}$ C$)(2.1 \times 10^7$ m/s$) =$ 1.61×10^{-3} T $= 16.1$ G. (b) For $\theta = 45°$, $B =$ $B_{min}/\sin 45° = \sqrt{2}\, B_{min} = 22.7$ G.

Problem

5. A particle carrying a 50-μC charge moves with velocity $\mathbf{v} = 5.0\hat{\imath} + 3.2\hat{k}$ m/s through a uniform magnetic field $\mathbf{B} = 9.4\hat{\imath} + 6.7\hat{\jmath}$ T. (a) What is the force on the particle? (b) Form the dot products $\mathbf{F} \cdot \mathbf{v}$ and $\mathbf{F} \cdot \mathbf{B}$ to show explicitly that the force is perpendicular to both $\mathbf{v}$ and $\mathbf{B}$.

Solution

(a) From Equation 27-2, $\mathbf{F} = q\mathbf{v} \times \mathbf{B} =$ $(50\ \mu$C$)(5\hat{\imath} + 3.2\hat{k}$ m/s$) \cdot 9.4\hat{\imath} + 6.7\hat{\jmath}$ T$) =$ $(50 \times 10^{-6}$ N$)(5 \times 6.7\hat{k} + 3.2 \times 9.4\hat{\jmath} - 3.2 \times 6.7\hat{\imath}) =$ $(-1.072\hat{\imath} + 1.504\hat{\jmath} + 1.675\hat{k}) \times 10^{-3}$ N. (The magnitude and direction can be found from the components, if desired.) (b) The dot products $\mathbf{F} \cdot \mathbf{v}$ and $\mathbf{F} \cdot \mathbf{B}$ are, respectively, proportional to $(-1.072)(5) + (1.675)(3.2) = 0$, and $(-1.072)(9.4) + (1.504)(6.7) = 0$, since the cross product of two vectors is perpendicular to each factor. (We did not round off the components of $\mathbf{F}$, so that the vanishing of the dot products could be exactly confirmed.)

Problem

9. An alpha particle (2 protons, 2 neutrons) is moving with velocity $\mathbf{v} = 150\hat{\imath} + 320\hat{\jmath} - 190\hat{k}$ km/s at a point where the magnetic field is $\mathbf{B} = 0.66\hat{\imath}$ T. Find the magnitude of the force on the particle.

Solution

The magnetic force on the alpha particle is (Equation 29-1a): $\mathbf{F}_B = 2e\mathbf{v} \times \mathbf{B} = 2(1.6 \times 10^{-19}$ C$)(150\hat{\imath} + 320\hat{\jmath} - 190\hat{k}) \times$ $(10^3$ m/s$) \times (0.66$ T$)\hat{\imath} = (2.11 \times 10^{-16}$ N$)(-320\hat{k} - 190\hat{\jmath})$. This has magnitude $F_B =$ 2.11×10^{-16} N$\sqrt{(-320)^2 + (-190)^2} = 7.86 \times 10^{-14}$ N.

Problem

13. A region contains an electric field $\mathbf{E} = 7.4\hat{\imath} + 2.8\hat{\jmath}$ kN/C and a magnetic field $\mathbf{B} = 15\hat{\jmath} + 36\hat{k}$ mT. Find the electromagnetic force on (a) a stationary proton, (b) an electron moving with velocity $\mathbf{v} = 6.1\hat{\imath}$ Mm/s.

Solution

The force on a moving charge is given by Equation 29-2 (called the Lorentz force) $\mathbf{F} = q(\mathbf{E} + \mathbf{v} \times \mathbf{B})$. (a) For a stationary proton, $q = e$ and $\mathbf{v} = 0$, so $\mathbf{F} = e\mathbf{E} =$ $(1.6 \times 10^{-19}$ C$)(7.4\hat{\imath} + 2.8\hat{\jmath})$ kN/C $= (1.18\hat{\imath} + 0.448\hat{\jmath})$ fN. (b) For the electron, $q = -e$ and $\mathbf{v} = 6.1\hat{\imath}$ Mm/s, so the electric force is the negative of the force in part (a) and the magnetic force is $-e\mathbf{v} \times \mathbf{B} = (-1.6 \times 10^{-19}$ C$)(6.1\hat{\imath}$ Mm/s$) \times$ $(15\hat{\jmath} + 36\hat{k})$ mT $= (-14.6\hat{k} + 35.1\hat{\jmath})$ fN. The total Lorentz force is the sum of these, or $(-1.18\hat{\imath} + 34.7\hat{\jmath} - 14.6\hat{k})$ fN.

Problem

17. Radio astronomers detect electromagnetic radiation at a frequency of 42 MHz from an interstellar gas cloud. If this radiation is caused by electrons spiraling in a magnetic field, what is the field strength in the gas cloud?

Solution

If this is electromagnetic radiation at the electron's cyclotron frequency, Equation 29-5 implies a field strength of $B = 2\pi f(m/e) = 2\pi(42$ MHz$)(9.11 \times 10^{-31}$ kg$/1.6 \times 10^{-19}$ C$) =$ 1.50×10^{-3} T $= 15.0$ G.

Problem

21. Typical particle energies in a nuclear fusion reactor are on the order of 10 keV. If the smallest dimension of the reactor is on the order of 1 m, estimate the minimum magnetic field strength needed to ensure that protons have orbits smaller than the size of the reactor. Will this field be sufficient for electrons of the same energy?

Solution

The radius of the proton orbits should be ≤ 0.5 m, so the result of Problem 19 (in atomic units) gives
$$B \geq \sqrt{2(10^{-2})(938)}/300(1)(0.5) = 2.89 \times 10^{-2} \text{ T, or}$$
roughly 300 G. This field is more than sufficient for electrons of the same energy, whose orbits are 43 times smaller.

Problem

25. A cyclotron is designed to accelerate deuterium nuclei. (Deuterium has one proton and one neutron in its nucleus.) (a) If the cyclotron uses a 2.0-T magnetic field, at what frequency should the dee voltage be alternated? (b) If the vacuum chamber has a diameter of 0.90 m, what is the maximum kinetic energy of the deuterons? (c) If the magnitude of the potential difference between the dees is 1500 V, how many orbits do the deuterons complete before achieving the energy of part (b)?

Solution

(a) The frequency of the accelerating voltage is the cyclotron frequency for deuterons (Equation 29-5), $f = eB/2\pi m \simeq$ $(1.6 \times 10^{-19} \text{ C})(2 \text{ T}) \div 2\pi(2 \times 1.67 \times 10^{-27} \text{ kg}) = 15.2$ MHz. (b) We can use the result of Problem 19 (expressed in atomic units), with the maximum orbital radius equal to the radius of the dees. Thus, $K_{max} = (300qBr_{max})^2/2m \simeq$ $(300 \times 1 \times 2 \times 0.45)^2/2(2 \times 938) = 19.4$ MeV. (c) If the deuterons start with essentially zero kinetic energy, and gain 1500 eV each half-orbit, they will make 19.4 MeV $\div$ $2(1500 \text{ eV}) = 6.48 \times 10^3$ orbits. (Of course, the same results follow in standard SI units.)

Problem

29. A mass spectrometer is used to separate the fissionable uranium isotope U-235 from the much more abundant isotope U-238. To within what percentage must the magnetic field be held constant if there is to be no overlap of these two isotopes? Both isotopes appear as constituents of uranium hexafluoride gas (UF_6), and the gas molecules are all singly ionized.

Solution

The separation of different uranium isotopes in UF_6 molecules can be found by differentiation of the result of Problem 27. Keeping the spectrometer parameters fixed, $x \sim m^{1/2}$, $dx \sim \frac{1}{2}m^{-1/2} dm$, and $dx/x = \frac{1}{2}(dm/m)$. The molecular masses of the two species are approximately 235 or 238 plus

6×19 which equals 349 or 352, respectively, so $\frac{1}{2}(dm/m) \approx$ $3/2 \times 350 = 0.43\%$. For a particular ion, $x \sim B^{-1}$ and $dx \sim B^{-2} dB$, therefore $dx/x = -dB/B$. Thus, variations in B should be less than 0.43% to separate these isotopes.

Section 29-4   The Magnetic Force on a Current

Problem

33. What is the magnitude of the force on a 50-cm-long wire carrying 15 A at right angles to a 500-G magnetic field?

Solution

The force on a straight current-carrying wire in a uniform magnetic field is (Equation 29-6) $\mathbf{F} = I\boldsymbol{\ell} \times \mathbf{B}$. Thus, $F = I\ell B \sin\theta = (15 \text{ A})(0.5 \text{ m})(0.05 \text{ T})\sin 90° = 0.375$ N. (The direction is given by the right-hand rule.)

Problem

37. In a high-magnetic-field experiment, a conducting bar carrying 7.5 kA passes through a 30-cm-long region containing a 22-T magnetic field. If the bar makes a 60° angle with the field direction, what force is necessary to hold it in place?

Solution

The magnitude of the force necessary to balance the magnetic force on the bar (Equation 29-6) is $F = |I\boldsymbol{\ell} \times \mathbf{B}| =$ $I\ell B \sin\theta = (7.5 \text{ kA})(0.3 \text{ m})(22 \text{ T})\sin 60° = 42.9$ kN (nearly 5 tons). The direction of this force is perpendicular to the plane of $\boldsymbol{\ell} \times \mathbf{B}$ in the opposite sense as the magnetic force.

Problem

41. A wire carrying 1.5 A passes through a region containing a 48-mT magnetic field. The wire is perpendicular to the field and makes a quarter-circle turn of radius 21 cm as it passes through the field region, as shown in Fig. 29-42. Find the magnitude and direction of the magnetic force on this section of wire.

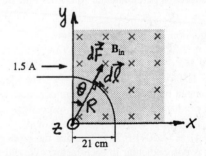

FIGURE 29-42   Problem 41 Solution.

Solution

Take the xy axes a shown on Fig. 29-50, with the z axis out of the page and the origin at the center of the quarter-circle arc. With θ measured clockwise from the y axis,

$d\ell = R\,d\theta(\hat{\imath}\cos\theta - \hat{\jmath}\sin\theta)$ as shown, and $\mathbf{B} = B(-\hat{\mathbf{k}})$. The magnetic force on the arc of wire is found from Equation 29-8.

$$\mathbf{F} = I\int_{arc} d\ell \times \mathbf{B} = I\int_0^{90°} R\,d\theta(\hat{\imath}\cos\theta - \hat{\jmath}\sin\theta) \times B(-\hat{\mathbf{k}})$$

$$= IRB\int_0^{90°} (\hat{\jmath}\cos\theta + \hat{\imath}\sin\theta)\,d\theta = IRB|-\hat{\imath}\cos\theta + \hat{\jmath}\sin\theta|_0^{90°}$$

$$= IRB(\hat{\imath} + \hat{\jmath}).$$

This has magnitude $\sqrt{2}\,IRB = \sqrt{2}(1.5\text{ A})(0.21\text{ m}) \times (48\text{ mT}) = 21.4$ mN and direction $45°$ between the positive x and y axes.

Problem

45. The probe in a Hall-effect magnetometer uses a semiconductor doped to a charge-carrier density of 7.5×10^{20} m^{-3}. The probe measures 0.35 mm thick in the direction of the magnetic field being measured, and carries a 2.5-mA current perpendicular to the field. If its Hall potential is 4.5 mV, what is the magnetic field strength?

Solution

If we assume the charge-carriers are of one type, with charge of magnitude e, then Equation 29-7 and the given data require $B = nqV_H t/I = (7.5\times10^{20}\text{ m}^{-3})(1.6\times10^{-19}\text{ C}) \times (4.5\text{ V})(0.35\text{ mm})/(2.5\text{ mA}) = 75.6$ mT.

Problem

49. A bar magnet experiences a 12-mN·m torque when it is oriented at $55°$ to a 100-mT magnetic field. What is the magnitude of its magnetic dipole moment?

Solution

Equation 29-11, solved for the magnitude of the dipole moment, gives $\mu = \tau/B\sin\theta = (12\times10^{-3}\text{ N·m})/(0.1\text{ T})\sin 55° = 0.146$ A·m^2.

Problem

53. Nuclear magnetic resonance (NMR) is a technique for analyzing chemical structures and is also the basis of magnetic resonance imaging used for medical diagnosis. The NMR technique relies on sensitive measurements of the energy needed to flip atomic nuclei upside-down in a given magnetic field. In an NMR apparatus with a 7.0-T magnetic field, how much energy is needed to flip a proton ($\mu = 1.41\times10^{-26}$ A·m^2) from parallel to antiparallel to the field?

Solution

From Equation 29-12, the energy required to reverse the orientation of a proton's magnetic moment from parallel to antiparallel to the applied magnetic field is $\Delta U = 2\mu B = 2(1.41\times10^{-26}\text{ A·m}^2)(7.0\text{ T}) = 1.97\times10^{-25}$ J $= 1.23\times10^{-6}$ eV. (This amount of energy is characteristic of radio waves of frequency 298 MHz, see Chapter 39.)

Problem

57. Proponents of space-based particle-beam weapons have to confront the effect of Earth's magnetic field on their beams. If a beam of protons with kinetic energy 100 MeV is aimed in a straight line perpendicular to Earth's magnetic field in a region where the field strength is 48 μT, what will be the radius of the protons' circular path?

Solution

It is simplest to use the result of Problem 22, in atomic units, to find the radius (see solution to Problem 19), $r = \sqrt{2Km} \div 300qB$, where r is in meters, K is in MeV, m in MeV/c^2, q in units of e, and B in teslas. Then $r = \sqrt{2(100)(938)}/300(0.48\times10^{-4}) = 30.1$ km. (The mass of a proton in atomic units is approximately 938 MeV/c^2.)

Problem

61. An old-fashioned analog meter uses a wire coil in a magnetic field to deflect the meter needle. If the coil is 2.0 cm in diameter and consists of 500 turns of wire, what should be the magnetic field strength if the maximum torque is to be 1.6 μN·m when the current in the coil is 1.0 mA?

Solution

The maximum torque on a flat coil, with magnetic moment $\mu = N\pi R^2 I$, is $\tau_{max} = \mu B$ (see Equations 29–10 and 11). Thus, $B = \tau_{max}/NI\pi R^3 = (1.6\ \mu\text{N·m})/(500)(1\text{ mA})\pi \times (1\text{ cm})^2 = 102$ G.

Problem

65. A conducting bar with mass 15.0 g and length 22.0 cm is suspended from a spring in a region where a 0.350-T magnetic field points into the page, as shown in Fig. 29-44. With no current in the bar, the spring length is 26.0 cm. The bar is supplied with current from outside the field region, using wires of negligible mass. When a 2.00-A current flows from left to right in the bar, it rises 1.2 cm from its equilibrium position. Find (a) the spring constant and (b) the unstretched length of the spring.

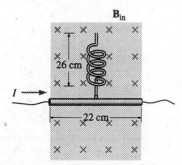

FIGURE 29-44   Problem 65.

Solution

In equilibrium, the vertical forces on the bar must sum to zero. These include the weight of the bar, $-mg$ (negative downward), the upward spring force $k \Delta \ell = k(\ell - \ell_0)$ (where ℓ is the length and ℓ_0 the unstretched length of the spring), and when current flows from left to right in the bar, an upward magnetic force of $F_B = ILB$ (L is the length of the bar). The conditions stated require that $k(26.0 \text{ cm} - \ell_0) - mg = 0$, and $k(24.8 \text{ cm} - \ell_0) + ILB - mg = 0$ or $k(26 \text{ cm} - \ell_0) = mg$ and $k(26 \text{ cm} - \ell_0 - 1.2 \text{ cm}) = mg - ILB$. These equations can be solved for k and ℓ_0 (subtract to eliminate ℓ_0, and divide to eliminate k) with the result $k = ILB/(1.2 \text{ cm}) = (2 \text{ A})(22 \text{ cm})(0.35 \text{ T})/(1.2 \text{ cm}) = 12.8 \text{ N/m}$, and $\ell_0 = 26 \text{ cm} - (mg/ILB)(1.2 \text{ cm}) = 26 \text{ cm} - (0.015 \times 9.8 \text{ N})/(12.8 \text{ N/m}) = 24.9 \text{ cm}$.

Problem

69. A 10-turn wire loop measuring 8.0 cm by 16 cm carrying 2.0 A lies in a horizontal plane but is free to rotate about the axis shown in Fig. 29-46. A 50-g mass hangs from one side of the loop, and a uniform magnetic field points horizontally, as shown. What magnetic field strength is required to hold the loop in its horizontal position?

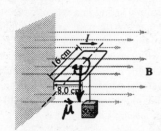

FIGURE 29-46   Problem 69.

Solution

For the direction of current shown in Fig. 29-46, the magnetic moment of the loop is downward, and the magnetic torque $\boldsymbol{\mu} \times \mathbf{B}$ is along the axis, out of the page. The gravitational torque $\mathbf{r} \times m\mathbf{g}$ is along the axis, into the page. The two torques cancel when $\mu B = mgr$, or $B = mgr/NIA = (0.05 \text{ kg})(9.8 \text{ m/s}^2)(0.04 \text{ m})/(10)(2.0 \text{ A})(0.8 \times 0.16 \text{ m}^2) = 76.6 \text{ mT}$.

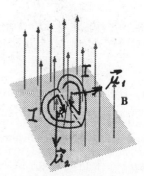

FIGURE 29-47   Problem 70 Solution.

Problem

73. Early models pictured the electron in a hydrogen atom as being in a circular orbit of radius 5.29×10^{-11} m about the stationary proton, held in orbit by the electric force. Find the magnetic dipole moment of such an atom. This quantity is called the *Bohr magnetron* and is typical of atomic-sized magnetic moments. *Hint:* The full electron charge passes any given point in the orbit once per orbital period. Use this fact to calculate the average current.

Solution

One electronic charge passes a given point on the orbit every period of revolution, so the magnitude of the average current corresponding to the electron's orbital motion is $I = \Delta q/\Delta t = e/(2\pi r/v)$. (The current circulates opposite to the orbital motion, since the electron is negatively charged.) In the simplest version of the Bohr model for the hydrogen atom, the electron moves in a circular orbit, around a fixed proton, under the influence of the Coulomb force, so that $mv^2/r = ke^2/r^2$, or $v/r = \sqrt{ke^2/mr^3}$. Thus, $I = (e/2\pi)(v/r) = (e^2/2\pi)\sqrt{k/mr^3}$. The magnetic dipole moment associated with this orbital atomic current (called a Bohr magneton) has magnitude $\mu_B = I\pi r^2 = \frac{1}{2}e^2\sqrt{kr/m} \approx \frac{1}{2}(1.6 \times 10^{-19} \text{ C})^2 \times \sqrt{(9 \times 10^9 \text{ N} \cdot \text{m}^2/\text{C}^2)(5.29 \times 10^{-11} \text{ m})(9.11 \times 10^{-31} \text{ kg})} \approx 9.25 \times 10^{-24} \text{ A} \cdot \text{m}^2$, and is typical of the size of atomic magnetic dipole moments in general.

● CHAPTER 30 SOURCES OF THE MAGNETIC FIELD

Section 30-1 The Biot-Savart Law

Problem

1. A wire carries 15 A. You form the wire into a single-turn circular loop with magnetic field 80 μT at the loop center. What is the loop radius?

Solution

Equation 30-3, with $x = 0$, gives the magnetic field at the center of a circular loop, $B = \mu_0 I/2a$ (with direction along the axis of the loop, consistent with the sense of circulation of the current and the right-hand rule). Thus, the radius is $a = \mu_0 I/2B = (2\pi\times10^{-7} \text{ N/A}^2)(15 \text{ A})/(80 \ \mu\text{T}) = 11.8$ cm.

Problem

5. Suppose Earth's magnetic field arose from a single loop of current at the outer edge of the planet's liquid core (core radius 3000 km). How large must the current be to give the observed magnetic dipole moment of 8.0×10^{22} A·m²?

Solution

The dipole moment of a circular loop is $\mu = I\pi R^2$ (Equation 29-10), so $I = (8.0\times10^{22} \text{ A·m}^2)/\pi(3\times10^6 \text{ m})^2 = 2.83\times10^9$ A.

Problem

9. You have a spool of thin wire that can handle a maximum current of 0.50 A. If you wind the wire into a loop-like coil 20 cm in diameter, how many turns should the coil have if the magnetic field at its center is to be 2.3 mT at this maximum current?

Solution

Equation 30-3 can be modified for N turns of wire (as in the solution to Problem 3), so at the center of a flat circular coil, $B = N\mu_0 I/2a$. Thus, $N = 2aB/\mu_0 I = (0.2 \text{ m})(2.3 \text{ mT}) \div (4\pi\times10^{-7} \text{ N/A}^2)(0.5 \text{ A}) = 732$.

Problem

13. A power line carries a 500-A current toward magnetic north and is suspended 10 m above the ground. The horizontal component of Earth's magetic field at the power line's latitude is 0.24 G. If a magnetic compass is placed on the ground directly below the power line, in what direction will it point?

Solution

A compass needle (small dipole magnet) is free to rotate in a horizontal plane until it is aligned with the direction of the total horizontal magnetic field (see Equation 29-11). A long, straight wire (the power line) carrying a current of 500 A parallel to the ground, in the direction of magnetic north, produces a magnetic field, at a distance 10 m below, to the west, with magnitude given by Equation 30-9 (see diagram):

$$B_y = \mu_0 I/2\pi r = (2\times10^{-7} \text{ N/A}^2)(500 \text{ A})/10 \text{ m}$$
$$= 10^{-5}\text{T} = 0.1 \text{ G}.$$

The horizontal component of the Earth's magnetic field is $B_x = 0.24$ G, so the compass needle will point $\theta = \tan^{-1}B_y/B_x = \tan^{-1}(0.1/0.24) = 22.6°$ west of magnetic north.

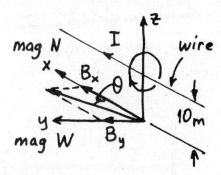

Problem 13 Solution.

Problem

17. Figure 30-48 shows a conducting loop formed from concentric semicircles of radii a and b. If the loop carries a current I as shown, find the magnetic field at point P, the common center.

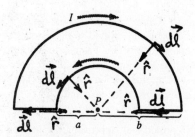

FIGURE 30-48 Problem 17 Solution.

Solution

The Biot-Savart law gives $\mathbf{B}(P) = (\mu_0/4\pi)\int_{\text{loop}} Id\boldsymbol{\ell} \times \hat{\mathbf{r}}/r^2$. $Id\boldsymbol{\ell} \times \hat{\mathbf{r}}/r^2$ on the inner semicircle has magnitude $I \ dl/a^2$ and direction out of the page, while on the outer semicircle, the magnitude is $I \ d\ell/b^2$ and the direction is into the page. On the straight segments, $d\ell \times \hat{\mathbf{r}} = 0$, so the total field at P is $(\mu_0/4\pi)[(I\pi a/a^2) - (I\pi b/b^2)] = \mu_0 I(b - a)/4ab$ out of the page. (Note: the length of each semicircle is $\int d\ell = \pi r$.)

Problem

21. The structure shown in Fig. 30-49 is made from conducting rods. The upper horizontal rod is free to slide vertically on the uprights, while maintaining electrical contact with them. The upper rod has mass 22 g and length 95 cm. A battery connected across the insulating gap at the bottom of the left-hand upright drives a 66-A current through the structure. At what height h will the upper wire be in equilibrium?

Solution

If h is small compared to the length of the rods, we can use Equation 30-6 for the repulsive magnetic force between the horizontal rods (upward on the top rod) $F = \mu_0 I^2 \ell/2\pi h$. The rod is in equilibrium when this equals its weight, $F = mg$, hence $h = \mu_0 I^2 \ell/2\pi mg =$ $(2\times10^{-7} \text{ N/A}^2)(66 \text{ A})^2(0.95 \text{ m})/(0.022 \times 9.8 \text{ N}) =$ 3.84 mm. (This is indeed small compared to 95 cm, as assumed.)

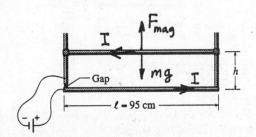

FIGURE 30-49   Problem 21 Solution.

Problem

25. A solenoid 10 cm in diameter is made with 2.1-mm-diameter copper wire wound so tightly that adjacent turns touch, separated only by enamel insulation of negligible thickness. The solenoid carries a 28-A current. In the long, straight wire approximation, what is the net force between two adjacent turns of the solenoid?

Solution

Equation 30-6 provides an approximate value of the force per unit length between adjacent turns of $F/\ell = \mu_0 I^2/2\pi d =$ $(2\times10^{-7} \text{ N/A}^2)(28 \text{ A})^2/(2.1 \text{ mm}) = 74.7 \text{ mN/m}$. The length of one turn is $2\pi r = 10\pi \text{ cm} = 0.314 \text{ m}$, so the force on one turn is approximately $(74.7 \text{ mN/m})(0.314 \text{ m}) = 23.5 \text{ mN}$.

Problem

29. The magnetic field shown in Fig. 30-53 has uniform magnitude 75 μT, but its direction reverses abruptly. How much current is encircled by the rectangular loop shown?

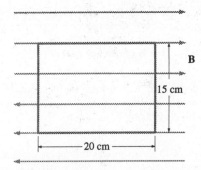

FIGURE 30-53   Problem 29.

Solution

Ampère's law applied to the loop shown in Fig. 30-53 (going clockwise) gives $\oint \mathbf{B} \cdot d\ell = 2B\ell = 2(75 \ \mu\text{T})(0.2 \text{ m}) =$ $\mu_0 I_{encircled} = (0.4\pi \mu\text{T·m/A})I_{encircled}$ (since the sides of the loop perpendicular to $\mathbf{B}$ give no contribution to the line integral). Thus, $I_{encircled} = (75/\pi)$ A $= 23.9$ A. As explained in the text, the current flows along the boundary surface between the regions of oppositely directed $\mathbf{B}$, positive into the page in Fig. 30-53, for clockwise circulation around the loop.

Problem

33. A solid wire 2.1 mm in diameter carries a 10-A current with uniform current density. What is the magnetic field strength (a) at the axis of the wire, (b) 0.20 mm from the axis, (c) at the surface of the wire, and (d) 4.0 mm from the wire axis?

Solution

The magnetic field strength is given by Equation 30-10 inside the wire ($r \leq R$) and Equation 30-9 outside ($r \geq R$) as shown in Fig. 30-21. (a) For $r = 0$, $B = 0$. (b) For $r = 0.2$ mm $< R = \frac{1}{2} \times 2.1$ mm $= 1.05$ mm, $B = \mu_0 Ir/2\pi R^2 =$ $(2\times10^{-7} \text{ N/A}^2)(10 \text{ A})(0.2 \text{ mm})/(1.05 \text{ mm})^2 = 3.63$ G. (c) For $r = R$, $B = \mu_0 I/2\pi R = (2\times10^{-7} \text{ N/A}^2)(10 \text{ A}) \div$ $(1.05 \text{ mm}) = 19.0$ G (d) For $r = 4$ mm $> R$, $B = \mu_0 I/2\pi r =$ $(2\times10^{-7} \text{ N/A}^2)(10 \text{ A})/(4 \text{ mm}) = 5$ G.

Problem

37. Typically, cylindrical wires made from yttrium-barium-copper-oxide superconductor can carry a maximum current density of 6.0 MA/m² at a temperature of 77 K, as long as the magnetic field at the conductor surface does not exceed 10 mT. Suppose such a wire is to carry the maximum current density. (a) At what wire diameter would the surface magnetic field equal the 10-mT limit? (b) Is this a maximum or minimum value for the diameter if the field is not to exceed the limit? (c) What current would a wire with this diameter carry?

Solution

The magnetic field strength at the surface of a wire with axial symmetry is $B = \mu_0 I/2\pi R$ (see Problem 34). If the current density in the wire is uniform over its circular cross-section, $I = J\pi R^2$ and $B = \frac{1}{2}\mu_0 JR$. (a) and (b) If $J = J_{\max}$, then $B \leq B_{\max}$ implies $2R \leq 4B_{\max}/\mu_0 J_{\max} = (10 \text{ mT}) \div (0.1\pi \ \mu\text{T·m/A})(6 \text{ MA/m}^2) = 5.31$ mm, which is the maximum diameter. (c) With the diameter above, the total current is $I = J_{\max}\pi R^2 = (6 \text{ MA/m}^2)\pi(5.31 \text{ mm}/2)^2 = 133$ A, not excessive for a superconductor.

Problem

41. Repeat the preceding problem for the case when one current flows in the $+x$ direction and the other in the $+y$ direction.

Solution

As in the preceding problem, combine the fields from each (approximately infinite) flat current sheet (see Equation 30-11 and Fig. 30-22) to obtain the total magnetic field. The field of the first plate ($J_{s,1}$ in the x direction) is $\mp\frac{1}{2}\mu_0 J_{s,1}\hat{\jmath}$ above and below the plate, respectively, whereas the field of the second ($J_{s,2}$ in the y direction) is $\pm\frac{1}{2}\mu_0 J_{s,2}\hat{\imath}$ above and below it. Superposition yields $\mathbf{B}_{\text{above}} = \frac{1}{2}\mu_0 J_s(\hat{\imath} - \hat{\jmath})$ above both plates, $\mathbf{B}_{\text{btw}} = \frac{1}{2}\mu_0 J_s(-\hat{\imath} - \hat{\jmath})$ between, and $\mathbf{B}_{\text{below}} = \frac{1}{2}\mu_0 J_s(-\hat{\imath} + \hat{\jmath})$ below both. The magnitude of the field in the three regions is the same, namely $\mu_0 J_s/\sqrt{2}$, but the directions are progressively 90° apart, namely $-45°$, $-135°$, and $-225°$ from the x axis, respectively.

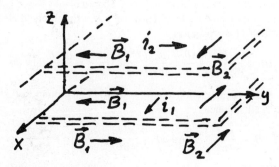

Problem 41 Solution.

Section 30-5 Solenoids and Toroids

Problem

45. A superconducting solenoid has 3300 turns per meter and can carry a maximum current of 4.1 kA. What is the magnetic field strength in the solenoid?

Solution

Equation 30-12 for a long thin solenoid (if applicable) gives $B = \mu_0 nI = (4\pi\times10^{-7} \text{ N/A}^2)(3300/\text{m})(4.1 \text{ kA}) = 17.0$ T.

Problem

49. A toroidal fusion reactor requires a magnetic field that varies by no more than 10% from its central value of 1.5 T. If the minor radius of the toroidal coil producing this field is 30 cm, what is the minimum value for the major radius of the device?

Solution

The central value of the toroidal field is given by Equation 30-13 with $r = R_{\text{maj}}$. At other values of r inside the toroid, the percent difference in field strength is $100(B - B_{\text{maj}})/B_{\text{maj}} = 100(R_{\text{maj}} - r)/r$. The extremes of r are $R_{\text{maj}} \pm R_{\min}$, so it is required that $10 \geq 100|(R_{\text{maj}} - (R_{\text{maj}} \pm R_{\min}))/(R_{\text{maj}} \pm R_{\min})|$, or $R_{\text{maj}} \geq (10 \mp 1)R_{\min}$. The minimum major radius (corresponding to the limit using the smallest value of r, or the lower sign above) is $R_{\text{maj}} \geq 11R_{\min} = 11\times30$ cm $= 3.30$ m. (Since 10% is not infinitesimal, differentiation w.r.t. r gives an alternative approximate limit: $|dB/B| = |-dr/r| = R_{\min}/R_{\text{maj}} \leq 10\%$, or $R_{\text{maj}} \geq 3$ m.)

Section 30-6 Magnetic Matter

Problem

53. When a sample of a certain substance is placed in a 250.0-mT magnetic field, the field inside the sample is 249.6 mT. Find the magnetic susceptibility of the substance. Is it ferromagnetic, paramagnetic, or diamagnetic?

Solution

Equation 30-14 gives the relation between the internal and applied magnetic fields in terms of the relative permeability or the magnetic susceptibility. For the sample described in this problem, the latter is $\chi_M = (B_{\text{int}} - B_{\text{app}})/B_{\text{app}} = (249.6 - 250.0)/250.0 = -1.6\times10^{-3}$. Since $\chi_M < 0$ (or $B_{\text{int}} < B_{\text{app}}$) the material is diamagnetic.

Paired Problems

Problem

57. Two concentric, coplanar circular current loops have radii a and $2a$. If the magnetic field is zero at their common center, how does the current in the outer loop compare with that in the inner loop?

Solution

The magnetic field strength at the center of a circular current loop is $\mu_0 I/2R$ (Equation 30-3). In order for the net field to cancel, the currents must be in opposite directions and have magnitudes such that $\mu_0 I_{\text{outer}}/2(2a) = \mu_0 I_{\text{inner}}/2a$. Then $I_{\text{outer}} = 2I_{\text{inner}}$.

Problem

61. The largest lightning strikes have peak currents around 250 kA, flowing in essentially cylindrical channels of ionized air. How far from such a flash would the resulting magnetic field be equal to Earth's magnetic field strength, about 50 μT?

Solution

Supposing that the cylindrical channel of ionized air acts like a long straight wire, we can use Equation 30-5 to estimate the distance: $y = \mu_0 I/2\pi B = (2 \times 10^{-7} \text{ N/A}^2)(250 \text{ kA}) \div (50 \ \mu T) = 1$ km.

Supplementary Problems

Problem

65. A circular wire loop of radius 15 cm and negligible thickness carries a 2.0-A current. Use suitable approximations to find the magnetic field of this loop (a) in the loop plane, 1.0 mm outside the loop, and (b) on the loop axis, 3.0 m from the loop center.

Solution

(a) As mentioned in the paragraph following Equation 30-4, on p. 756, the calculation of the magnetic field from a circular current loop, at points not on its axis, is difficult. Fortunately, at a distance of 1 mm from a loop of radius 15 cm, the field is approximately that of a long, straight wire, Equation 30-5 gives $B \approx \mu_0 I/2\pi y = (2 \times 10^{-7} \text{ N/A}^2)(2A)/(10^{-3} \text{ m}) = 4$ G. (b) Since 3m $\gg$ 15 cm, the approximation $x \gg a$ in Equation 30-4 is justified. Then $B \approx \mu_0 Ia^2/2|x|^3 = (2\pi \times 10^{-7} \text{ N/A}^2)(2A)(0.15 \text{ m})^2/(3 \text{ m})^3 = 1.05 \times 10^{-5}$ G.

Problem

69. A wide, flat conducting spring of spring constant $k = 20$ N/m and negligible mass consists of two 6.0-cm-diameter turns, as shown in Fig. 30-63. In its unstretched configuration the coils are nearly touching. A 10-g mass is hung from the spring, and at the same time a current I is passed through it. The spring stretches 2.0 mm. Find I, assuming the coils remain close enough to be treated as parallel wires.

FIGURE 30-63 Problem 69.

Solution

The length of one turn, $\ell = 6\pi$ cm, is large compared to the separation of the two turns, $d = 2$ mm, so that Equation 30-6 can be used to find the magnetic force in the spring. $F_{mag} = \mu_0 I^2 \ell/2\pi d = (2 \times 10^{-7} \text{ N/A}^2)I^2 (6\pi \text{ cm})/2 \text{ mm} = 1.88 I^2 \times 10^{-5}$ N/A^2. The elastic force in the spring, in the same direction as F_{mag}, is $F_{el} = kd = (20 \text{ N/m})(2 \text{ mm}) = 4 \times 10^{-2}$ N. At equilibrium, $F_{el} + F_{mag} = mg = (10^{-2} \text{ kg})(9.8 \text{ m/s}^2) = 9.8 \times 10^{-2}$ N. Thus, $1.88 I^2 \times 10^{-5}$ N/A$^2 = 5.8 \times 10^{-2}$ N, or $I = 55.5$ A.

Problem

73. Work Example 30-2 by expressing all variables in terms of the angle θ and integrating over the appropriate range in θ.

Solution

In Example 10-2 and Fig. 30-9, $\cos\theta = -x/r$, $\tan\theta = -y/x$, so $d(\tan\theta) = d\theta/\cos^2\theta = y \, dx/x^2$. Then $dx = (x^2/\cos^2\theta)d\theta/y = r^2 \, d\theta/y$. Thus, we can write the field element (out of the page) as

$$dB = \frac{\mu_0 I}{4\pi} \frac{dx \sin\theta}{r^2} = \frac{\mu_0 I}{4\pi}\left(\frac{r^2 \, d\theta}{y}\right)\frac{\sin\theta}{r^2} = \frac{\mu_0 I}{4\pi y}\sin\theta \, d\theta.$$

The limits of integration $x = -\infty$ to $+\infty$ correspond to $\theta = 0$ to π (180°), hence

$$B = \frac{\mu_0 I}{4\pi y}\int_0^\pi \sin\theta \, d\theta = \frac{\mu_0 I}{4\pi y}\left.-\cos\theta\right|_0^\pi = \frac{\mu_0 I}{2\pi y},$$

which is the same as Equation 30-5. ●

● CHAPTER 31   ELECTROMAGNETIC INDUCTION

Sections 31-2 and 31-3   Faraday's Law and Induction and the Conservation of Energy

Problem

1. Show that the volt is the correct SI unit for the rate of change of magnetic flux, making Faraday's law dimensionally correct.

Solution

The units of $d\phi_B/dt$ are T·m^2/s = (N/A·m)(m^2/s) = (N·m/A·s) = J/C = V.

Problem

5. A conducting loop of area 240 cm² and resistance 12 Ω lies at right angles to a spatially uniform magnetic field. The loop carries an induced current of 320 mA. At what rate is the magnetic field changing?

Solution

The flux through a stationary loop perpendicular to a magnetic field is $\phi_B = BA$ (see Problem 3), so Faraday's law (Equation 31-2) and Ohm's law (Equation 27-7) relate this to the magnitude of the induced current: $I = |\mathscr{E}/R| = |d\phi_B/dt|/R = A|dB/dt|/R$. Therefore $|dB/dt| = IR/A = (320 \text{ mA})(12 \text{ Ω})/(240 \text{ cm}^2) = 160 \text{ T/s}$.

Problem

9. A square wire loop of side ℓ and resistance R is pulled with constant speed v from a region of no magnetic field until it is fully inside a region of constant, uniform magnetic field **B** perpendicular to the loop plane. The boundary of the field region is parallel to one side of the loop. Find an expression for the total work done by the agent pulling the loop.

Solution

The loop can be treated analogously to the situation analysed in Section 31-3, under the heading "Motional EMF and Lenz's Law"; instead of exiting, the loop is entering the field region at constant velocity. All quantities have the same magnitudes, except the current in the loop is CCW instead of CW, as in Fig. 31-13. Since the applied force acts over a displacement equal to the side-length of the loop, the work done can be calculated directly: $W_{app} = \mathbf{F}_{app} \cdot \boldsymbol{\ell} = (I\ell B)\ell = I\ell^2 B$. But, $I = \mathscr{E}/R = |d\phi_B/dt|/R = d/dt \, (B\ell x)/R = B\ell v/R$, as before, so $W_{app} = B^2 \ell^3 v/R$. (Alternatively, the work can be calculated from the conservation of energy: $I = B\ell v/R$, $\mathscr{P}_{diss} = I^2 R = (B\ell v)^2/R$, and $W_{app} = \mathscr{P}_{diss} t = [(B\ell v)^2/R](\ell/v)$.)

Problem

13. The wingspan of a 747 jetliner is 60 m. If the plane is flying at 960 km/h in a region where the vertical component of Earth's magnetic field is 0.20 G, what emf develops between the plane's wingtips?

Solution

A motional emf causes electrons in the wing to drift until equilibrium with the electrostatic field from accumulated wingtip charges is achieved. The magnetic and electric forces on an electron have magnitudes $F_{mag} = |-e\mathbf{v} \times \mathbf{B}| = evB_\perp$ (we suppose that the 747 is flying horizontally so only the vertical magnetic field gives a force parallel to the wingspan), and $F_{el} = |-e\mathbf{E}_s| = eV/\ell$, where V is the potential difference between the wingtips. At equilibrium, $evB = eV/\ell$, or

$V = B\ell v = (2 \times 10^{-5} \text{ T})(60 \text{ m})(960 \text{ m}/3.6 \text{ s}) = 320 \text{ mV}$. (Motional emf's like this need to be considered in rocket measurements of ionospheric electric fields.)

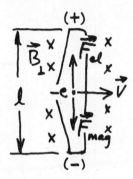

Problem 13 Solution.

Problem

17. A square conducting loop of side $s = 0.50$ m and resistance $R = 5.0$ Ω moves to the right with speed $v = 0.25$ m/s. At time $t = 0$ its rightmost edge enters a uniform magnetic field $B = 1.0$ T pointing into the page, as shown in Fig. 31-43. The magnetic field covers a region of width $w = 0.75$ m. Plot (a) the current and (b) the power dissipation in the loop as functions of time, taking a clockwise current as positive and covering the time until the entire loop has exited the field region.

Solution

Let x be the distance between the right side of the loop and the left edge of the field region. Take $t = 0$ when $x = 0$, so that $x = vt$. The loop enters the field region at $t = 0$, is completely within the region for t between $\ell/v = 2$ s and $w/v = 3$ s, and is out of the region for $t \geq (w + \ell)/v = 5$ s. The

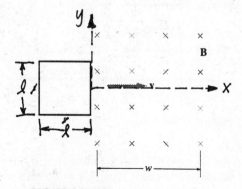

FIGURE 31-43 Problem 17 Solution.

area of loop overlapping the field region increases linearly from 0 to ℓ^2, stays constant at ℓ^2, then decreases to 0 between these times. (We use ℓ for side length to avoid confusion with time units.) Thus,

$$\phi_B = BA = B\ell^2 \begin{cases} 0 \\ vt/\ell \\ 1 \\ (w + \ell - vt)/\ell \\ 0 \end{cases}$$

$$= 0.25 \text{ Wb} \begin{cases} 0, & t \le 0 \\ 0.5t, & 0 \le t \le 2 \\ 1, & 2 \le t \le 3 \\ 0.5(5 - t), & 3 \le t \le 5 \\ 0, & 5 \le t \end{cases}$$

(We substituted the given numerical values and used SI units for flux, with time t in seconds, see solution to Problem 3.)
(a) The induced current (positive clockwise) is given by Faraday's and Ohm's laws:

$$I = -\frac{1}{R}\frac{d\phi_B}{dt} = 25 \text{ mA} \begin{cases} 0, & t \le 0 \\ -1, & 0 \le t \le 2 \\ 0, & 2 \le t \le 3 \\ +1, & 3 \le t \le 5 \\ 0, & 5 \le t \end{cases}$$

(b) The power dissipated, I^2R, is $(\pm 25 \text{ mA})^2(5 \ \Omega) = 3.13$ mW when the current is not zero.

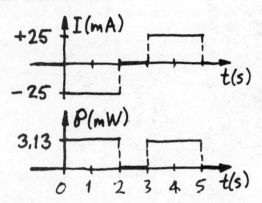

Problem 17 Solution.

Problem

21. (a) Find an expression for the resistor current in Problem 19 if the solenoid current is given by $I = I_0 \sin \omega t$, where $I_0 = 85$ A and $\omega = 210 \text{ s}^{-1}$. (b) What is the peak current in the resistor? (c) What is the resistor current when the solenoid current is a maximum?

Solution

(a) In Problem 19, $I_c = \mu_0(N_cN_s/\ell)\frac{1}{4}\pi(D_s^2/R)(-dI_s/dt)$ was the expression for the current in the coil, taken as positive from left to right in the resistor. When the current in the solenoid is $I_s = I_0 \sin \omega t$, $dI_s/dt = \omega I_0 \cos \omega t$, so $I_c = -(\mu_0\pi/4)(N_cN_s/\ell)(D_s^2/R)\omega I_0 \cos \omega t \equiv -I_{\text{peak}}\cos \omega t$.
(b) Numerically, $I_{\text{peak}} = (\pi^2 \times 10^{-7} \text{N/A}^2)(5 \times 5000/2 \text{ m}) \times (0.3 \text{ m})^2(210 \text{ s}^{-1})(85 \text{ A})/(180 \ \Omega) = 110$ mA. (c) When $\sin \omega t$ is a maximum, $\cos \omega t$ is zero.

Problem

25. A credit-card reader extracts information from the card's magnetic stripe as it is pulled past the reader's head. At some instant the card motion results in a magnetic field at the head that is changing at the rate of 450 μT/ms. If this field passes perpendicularly through a 5000-turn head coil 2.0 mm in diameter, what will be the induced emf?

Solution

The magnetic flux through the coil in the reader's head is changing at a rate of $NA \ dB/dt = (5000)\pi(1 \text{ mm})^2 \times (450 \ \mu\text{T/ms}) = 7.07$ mV. According to Faraday's law, this is equal to the magnitude of the induced emf.

Problem

29. A battery of emf $\mathcal{E}$ is inserted in series with the resistor in Fig. 31-46, with its positive terminal toward the top rail. The bar is initially at rest, and now no agent pulls it. (a) Describe the bar's subsequent motion. (b) The bar eventually reaches a constant speed. Why? (c) What is that constant speed? Express in terms of the magnetic field, the battery emf, and the rail spacing ℓ. Does the resistance R affect the final speed? If not, what role does it play?

Solution

(a) The battery causes a CW current (downward in the bar) to flow in the circuit composed of the bar, resistor, and rails. (For positive circulation CW, the right-hand rule gives a positive normal to the area bounded by the circuit into the page, so that the flux $\phi_m = \mathbf{B} \cdot \mathbf{A} = B\ell x$ is positive. The length of the circuit is x, as in Example 31-4.) Thus, there is a magnetic force $\mathbf{F}_{\text{mag}} = I\ell \times \mathbf{B} = I\ell B$ to the right, which accelerates the bar in that direction. (Any other forces on the bar are assumed to cancel, or be negligible.) An induced emf opposes the battery ($\mathcal{E}_i = -d\phi_M/dt = -B\ell v$, as in Example 31-4, the negative sign indicating a CCW sense in the circuit) so the instantaneous current is $I(t) = (\mathcal{E} + \mathcal{E}_i)/R = (\mathcal{E} - B\ell v)/R$. Thus, as v increases, I (and the accelerating force) decreases. (b) Eventually $(t \to \infty)$, $I(\infty) = 0$, $F_{\text{mag}} = I(\infty)\ell B = 0$, and the velocity v_∞ stays constant. (c) When $I(\infty) = 0$, $\mathcal{E} - B\ell v_\infty = 0$, so $v_\infty = \mathcal{E}/B\ell$. Although v_∞ doesn't depend on the resistance, the value of R does affect how rapidly v approaches v_∞. For large R, I charges slowly and v takes a long time to reach v_∞. (The equation of motion of the bar (mass m) is $m(dv/dt) = I\ell B = (\mathcal{E} - B\ell v)\ell B/R$, which can be separated: $(dv/(v_\infty - v)) = (\ell^2 B^2/mR)dt$. For $v_0 = 0$, this integrates to $\ln(1 - v/v_\infty) = -\ell^2 B^2 t/mR$, or $v = v_\infty(1 - e^{-\ell^2 B^2 t/mR})$. The time constant, $\tau = mR/\ell^2 B^2$, depends on the resistance.)

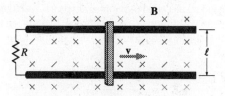

FIGURE 31-46 Problem 29 Solution.

Problem

33. In Fig. 31-46, take $\ell = 10$ cm, $B = 0.50$ T, $R = 4.0\ \Omega$, and $v = 2.0$ m/s. Find (a) the current in the resistor, (b) the magnetic force on the bar, (c) the power dissipation in the resistor, and (d) the mechanical work done by the agent pulling the bar. Compare your answers to (c) and (d).

Solution

The situation is like that described in Example 31-4 and the solution to Problem 27. (a) $I = \mathcal{E}/R = B\ell v/R = (0.5\text{ T})(0.1\text{ m})(2\text{ m/s})/4\ \Omega = 25$ mA. (Neglect the resistance of the bar and rails.) (b) $F_{mag} = I\ell B = (25\text{ mA})(0.1\text{ m}) \times (0.5\text{ T}) = 1.25\times10^{-3}$N. (c) $\mathcal{P}_J = I^2 R = (25\text{ mA})^2(4\ \Omega) = 2.5$ mW. (d) The agent pulling the bar must exert a force equal in magnitude to F_{mag} and parallel to **v**. Therefore, it does work at a rate $Fv = (1.25\times10^{-3}N)(2\text{ m/s}) = 2.5$ mW. The conservation of energy requires the answers to parts (c) and (d) to be equal.

Section 31-4 Induced Electric Fields

Problem

37. The induced electric field 12 cm from the axis of a solenoid with 10 cm radius is 45 V/m. Find the rate of change of the solenoid's magnetic field.

Solution

The geometry of the induced electric field from the solenoid is described in Example 31-9, where $2\pi r|E| = |-d(\pi R^2 B)/dt| = \pi R^2|dB/dt|$. Thus, $|dB/dt| = 2r|E|/R^2 = 2(12\text{ cm})(45\text{ V/m})/(10\text{ cm})^2 = 1.08\times10^3$ T/s = 1.08 T/ms. (The sign of dB/dt and the direction of the induced electric field are related by Lenz's law.)

Problem

41. In Example 31-9, take the solenoid radius $R = 10$ cm and suppose the magnetic field inside the solenoid is given by $B = 0.10t^3 - 1.1t^2 + 2.8t$, with B in T and t in ms. (a) Find the electric field strength 14 cm from the solenoid axis at $t = 1.0$ ms. (b) Find a time when the induced electric field in and around the solenoid is zero.

Solution

The induced electric field strength outside the solenoid was found in Example 31-9 to be $|E| = (R^2/2r)|dB/dt|$. (a) When $t = 1$ ms, the derivative of the given B is $dB/dt = [3(0.1)(1)^2 - 2(1.1)(1) + 2.8]$ T/ms = 900 T/s, so $|E| = (10\text{ cm})^2(900\text{ T/s})/(2 \times 14\text{ cm}) = 32.1$ V/m. (b) $dB/dt = 0$ implies $0.3t^2 - 2.2t + 2.8 = 0$ (for t in ms), hence

$$t = [2.2 \pm \sqrt{(2.2)^2 - 4(0.3)(2.8)}]/0.6$$
$$= 1.64\text{ ms or } 5.69\text{ ms}.$$

Paired Problems

Problem

45. A magnetic field is given by $\mathbf{B} = B_0(x/x_0)^2\hat{\mathbf{k}}$, where B_0 and x_0 are constants. Find an expression for the magnetic flux through a square of side $2x_0$ that lies in the x-y plane with one corner at the origin and two sides coinciding with the positive x and y axes.

Solution

Take elements of area, $d\mathbf{A} = 2x_0\,dx\hat{\mathbf{k}}$, which are rectangular strips parallel to the y axis. Then

$$\phi_B = \int_{square} \mathbf{B} \cdot d\mathbf{A} = \left(\frac{B_0}{x_0^2}\right)2x_0 \int_0^{2x_0} x^2\,dx$$
$$= (2B_0/x_0)(2x_0)^3/3 = 16B_0 x_0^2/3.$$

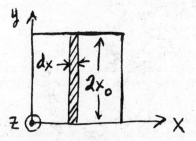

Problem 45 Solution.

Problem

49. A pair of vertical conducting rods are a distance ℓ apart and are connected at the bottom by a resistance R. A conducting bar of mass m runs horizontally between the rods and can slide freely down them while maintaining electrical contact. The whole apparatus is in a uniform magnetic field **B** pointing horizontally and perpendicular to the bar. When the bar is released from rest it soon reaches a constant speed. Find this speed.

Solution

When the bar is falling, a motional emf causes an induced current to flow in the bar (in the direction of **v** × **B**) as shown. When we are looking horizontally in the direction of **B** (into the page), the forces on the bar are gravity, mg downward, and the magnetic force, $I\ell B$ upward, where the induced current opposes the decrease in flux. The velocity is constant when $I\ell B = mg$. Now, $\phi_m = B\ell y$ and $I = \mathcal{E}/R = -(d\phi_B/dt)/R = -(B\ell/R)(dy/dt) = B\ell v/R$ (where $dy/dt = -v$ is the speed downward). Therefore, $v = IR/B\ell = mgR/B^2\ell^2$.

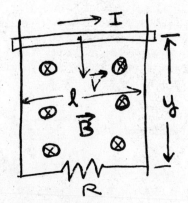

Problem 49 Solution.

Problem

53. An electron is inside a solenoid, 28 cm from the solenoid axis. It experiences an electric force of magnitude 1.3 fN. At what rate is the solenoid's magnetic field changing?

Solution

The electric field has magnitude $E = F/e$. If we suppose that this is the electric field induced by the changing magnetic field in the solenoid, and use the axial symmetric approximation in Example 31-9 (modified as in the accompanying exercise, $\phi_B = \pi r^2 B$ inside the solenoid, and $2\pi r E = -\pi r^2(dB/dt)$), then $E = \frac{1}{2}r|dB/dt| = F/e$, or $|dB/dt| = 2F/re = 2(1.3 \text{ fN})/(28 \text{ cm}\times1.6\times10^{-19} \text{ C}) = 58.0$ T/ms.

Problem

57. A conducting rod of length ℓ moves at speed v in a plane perpendicular to a uniform magnetic field **B**, as shown in Fig. 31-56. The magnetic force on charge carriers in the rod causes charge separation, which creates an electric field. Charge motion stops when the electric and magnetic forces on the charge carriers are equal. Show that this condition results in an electric field of magnitude vB and, therefore, in a potential difference $B\ell v$ between the rod ends, as we found from flux considerations in Example 31-8.

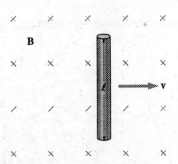

FIGURE 31-56 Problem 57.

Solution

The force on a charge carrier in the rod is $q\mathbf{v} \times \mathbf{B}$, with magnitude qvB and direction upward in Fig. 31-56 for positive q. The motional equivalent electric field is $F/q = vB = E$, leading to a potential difference of $\int \mathbf{E} \cdot d\boldsymbol{\ell} = vB\ell$ between the ends of the rod. (The motional field is canceled by the field of separated charge at the ends of the rod, when the current is zero.)

Problem

61. A generator like that shown in Fig. 31-14 has an N-turn coil of area A spinning with angular speed ω in a uniform magnetic field B. A resistor R is connected across the generator. (a) Find an expression for the power dissipated in the resistor as a function of time. (b) Find an expression for the magnetic torque on the generator coil as a function of time. (c) Study the discussion associated with Equations 12-28a and b, and use it to show that the rate at which the agent turning the generator at constant angular speed does work is equal to the power dissipation in the resistor. Assume the coil's magnetic moment is aligned with the field a time $t = 0$.

Solution

(a) For a generator like the one considered in Example 31-6, $\phi_B = NAB\cos\omega t$, and $\mathcal{E} = -d\phi_B/dt = \omega NAB\sin\omega t$, where $\omega = 2\pi f = d\theta/dt$. The instantaneous power dissipated in a resistor attached to this emf is $\mathcal{P}_R = I^2R = \mathcal{E}^2/R = (\omega NAB \sin \omega t)^2/R$. (b) The torque could be found by considering the forces on the wires of the coil (as in Section 29-5); the result, expressed in terms of the induced magnetic dipole moment ($\mu = NIA$) is $\tau = |\boldsymbol{\mu} \times \mathbf{B}| = \mu B \sin \theta = NIAB \sin \omega t = \omega (NAB \sin \omega t)^2/R$, where $I = \mathcal{E}/R$ from part (a). (c) The work done turning the coil is $dW = \tau \, d\theta$, so the power expended operating the generator is $dW/dt = \tau \, d\theta/dt = \tau \omega$. Conservation of energy requires that $dW/dt = \mathcal{P}_R$, which is entirely consistent with the results in parts (a) and (b).

Problem

65. A pendulum consists of a mass m suspended from two identical copper wires of negligible mass. At equilibrium the mass is a vertical distance ℓ below its supports, and the wires make 45° angles with the vertical, as shown in Fig. 31-58. A uniform magnetic field $\mathbf{B}$ points into the page. The pendulum is displaced from the plane of the page by a small angle θ_0, and at time $t = 0$ it is released. Find an expression for the voltmeter reading as a function of time.

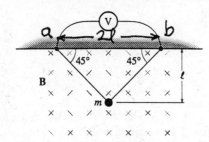

FIGURE 31-58 Problem 65.

Solution

The pendulum is equivalent to a simple pendulum with angular frequency $\omega = \sqrt{g/\ell}$. For small oscillations, $\theta = \theta_0 \cos \omega t$, where θ is the angle between the plane of the triangle, amb, formed by the mass and its supports, and the plane of Fig. 31-58. As the wires swing through the magnetic field,

motional emf's cause charges to separate at the ends a and b, and a potential difference is measured by the voltmeter, $V = V_b - V_a$. We assume that the accumulation of charge at a and b takes place so rapidly, compared to the pendulum's motion, that the induced electric field in the wires is in equilibrium with the electrostatic field from the charge separation, at every instant. Then, Faraday's law may be written as $V = -d\phi_B/dt$, where ϕ_B is the flux linking triangle amb. (Although the interpretation of Faraday's law in this case is subtle, its application is straightforward, as in Example 30-8.) With reference to Fig. 31-58 and the side view sketch, $\phi_B = BA \cos \theta = B\ell^2 \cos(\theta_0 \cos \omega t)$, where the area of the triangle is $A = \frac{1}{2}(\text{base})(\text{altitude}) = \frac{1}{2}(2\ell)(\ell) = \ell^2$, and the angle between $\mathbf{B}$ and $\mathbf{A}$ is θ, the pendulum angle. Therefore,

$$V = -\frac{d}{dt}[B\ell^2 \cos(\theta_0 \cos \omega t)]$$
$$= -B\ell^2[-\sin(\theta_0 \cos \omega t)](-\omega \theta_0 \sin \omega t).$$

For small angles, $\sin \theta \approx \theta$, so

$$V \approx -B\ell^2 \omega \theta_0^2 \cos \omega t \sin \omega t = -\frac{1}{2} B\ell^2 \theta_0^2 \sqrt{\frac{g}{\ell}} \sin\left(2\sqrt{\frac{g}{\ell}} t\right).$$

(We used a trigonometric identity for $\sin 2\omega t$, and substituted for ω.)

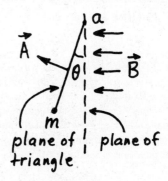

Problem 65 Solution.

● **CHAPTER 32** INDUCTANCE AND MAGNETIC ENERGY

Section 32-1 Mutual Inductance

Problem

1. Two coils have a mutual inductance of 2.0 H. If current in the first coil is changing at the rate of 60 A/s, what is the emf in the second coil?

Solution

From Equation 32-2, $\mathcal{E}_2 = -M(dI_1/dt) = -(2 \text{ H})(60 \text{ A/s}) = -120$ V. (The minus sign, Lenz's law, signifies that an induced emf opposes the process which creates it.)

Problem

5. An alternating current given by $I_p \sin 2\pi ft$ is supplied to one of two coils whose mutual inductance is M. (a) Find an expression for the emf in the second coil. (b) When $I_p = 1.0$ A and $f = 60$ Hz, the peak emf in the second coil is measured at 50 V. What is the mutual inductance?

Solution

(a) From Equation 32-2, $\mathcal{E}_2 = -M \, dI_1/dt = -M 2\pi f I_p \cos(2\pi ft)$. (b) The peak value of the cosine is 1, so $|M| = \mathcal{E}_{2p}/2\pi f I_p = 50$ V$/(2\pi \times 60$ Hz$)(1$ A$) = 133$ mH. (From the information given, only the magnitude of M can be determined; its sign depends on how the coils are coupled.)

Problem

9. A rectangular loop of length ℓ and width w is located a distance a from a long, straight, wire, as shown in Fig. 32-18. What is the mutual inductance of this arrangement?

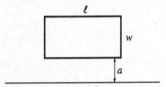

FIGURE 32-18   Problem 9.

Solution

When current I_1 flows to the left in the wire, the flux through the loop is $\phi_{B,2} = (\mu_0 I_1/2\pi) \int_a^{a+w} \ell \, dr/r = (\mu_0 I_1 \ell/2\pi) \times \ln(1 + w/a)$ (see Example 31-2). Then Equation 32-1 gives $M = \phi_{B,2}/I_1 = (\mu_0 \ell/2\pi)\ln(1 + w/a)$. (In calculating the flux, the normal to the loop area was taken into the page, so the positive sense of circulation around the loop is CW. This determines the direction of the induced emf $\mathcal{E}_2$ in Equation 32-2.)

Section 32-2   Self-Inductance

Problem

13. A 2.0-A current is flowing in a 20-H inductor. A switch is opened, interrupting the current in 1.0 ms. What emf is induced in the inductor?

Solution

Assume that the current changes uniformly from 2 A to zero in 1 ms (or consider average values). Then $dI/dt = -2$ A/ms, and Equation 32-5 gives $\mathcal{E} = -(20$ H$)(-20$ A/ms$) = 40$ kV. (The emf opposes the decreasing current.)

Problem

17. The emf in a 50-mH inductor has magnitude $|\mathcal{E}| = 0.020t$, with t in seconds and $\mathcal{E}$ in volts. At $t = 0$ the inductor current is 300 mA. (a) If the current is increasing, what will be its value at $t = 3.0$ s? (b) Repeat for the case when the current is decreasing.

Solution

(a) $\mathcal{E}$ has the opposite sign to dI/dt in Equation 32-5. When I is increasing, $dI/dt > 0$, $\mathcal{E}$ is negative, $\mathcal{E} = -|\mathcal{E}|$. Thus,

$$\frac{dI}{dt} = -\frac{\mathcal{E}}{L} = \frac{(0.02 \text{ V/s})t}{(0.05 \text{ H})} = \left(0.4 \frac{\text{A}}{\text{s}^2}\right)t, \quad \text{and}$$

$$I = \frac{1}{2}\left(0.4 \frac{\text{A}}{\text{s}^2}\right)t^2 + I_0.$$

At $t = 3$ s, $I = (0.2$ A/s$^2)(3$ s$)^2 + 0.3$ A $= 2.1$ A. (b) For $dI/dt < 0$, $\mathcal{E} = |\mathcal{E}| > 0$, so $dI/dt = -(0.4$ A/s$^2)t$, and $I = -(0.2$ A/s$^2)t^2 + I_0$. At $t = 3$ s in this case, $I = -1.5$ A. (After $t = \sqrt{3/2}$ s, the current reverses direction and begins increasing in absolute value.)

Problem

21. The emf in a 50-mH inductor is given by $\mathcal{E} = \mathcal{E}_p \sin \omega t$, where $\mathcal{E}_p = 75$ V and $\omega = 140$ s^{-1}. What is the peak current in the inductor? (Assume the current swings symmetrically about zero.)

Solution

From Equation 32-5, $dI/dt = -(\mathcal{E}_p/L)\sin \omega t$, so integration yields $I(t) = (\mathcal{E}_p/\omega L)\cos \omega t$. (Since $I(t)$ is symmetric about $I = 0$, there is no constant of integration.) The peak current is $I_p = \mathcal{E}_p/\omega L = 75$ V$/(140$ s$^{-1} \times 50$ mH$) = 10.7$ A.

Section 32-3   Inductors in Circuits

Problem

25. The current in a series RL circuit rises to 20% of its final value in 3.1 μs. If $L = 1.8$ mH, what is the resistance R?

Solution

The buildup of current in an RL circuit with a battery is given by Equation 32-8, $I(t) = I_\infty(1 - e^{-Rt/L})$, where $I_\infty = \mathcal{E}_0/R$ is the final current. Solving for R, one finds $R = -(L/t) \times \ln(1 - I/I_\infty) = -(1.8$ mH/3.1 μs$)\ln(1 - 20\%) = 130$ Ω.

Problem

29. In Fig. 32-7a, take $R = 2.5$ kΩ and $\mathcal{E}_0 = 50$ V. When the switch is closed, the current through the inductor rises to 10 mA in 30 μs. (a) What is the inductance? (b) What will be the current in the circuit after many time constants?

Solution

(b) After a long time ($t \to \infty$), the exponential term in Equation 32-8 is negligible. Thus, $I_\infty = \mathcal{E}_0/R = 50 \text{ V}/2.5 \text{ k}\Omega = 20$ mA. (a) The current has risen to half its final value in 30 μs. Thus (Equation 32-8 again), $\frac{1}{2} = 1 - e^{-Rt/L}$, or $L = Rt/\ln 2 = (2.5 \text{ k}\Omega)(30 \ \mu\text{s})/\ln 2 = 108$ mH.

Problem

33. Resistor R_2 in Fig. 32-20 is to limit the emf that develops when the switch is opened. What should be its value in order that the inductor emf not exceed 100 V?

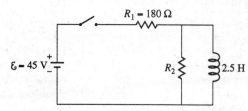

FIGURE 32-20 Problem 33.

Solution

As explained in Example 32-6, when the switch is opened (after having been closed a long time), the voltage across R_2 (which equals the inductor emf) is $V_2 = I_2 R_2 = \mathcal{E}_0 R_2/R_1$. If we choose to limit this to no more than 100 V, then $R_2 \leq (100 \text{ V})(180 \ \Omega)/45 \text{ V} = 400 \ \Omega$.

Problem

37. In Fig. 32-22, take $\mathcal{E}_0 = 20$ V, $R_1 = 10 \ \Omega$, $R_2 = 5.0 \ \Omega$, and assume the switch has been open for a long time. (a) What is the inductor current immediately after the switch is closed? (b) What is the inductor current a long time after the switch is closed? (c) If after a long time the switch is again opened, what will be the voltage across R_1 immediately afterwards?

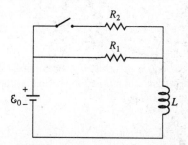

FIGURE 32-22 Problem 37.

Solution

(a) If the switch has been open a long time, a steady current flows through the inductance ($dI_L/dt = 0$). When the switch is closed (at $t = 0$), I_L cannot change instantaneously, so $I_L(0) = \mathcal{E}/R_1 = 20 \text{ V}/10 \ \Omega = 2$ A. (Of course, $I_1(0) = I_L(0)$, and $I_2(0) = 0$.) (b) After another long time ($t \to \infty$), the currents are steady again and $\mathcal{E}_L = 0$ (the inductance behaves like a short circuit). The resistors are in parallel, therefore $I_L(\infty) = \mathcal{E}(1/R_1 + 1/R_2) = 20 \text{ V}(\frac{1}{5} + \frac{1}{10}) \ \Omega^{-1} = 6$ A. (c) When the switch is again opened, the current through R_2 is zero, but I_L cannot change instantly, so $I_L = I_1 = I_L(\infty) = 6$ A. Thus, the voltage across R_1 is $V_1 = I_1 R_1 = (6 \text{ A})(10 \ \Omega) = 60$ V.

Section 32-4 Magnetic Energy

Problem

41. A 12-V battery, 5.0-Ω resistor, and 18-H inductor are connected in series and allowed to reach a steady state. (a) What is the energy stored in the inductor? (b) Once in the steady state, over what time interval is the energy dissipated in the resistor equal to that stored in the inductor?

Solution

(a) The steady state (i.e., final) current in an RL circuit with emf $\mathcal{E}_0$ is $I = \mathcal{E}_0/R$, so the energy stored in the inductor (Equation 32-10) is $U_L = \frac{1}{2}LI^2 = \frac{1}{2}(18 \text{ H})(12 \text{ V}/5 \ \Omega)^2 = 51.8$ J. (b) Energy is dissipated in the resistor at the rate $\mathcal{P}_R = I^2 R$ (the joule heat), so the time interval querried is $\Delta t = U_L/\mathcal{P}_R = \frac{1}{2}LI^2/I^2 R = L/2R = 18 \text{ H}/(2 \times 5 \ \Omega) = 1.8$ s.

Problem

45. The current in a 2.0-H inductor is increasing. At some instant, the current is 3.0 A and the inductor emf is 5.0 V. At what rate is the inductor's magnetic energy increasing at this instant?

Solution

The rate at which energy is stored in an inductor is $\mathcal{P}_L = LI(dI/dt)$ (see the discussion of "Magnetic Energy in an Inductor" leading to Equation 32-10). When the current is increasing, as for this inductor, $L(dI/dt) = |\mathcal{E}_L|$, and $\mathcal{P}_L = I|\mathcal{E}_L| = (3 \text{ A})(5 \text{ V}) = 15$ W.

Problem

49. The Alcator fusion experiment at MIT has a 50-T magnetic field. What is the magnetic energy density in Alcator?

Solution

From Equation 32-11, $u_B = (50 \text{ T})^2/(8\pi \times 10^{-7} \text{ N/A}^2) = 995 \text{ MJ/m}^3$. (This is about 2.8% of the energy density content of gasoline, see Appendix C.)

Problem

53. A single-turn loop of radius R carries current I. How does the magnetic energy density at the loop center compare with that of a long solenoid of the same radius, carrying the same current, and consisting of n turns per unit length?

Solution

The energy density at the center of the loop is $u_B^{(loop)} = B^2/2\mu_0 = (\mu_0 I/2R)^2/2\mu_0 = \mu_0 I^2/8R^2$ (see Equations 30-3 and 32-11). In a long thin solenoid of the same radius, $u_B^{(solenoid)} = (\mu_0 nI)^2/2\mu_0 = \mu_0 n^2 I^2/2$, so the ratio of $u_B^{(loop)}$ to $u_B^{(solenoid)}$ is $1/4n^2 R^2$.

Paired Problems

Problem

57. Two coils have mutual inductance M. The current supplied to coil A is given by $I = bt^2$. Find an expression for the magnitude of the induced emf in coil B.

Solution

From Equation 32-2, $\mathcal{E}_B = -M(dI_A/dt) = -M(2bt)$, where the negative value means the emf opposes the change in current.

Problem

61. In Fig. 32-11a, take $\mathcal{E}_0 = 25$ V, $R_1 = 1.5$ Ω, and $R_2 = 4.2$ Ω. What is the voltage across R_2 (a) immediately after the switch is first closed and (b) a long time after the switch is closed? (c) Long after the switch is closed it is again opened. Now what is the voltage across R_2?

Solution

The circuit is analyzed in Example 32-6 at the instants of time mentioned in this problem, so all that is necessary here is to find numerical values. (a) At the moment the switch is first closed, $I = \mathcal{E}_0/(R_1 + R_2) = 25$ V/(1.5 + 4.2) Ω = 4.39 A. (b) When the current is steady, the current through R_2 is zero (since the inductor is an ideal one). (c) When the switch is reopened, $V_2 = \mathcal{E}_0 R_2/R_1 = (25$ V)(4.2/1.5) = 70.0 V, momentarily.

Supplementary Problems

Problem

65. (a) Use the result of Example 32-9 to determine the inductance of a toroid. (b) Show that your result reduces to the inductance of a long solenoid when $R \gg \ell$.

Solution

(a) Since $U = \frac{1}{2}LI^2$, dividing the expression for U in Example 32-9 by $\frac{1}{2}I^2$, we find $L = (\mu_0 N^2/2\pi)\ell \ln(1 + \ell/R)$. (b) For $\ell \ll R$, $\ln(1 + \ell/R) \approx \ell/R$, so $L \approx \mu_0 N^2\ell^2/2\pi R$. Since $\ell^2 = A$ is the cross-sectional area, $2\pi R = \ell_{Solenoid}$ is the length, and $n = N/\ell_{Solenoid}$ is the number of turns per unit length of the toroidal solenoid, this result is approximately the same as Equation 32-4.

Problem

69. An electric field and a magnetic field have the same energy density. Obtain an expression for the ratio E/B, and evaluate this ratio numerically. What are its units? Is your answer close to any of the fundamental constants listed inside the front cover?

Solution

εThe combination of Equations 26-3 and 32-11 implies that if $u_E = \frac{1}{2}\varepsilon_0 E^2 = u_B = B^2/2\mu_0$, then $E/B = 1/\sqrt{\mu_0\varepsilon_0}$. Numerically, $\mu_0 = 4\pi\times10^{-7}$ N/A^2 and $(1/4\pi\varepsilon_0) \approx 9\times10^9$ N·m^2/C^2, so $1/\sqrt{\mu_0\varepsilon_0} \approx \sqrt{(9\times10^9\text{ N·m}^2/\text{C}^2)/(10^{-7}\text{ N/A}^2)} = 3\times10^8$ m/s, which is, in fact, the speed of light (see Section 34-5). ●

● **CHAPTER 33** ALTERNATING-CURRENT CIRCUITS

Section 33-1 Alternating Current

Problem

1. Much of Europe uses AC power at 230 V rms and 50 Hz. Express this AC voltage in the form of Equation 33-3, taking $\phi = 0$.

Solution

Use of Equations 33-1 and 2 allows us to write $V_p = \sqrt{2}\,V_{rms} = \sqrt{2}(230\text{ V}) = 325$ V, and $\omega = 2\pi f = 2\pi(50\text{ Hz}) = 314$ s^{-1}. Then the voltage expressed in the form of Equation 33-3 is $V(t) = (325\text{ V})\sin[(314\text{ s}^{-1})t]$.

Problem

5. An AC current is given by $I = 495\sin(9.43t)$, with I in milliamperes and t in milliseconds. Find (a) the rms current and (b) the frequency in Hz.

Solution

Comparison of the current with Equation 33-3 shows that its amplitude and angular frequency are $I_p = 495$ mA and $\omega = 9.43$ (ms)$^{-1}$. Application of Equations 33-1 and 2 give (a) $I_{rms} = 495$ mA/$\sqrt{2} = 350$ mA, and (b) $f = 9.43/2\pi$(ms) = 1.50 kHz.

Problem

9. How are the rms and peak voltages related for the triangle wave in Fig. 33-30? See Problem 7.

Solution

Define the zero of time to coincide with a positive apex of the waveform, which has period T, as drawn upon Fig. 33-30. The analytic form of the voltage signal (for $0 \le t \le T$) is

$$V(t) = V_p \begin{cases} -(4t/T) + 1, & 0 \le t \le \tfrac{1}{2}T \\ (4t/T) - 3, & \tfrac{1}{2}T \le t \le T \end{cases}.$$

Because the negative part of the waveform is the reflection of the positive part, the square of the waveform has period $\tfrac{1}{2}T$. Thus, the average over T equals the average over $\tfrac{1}{2}T$, or

$$\langle V^2 \rangle = \frac{1}{T} \int_0^T V^2 \, dt = \frac{1}{\tfrac{1}{2}T} \int_0^{T/2} V^2 \, dt$$

$$= \frac{2V_p^2}{T} \int_0^{T/2} \left(\frac{16t^2}{T^2} - \frac{8t}{T} + 1 \right) dt$$

$$= \frac{2V_p^2}{T} \left[\frac{16}{3T^2} \left(\frac{T}{2} \right)^3 - \frac{8}{2T} \left(\frac{T}{2} \right)^2 + \frac{T}{2} \right]$$

$$= 2V_p^2 \left(\frac{2}{3} - 1 + \frac{1}{2} \right) = \frac{V_p^2}{3}.$$

Consequently, $V_{\text{rms}} = \sqrt{\langle V^2 \rangle} = V_p/\sqrt{3}$. (We used a symmetry of the waveform to reduce the number of integrations by half.)

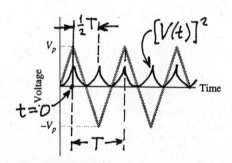

FIGURE 33-30 Problem 9.

Section 33-2 Circuit Elements in AC Circuits

Problem

13. What is the rms current in a $1.0\text{-}\mu\text{F}$ capacitor connected across the 120-V rms, 60-Hz AC line?

Solution

Equation 33-5 can be used with the rms current and voltage, since both are $1/\sqrt{2}$ times their peak values. Thus, $I_{\text{rms}} = \omega C V_{\text{rms}} = (2\pi \times 60 \text{ Hz})(1 \ \mu\text{F})(120 \text{ V}) = 45.2$ mA.

Problem

17. A capacitor and a 1.8-kΩ resistor pass the same current when each is separately connected across a 60-Hz power line. What is the capacitance?

Solution

The currents (rms or peak) are the same if $X_C = R = 1/\omega C$, so $C = 1/\omega R = [2\pi(60 \text{ Hz})(1.8 \text{ k}\Omega)]^{-1} = 1.47 \ \mu\text{F}$.

Problem

21. A $1.2\text{-}\mu\text{F}$ capacitor is connected across a generator whose output is given by $V = V_p \sin 2\pi ft$, where $V_p = 22$ V, $f = 60$ Hz, and t is in seconds. (a) What is the peak current? (b) What are the magnitudes of the voltage and (c) the current at $t = 6.5$ ms?

Solution

(a) $I_p = V_p \omega C = (22 \text{ V})(377 \text{ s}^{-1})(1.2 \ \mu\text{F}) = 9.95$ mA (see Equation 33-5 and Example 33-1). (b) $V(6.5 \text{ ms}) = V_p \sin \omega t = (22 \text{ V}) \sin[(377 \text{ s}^{-1}) \times (6.5 \text{ ms})] = 14.0$ V. (Remember that ωt is in radians.) (c) In a capacitor, the current leads the voltage by 90°, so $I(6.5 \text{ ms}) = I_p \sin(\omega t + \pi/2) = (9.95 \text{ mA}) \cos[(377 \text{ s}^{-1})(6.5 \text{ ms})] = -7.67$ mA (the magnitude is 7.67 mA).

Problem

25. A 0.75-H inductor is in series with a flourescent lamp, and the series combination is across the 120-V rms, 60-Hz power line. If the rms inductor voltage is 90 V, what is the rms lamp current?

Solution

In a series circuit, the same current flows through the inductor and lamp. Therefore, since the ratio of the rms quantities for a given circuit element equals that of the peak values, Equation 33-7 gives $I_{\text{rms}} = V_{\text{rms}}/\omega L = 90 \text{ V}/(2\pi \times 60 \text{ Hz})(0.75 \text{ H}) = 318$ mA.

Section 33-3 *LC* Circuits

Problem

29. You have a 2.0-mH inductor and wish to make an *LC* circuit whose resonant frequency spans the AM radio band (550 kHz to 1600 kHz). What range of capacitance should your variable capacitor cover?

Solution

The resonant frequency of an *LC* circuit is (Equation 33-11) $\omega = 1/\sqrt{LC}$, so the capacitance should cover a range from $C = 1/\omega^2 L = 1/(2\pi \times 550 \text{ kHz})^2(2 \text{ mH}) = 41.9$ pF down to $C = 1/(2\pi \times 1.6 \text{ MHz})^2(2 \text{ mH}) = 4.95$ pF.

Problem

33. An LC circuit includes a 0.025-μF capacitor and a 340-μH inductor. (a) If the peak voltage on the capacitor is 190 V, what is the peak current in the inductor? (b) How long after the voltage peak does the current peak occur?

Solution

(a) In an LC circuit, the peak current and voltage are related by $I_p = \omega q_p = \omega C V_p = C V_p / \sqrt{LC} = V_p \sqrt{C/L}$ (see Example 33-3). Thus, $I_p = (190 \text{ V}) \sqrt{0.025 \ \mu\text{F}/340 \ \mu\text{H}} = 1.63$ A.
(b) The current peaks one quarter of a period after the voltage, or $\Delta t = \frac{1}{4} T = \frac{1}{4}(2\pi/\omega) = \frac{1}{2} \pi \sqrt{LC} = (\pi/12)\sqrt{0.025 \ \mu\text{F} \times 340 \ \mu\text{H}} = 4.58 \ \mu$s.

Problem

37. One-eighth of a cycle after the capacitor in an LC circuit is fully charged, what are each of the following as fractions of their peak values: (a) capacitor charge, (b) energy in the capacitor, (c) inductor current, (d) energy in the inductor?

Solution

The equations in Section 33-3 give the desired quantities, which we evaluate when $\omega t = \omega(T/8) = 2\pi/8 = \frac{1}{4}\pi = 45°$ (i.e., $\frac{1}{8}$ cycle). (Note that phase constant zero corresponds to a fully charged capacitor at $t = 0$.) (a) From Equation 33-10, $q/q_p = \cos 45° = 1/\sqrt{2}$. (b) From the equation for electric energy, $U_E/U_{Ep} = \cos^2 45° = 1/2$. (c) From Equation 33-12, $I/I_p = -\sin 45° = -1/\sqrt{2}$. (The direction of the current is away from the positive capacitor plate at $t = 0$.) (d) From the equation for magnetic energy, $U_B/U_{Bp} = \sin^2 45° = 1/2$.

Problem

41. A damped RLC circuit includes a 5.0-Ω resistor and a 100-mH inductor. If half the initial energy is lost after 15 cycles, what is the capacitance?

Solution

If only half the energy is lost after 15 cycles, the damping is small and the energy varies like the square of Equation 33-13, namely $U_{tot} = U_p e^{-Rt/L} \cos^2 \omega t$. (The energy time constant is L/R, one half the charge time constant.) After 15 cycles, $t = 15\,T = 15(2\pi/\omega)$, the fraction of energy remaining is $\frac{1}{2} = e^{-15RT/L}\cos^2 30\pi = e^{-15RT/L}$. Take logarithms and use $\omega = 1/\sqrt{LC}$ to get $L\ln 2 = 15RT = 30\pi R\sqrt{LC}$, from which we find

$$C = \left(\frac{\ln 2}{30\pi R}\right)^2 L = \left(\frac{\ln 2}{30\pi \times 5 \ \Omega}\right)^2 (100 \text{ mH}) = 0.216 \ \mu\text{F}.$$

Section 33-4 Driven RLC Circuits and Resonance

Problem

45. TV channel 2 occupies the frequency range from 54 MHz to 60 MHz. A series RLC tuning circuit in a TV receiver includes an 18-pF capacitor and resonates in the middle of the channel 2 band. (a) What is the inductance? (b) To let the whole signal in, the resonance curve must be broad enough that the current throughout the band be no less than 70% of the current at the resonant frequency. What constraint does this place on the circuit resistance?

Solution

(a) The condition for resonance in a series RLC circuit requires $L = 1/\omega_r^2 C = 1/(2\pi \times 57 \text{ MHz})^2(18 \text{ pF}) = 0.433 \ \mu$H. (b) The peak current at any frequency is $I_p = V_p/Z$, while at resonance, $I_{res} = V_p/R$. Thus, $I_p/I_{res} = R/Z = 1/\sqrt{1 + (X_C - X_L)^2/R^2}$ (see Equation 33-14). If this ratio is required to be not less than 70%, then

$$\left(\frac{X_C - X_L}{R}\right)^2 \le \left(\frac{1}{0.7}\right)^2 - 1, \quad \text{or} \quad R \ge \sqrt{\frac{49}{51}}\,|X_C - X_L|.$$

The reactance can be expressed in several equivalent forms, one being in terms of the given data and the resonant frequency:

$$|X_C - X_L| = \left|\frac{1}{\omega C} - \omega L\right| = \sqrt{\frac{L}{C}}\left|\frac{1}{\omega\sqrt{LC}} - \omega\sqrt{LC}\right|$$
$$= \frac{1}{\omega_r C}\left|\frac{\omega_r}{\omega} - \frac{\omega}{\omega_r}\right|.$$

If this is evaluated at the lower band edge (54 MHz), one gets

$$R \ge \frac{\sqrt{49/51}}{(2\pi \times 57 \text{ MHz})(18 \text{ pF})}\left|\frac{57}{54} - \frac{54}{57}\right| = 16.4 \ \Omega.$$

(The upper band edge, 60 MHz, gives a weaker limit, $R \ge 15.6 \ \Omega$.)

Problem

49. Figure 33-32 shows the phasor diagram for an RLC circuit. (a) Is the driving frequency above or below resonance? (b) Complete the diagram by adding the applied voltage phasor, and from your diagram determine the phase difference between applied voltage and current.

Solution

(a) From the observation that $V_{L,p} = I_p\omega L > V_{C,p} = I_p/\omega C$, we conclude that the frequency is above resonance, $\omega^2 > 1/LC$. (b) The applied voltage phasor is the vector sum of the resistor, capacitor, and inductor voltage phasors, as drawn on Fig. 33-32. The current is in phase with the voltage across the resistor, in this case lagging the applied voltage (since $\phi = \tan^{-1}[(V_{C,p} - V_{L,p})/V_{R,p}] < 0$) by approximately 50° (as estimated from Fig. 33-32).

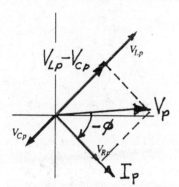

FIGURE 33-32 Problem 49 Solution.

Section 33-5 Power in AC Circuits

Problem

53. An electric drill draws 4.6 A rms at 120 V rms. If the current lags the voltage by 25°, what is the drill's power consumption?

Solution

The average power consumed by an AC circuit is given by Equation 33-17, $\mathcal{P}_{av} = V_{rms} I_{rms} \cos \phi = (120 \text{ V})(4.6 \text{ A})\cos(-25°) = 500 \text{ W}$.

Problem

57. A power plant produces 60-Hz power at 365 kV rms and 200 A rms. The plant is connected to a small city by a transmission line with total resistance 100 Ω. What fraction of the power is lost in transmission if the city's power factor is (a) 1.0 or (b) 0.60? (c) Is it more economical for the power company if the load has a large power factor or a small one? Explain.

Solution

(a) We assume that the average power supplied to the city is $\mathcal{P}_{av} = I_{rms} V_{rms} \cos \phi = (200 \text{ A})(365 \text{ kV})(1) = 73.0 \text{ MW}$, and that the average power lost in the transmission line is $\Delta \mathcal{P} = I_{rms}^2 R = (200 \text{ A})^2 (100 \text{ Ω}) = 4 \text{ MW}$. Thus, the percent lost is $(4/73) \times 100 \approx 5.5\%$. (b) If the power factor in part (a) were 0.6 instead of 1.0, the percent lost would be

5.5%/0.6 = 9.1%. (c) Since $\Delta \mathcal{P}$ is a constant, the larger $\mathcal{P}_{av}$ (which is proportional to $\cos \phi$), the smaller the fraction of power lost, $\Delta \mathcal{P}/\mathcal{P}_{av}$. A large power factor is better for the power plant owners.

Section 33-6 Transformers and Power Supplies

Problem

61. The transformer in the power supply of Fig. 33-27a has an output voltage of 6.3 V rms at 60 Hz, and the capacitance is 1200 μF. (a) With an infinite load resistance, what would be the output voltage of the power supply? (b) What is the minimum load resistance for which the output would not drop more than 1% from this value? Assume that the discharge time in Fig. 33-27b is essentially a full cycle.

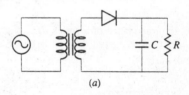

(a)

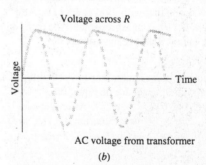

Voltage across R

AC voltage from transformer

(b)

FIGURE 33-27 For reference.

Solution

(a) With no load present, the capacitor cannot discharge through the transformer because the diode blocks such a current. Therefore, it remains charged to the peak AC voltage, $V_p = \sqrt{2} \, V_{rms} = \sqrt{2}(6.3 \text{ V}) = 8.91 \text{ V}$. (b) With the load resistance present, the capacitor voltage must not decay to less than 99% of its maximum value over one period of AC ($T = \frac{1}{60}$ s) (see Fig. 33-27b). Therefore, $e^{-T/RC} \geq 0.99$, or $R \geq -T/C \ln(0.99) = [-(60 \text{ Hz})(1200 \text{ μF})\ln(0.99)]^{-1} = 1.38 \text{ kΩ}$. (Note: The voltage for a discharging RC circuit is given by Equation 28-8.)

Paired Problems

Problem

65. The peak current in an oscillating LC circuit is 850 mA. If $L = 1.2$ mH and $C = 5.0$ μF, what is the peak voltage?

Solution

In an LC circuit, the peak electric and magnetic energies are equal (see the analysis in the text), so $\frac{1}{2}CV_p^2 = \frac{1}{2}LI_p^2$. Thus, $V_p = \sqrt{L/C}\, I_p = \sqrt{1.2 \text{ mH}/5.0 \text{ } \mu\text{F}}\, (850 \text{ mA}) = 13.2$ V.

Problem

69. A series RLC circuit with $R = 5.5$ Ω, $L = 180$ mH, and $C = 0.12$ μF is connected across a sine-wave generator. If the inductor can handle a maximum current of 1.5 A, what is the maximum safe value for the generator's peak output voltage when it is tuned to resonance?

Solution

At resonance, the impedance of a series RLC circuit is $Z = R$, so $V_p = ZI_p = RI_p = (5.5 \text{ }\Omega)(1.5 \text{ A}) = 8.25$ V at the maximum safe peak current.

Supplementary Problems

Problem

73. An undriven RLC circuit with inductance L and resistance R starts oscillating with total energy U_0. After N cycles the energy is U_1. Find an expression for the capacitance, assuming the circuit is not heavily damped.

Solution

The total energy in an underdamped oscillating RLC circuit decays like the square of Equation 33-13 $(U = q^2/2C)$, or $U_{tot} = U_0 e^{-Rt/L}$. After N cycles, $t = N(2\pi/\omega) = 2\pi N\sqrt{LC}$ (see Equation 33-11), so $U_1 = U_0 e^{-2\pi RN\sqrt{C/L}}$, and $C = L[\ln(U_0/U_1)/2\pi RN]^2$.

Problem

77. A sine-wave generator with peak output voltage of 20 V is applied across a series RLC circuit. At the resonant frequency of 2.0 kHz the peak current is 50 mA, while at 1.0 kHz it is 15 mA. Find R, L, and C.

Solution

At resonance, $I_p = V_p/R$, so $R = 20$ V/50 mA $= 400$ Ω. The impedance at resonance is $Z = R$ (i.e., $X = X_C - X_L = 0$), while at half the resonant frequency (1 kHz $= \frac{1}{2}(2$ kHz)) $Z = V_p/I_p = 20$ V/15 mA $= 1.33$ kΩ $= (10/3)R$ (i.e., $X = \sqrt{Z^2 - R^2} = \sqrt{(10/3)^2 - 1}\, R = \sqrt{91}\, R/3$.) Therefore,

$$\frac{1}{\omega_r C} - \omega_r L = 0, \quad \text{and} \quad \frac{1}{\frac{1}{2}\omega_r C} - \frac{1}{2}\omega_r L = \sqrt{91}\, R/3.$$

These equations can be solved for C and L, with the result:

$$L = \frac{2\sqrt{91}\, R}{9\omega r} = \frac{2\sqrt{91}\,(400 \text{ }\Omega)}{(9 \times 2\pi \times 2 \text{ kHz})} = 67.5 \text{ mH, and}$$

$C = (\omega_r^2 L)^{-1} = (2\pi \times 2 \text{ kHz})^{-2}(67.5 \text{ mH})^{-1} = 93.8$ nF.

Problem

81. For RLC circuits in which the resistance is not too large, the Q factor may be defined as the ratio of the resonant frequency to the difference between the two frequencies where the power dissipated in the circuit is half that dissipated at resonance. Show, using suitable approximations, that this definition leads to the expression $Q = \omega_r L/R$, with ω_r the resonant frequency.

Solution

From Equations 33-14 (with rms values), 33-17, and the result in the solution to Problem 55(a), the average power in a series RLC circuit becomes $\langle \mathcal{P} \rangle = I_{rms} V_{rms} \cos \phi = (V_{rms}/Z)V_{rms}(R/Z) = V_{rms}^2 R/Z^2$. From Equation 33-15, one sees that the power falls to half its resonance value (V_{rms}^2/R) when $Z = \sqrt{2}R$, or when $|X_C - X_L| = R$. In terms of the resonant frequency, $\omega_r = 1/\sqrt{LC}$, this condition becomes

$$\left|\frac{1}{\omega C} - \omega L\right| = L\left|\frac{\omega_r^2}{\omega} - \omega\right| = R, \quad \text{or} \quad \omega_r^2 - \omega^2 = \pm\frac{R}{L}\omega.$$

The solutions of these quadratics, with $\omega > 0$, are

$$\omega = \frac{1}{2}\left[\pm\frac{R}{L} + \sqrt{\frac{R^2}{L^2} + 4\omega_r^2}\right].$$

If $\frac{R}{L} \ll \omega_r$ (equivalent to $R \ll \sqrt{\frac{L}{C}}$) we can neglect the first term under the square root sign, compared to the second, obtaining $\omega \approx \omega_r \pm R/2L$. The difference between these two values of ω is $\Delta\omega = R/L$, from which the desired expression for Q follows. ●

● CHAPTER 34 MAXWELL'S EQUATIONS AND ELECTROMAGNETIC WAVES

Section 34-2 Ambiguity in Ampère's Law

Problem

1. A uniform electric field is increasing at the rate of 1.5 V/m·μs. What is the displacement current through an area of 1.0 cm^2 at right angles to the field?

Solution

Maxwell's displacement current is $\varepsilon_0 \partial \phi_E / \partial t = (8.85 \times 10^{-12}$ F/m$)(1.5$ V/m·μs$)(1$ cm$^2) = 1.33$ nA. (See Equations 34-1 and 24-2.)

Problem

5. A parallel-plate capacitor has circular plates with radius 50 cm and spacing 1.0 mm. A uniform electric field between the plates is changing at the rate 1.0 MV/m·s. What is the magnetic field between the plates (a) on the symmetry axis, (b) 15 cm from the axis, and (c) 150 cm from the axis?

Solution

(a) As explained in Example 34-1, cylindrical symmetry and Gauss's law for magnetism require that the **B**-field lines are circles around the symmetry axis, as in Fig. 34-5. For a radius, r, less than the radius of the plates, R, the displacement current is $I_D = \varepsilon_0 d\phi_E / dt = \varepsilon_0 (d/dt) \int \mathbf{E} \cdot d\mathbf{A} = \varepsilon_0 \pi r^2 (dE/dt)$, where the integral is over disk of radius r centered between the plates. Maxwell's form of Ampère's law gives $\oint \mathbf{B} \cdot d\boldsymbol{\ell} = 2\pi r B = \mu_0 I_D$, where the line integral is around the circumference of the disk. Thus, $B = \frac{1}{2}\mu_0 \varepsilon_0 r(dE/dt) = r(dE/dt)/2c^2$, where c is the speed of light (Equation 34-16). On the symmetry axis, $r = 0$, so $B = 0$. (b) For $r = 15$ cm $< R$, $B = \frac{1}{2}(0.15$ m$)(10^6$ V/m·s$) \div (3 \times 10^8$ m/s$)^2 = 8.33 \times 10^{-13}$ T. (c) For $r > R$, the displacement current is $I_D = \varepsilon_0 \pi R^2 (dE/dt)$, so $B = (dE/dt)R^2/2c^2 r$. At $r = 150$ cm, $B = (10^6$ V/m·s$) \times (50$ cm$)^2/2(3 \times 10^8$ m/s$)^2(150$ cm$) = 9.26 \times 10^{-13}$ T.

Section 34-4 Electromagnetic Waves

Problem

9. The electric field of a radio wave is given by $\mathbf{E} = E\sin(kz - \omega t)(\hat{\mathbf{i}} + \hat{\mathbf{j}})$. (a) What is the peak amplitude of the electric field? (b) Give a unit vector in the direction of the magnetic field at a place and time where $\sin(kz - \omega t)$ is positive.

Solution

(a) The peak amplitude is the magnitude of $E(\hat{\mathbf{i}} + \hat{\mathbf{j}})$, which is $E\sqrt{2}$. Note that $\hat{\mathbf{i}} + \hat{\mathbf{j}} = \sqrt{2}\,\hat{\mathbf{n}}$, where $\hat{\mathbf{n}}$ is a unit vector 45° between the positive x and y axes. (b) When **E** is parallel to $\hat{\mathbf{n}}$

(for $\sin(kz - \omega t)$ positive) **B** points 45° into the second quadrant (so that $\mathbf{E} \perp \mathbf{B}$ and $\mathbf{E} \times \mathbf{B}$ is in the $+z$ direction). Thus, **B** is parallel to the unit vector $(-\hat{\mathbf{i}} + \hat{\mathbf{j}})/\sqrt{2}$.

Section 34-5 The Speed of Electromagnetic Waves

Problem

13. Your intercontinental telephone call is carried by electromagnetic waves routed via a satellite in geosynchronous orbit at an altitude of 36,000 km. Approximately how long does it take before your voice is heard at the other end?

Solution

Assuming the satellite is approximately overhead, we can estimate the round-trip travel time by $\Delta t = \Delta r/c = (2 \times 36{,}000$ km$)/(3 \times 10^5$ km/s$) = 0.24$ s.

Problem

17. "Ghosts" on a TV screen occur when part of the signal goes directly from transmitter to receiver, while part takes a longer route, reflecting off mountains or buildings (Fig. 34-29). The electron beam in a 50-cm-wide TV tube "paints" the picture by scanning the beam from left to right across the screen in about 10^{-4} s. If a "ghost" image appears displaced about 1 cm from the main image, what is the difference in path lengths of the direct and indirect signals?

Solution

The time it takes for the electron beam to sweep across 1 cm of the TV screen is 1 cm/(50 cm/10^{-4} s$) = 2 \times 10^{-6}$ s, which equals the time delay of the ghost signal. Therefore, the path difference is $\Delta r = c\,\Delta t = (3 \times 10^8$ m/s$)(2 \times 10^{-6}$ s$) = 600$ m.

Section 34-6 Properties of Electromagnetic Waves

Problem

21. A 60-Hz power line emits electromagnetic radiation. What is the wavelength?

Solution

The wavelength in vacuum (or air) is $\lambda = c/f = (3 \times 10^8$ m/s$)/(60$ Hz$) = 5 \times 10^6$ m, almost as large as the radius of the Earth.

Problem

25. What would be the electric field strength in an electromagnetic wave whose magnetic field equalled that of Earth, about 50 μT?

Solution

For a wave in free space, Equation 34-18 gives $E = cB = (3 \times 10^8$ m/s$)(0.5 \times 10^{-4}$ T$) = 15$ kV/m.

Section 34-8   Polarization

Problem

29. Polarized light is incident on a sheet of polarizing material, and only 20% of the light gets through. What is the angle between the electric field and the polarization axis of the material?

Solution

From the law of Malus (Equation 34-19), $S/S_0 = \cos^2\theta = 20\%$, or $\theta = \cos^{-1}(\sqrt{0.2}) = 63.4°$.

Problem

33. Unpolarized light of intensity S_0 passes first through a polarizer with its polarization axis vertical, then through one with its axis at 35° to the vertical. What is the light intensity after the second polarizer?

Solution

Only 50% (one half the intensity) of the unpolarized light is transmitted through the first polarizer, and the second cuts this down by $\cos^2 35°$. Therefore $\frac{1}{2}\cos^2 35° = 33.6\%$ of the unpolarized intensity gets through both polarizers.

Problem

37. Polarized light with average intensity S_0 passes through a sheet of polarizing material which is rotating at 10 rev/s. At time $t = 0$ the polarization axis is aligned with the incident polarization. Write an expression for the transmitted intensity as a function of time.

Solution

Because the frequency of light is much greater than that of the rotating polarizer (5×10^{14} Hz $\gg$ 10 Hz), the law of Malus relates the average light intensities (see discussion leading to Equation 34-21a). Thus, $S = S_0\cos^2\theta$. For $\theta = \omega t$, where $\omega = 2\pi\times10$ s^{-1}, $S = S_0\cos^2(20\pi s^{-1})t = \frac{1}{2}S_0[1 + \cos(40\pi s^{-1})t]$.

Section 34-10   Energy in Electromagnetic Waves

Problem

41. A radio receiver can pick up signals with peak electric fields as low as 450 μV/m. What is the average intensity of such a signal?

Solution

From Equation 34-21b, $\bar{S} = E_p^2/2\mu_0 c = (450\ \mu V/m)^2/(8\pi\times10^{-7}$ H/m$)(3\times10^8$ m/s$) = 2.69\times10^{-10}$ W/m^2.

Problem

45. The United States' safety standard for continuous exposure to microwave radiation is 10 mW/cm^2. The glass door of a microwave oven measures 40 cm by 17 cm and is covered with a metal screen that blocks microwaves. What fraction of the oven's 625-W microwave power can leak through the door window without exceeding the safe exposure to someone right outside the door? Assume the power leaks uniformly through the window area.

Solution

The power corresponding to the safety standard of intensity, uniformly distributed over the window area, is (10 mW/cm^2)(40 $\times$ 17 cm^2) = 6.8 W, which is 1.09% of the microwave's 625 W power output.

Problem

49. During its 1989 encounter with Neptune, the Voyager 2 spacecraft was 4.5×10^9 km from Earth (Fig. 34-30). Its images of Neptune were broadcast by a radio transmitter with a mere 21-W average power output. What would be (a) the average intensity and (b) the peak electric field received at Earth if the transmitter broadcast equally in all directions? (The received signal was actually somewhat stronger because Voyager used a directional antenna.)

Solution

(a) The average intensity at a distance r from an isotropic emitter is (Equation 34-22) $\bar{S} = \mathcal{P}/4\pi r^2 = 21$ W/$4\pi(4.5\times10^{12}$ m$)^2 = 8.25\times10^{-26}$ W/m^2. (b) This corresponds to a peak electric field of only (Equation 34-21b) $E_p = \sqrt{2\mu_0 c\bar{S}} = 7.89\times10^{-12}$ V/m. (The one-way travel time querried in the caption of Fig. 34-30 is $r/c = 4$h, 10 min.)

Problem

53. A typical fluorescent lamp is a little over 1 m long and a few cm in diameter. How do you expect the light intensity to vary with distance (a) near the lamp but not near either end and (b) far from the lamp?

Solution

(a) Near the lamp, but far from its ends, light waves travel approximately radially outwards from the tube axis. The power crossing two co-axial cylindrical patches is the same, but the area of each patch is proportional to the radius. Therefore, the intensity varies like $1/r$. ($S_1 A_1 = S_2 A_2 = S_1\theta r_1\ell = S_2\theta r_2\ell$.) This is the same relation as depicated in Fig. 16-18c. (b) Very far away, the lamp appears as a point source and Equation 34-22 holds, so the intensity varies like $1/r^2$.

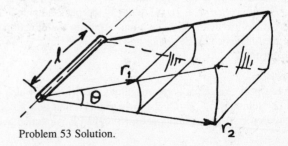

Problem 53 Solution.

Section 34-11   Wave Momentum and Radiation Pressure

Problem

57. The average intensity of noonday sunlight is about 1 kW/m². What is the radiation force on a solar collector measuring 60 cm by 2.5 m if it is oriented at right angles to the incident light and absorbs all the light?

Solution

For sunlight incident normally on a perfect absorber, $P_{rad} = \bar{S}/c$ (Equation 34-24). Therefore, the force on the solar collector is $P_{rad}A = (1 \text{ kW/m}^2)(0.6 \times 2.5 \text{ m}^2)/(3\times10^8 \text{ m/s}) = 5 \ \mu\text{N}$.

Paired Problems

Problem

61. Find the peak electric and magnetic fields 1.5 m from a 60-W light bulb that radiates equally in all directions.

Solution

For an isotropic source of electromagnetic waves (in a medium with vacuum permittivity and permeability) Equations 34-21b and 22 give $\bar{S} = \mathcal{P}/4\pi r^2 = E_p^2/2\mu_0 c$, therefore $E_p = \sqrt{2\mu_0 c\mathcal{P}/4\pi r^2} = (2\times10^{-7} \times 3\times10^8 \times 60)^{1/2}(\text{V/m}) \div (1.5 \text{ m}) = 40 \text{ V/m}$. Then Equation 34-18 gives $B_p = E_p/c = 133 \text{ nT}$.

Problem

65. What is the radiation force on the door of a microwave oven if 625 W of microwave power hits the door at right angles and is reflected?

Solution

The radiation pressure for normally incident, perfectly reflected electromagnetic waves is $2\bar{S}/c$. We suppose that the microwave power is uniformly spread over the area of the door (as for plane waves), so $\bar{S} = \mathcal{P}/A$. Then the force on the door is simply $P_{rad}A = (2\bar{S}/c)A = 2\mathcal{P}/c = 2(625 \text{ W})/(3\times10^8 \text{ m/s}) = 4.17 \ \mu\text{N}$.

Supplementary Problems

Problem

69. Maxwell's equations in a dielectric resemble those in vacuum (Equations 34-6 through 34-9), but with ϕ_E in Ampère's law replaced by $\kappa\phi_E$, where κ is the dieletric constant introduced in Chapter 26. Show that the speed of electromagnetic waves in such a dielectric is $c/\sqrt{\kappa}$.

Solution

The effect of a linear, isotropic, homogeneous dielectric medium in Gauss's law is to replace ε_0 by $\kappa\varepsilon_0$. Maxwell defined the displacement current in a dielectric analogously, as $\kappa\varepsilon_0 d\phi_E/dt$. Therefore, Maxwell's equations in a dielectric medium (containing no free charge or conduction currents) are just those in Table 34-2 with $\kappa\varepsilon_0$ replacing ε_0. The discussion in Sections 34-4 through 6 applies to waves in a dielectric medium, with the same replacement. In particular, the wave speed (Equation 34-16) becomes $1/\sqrt{\kappa\varepsilon_0\mu_0} = c/\sqrt{\kappa}$. (In Section 35-3, $\sqrt{\kappa}$ is defined as the index of refraction of the medium.)

Problem

73. The peak electric field measured at 8.0 cm from a light source is 150 W/m², while at 12 cm it measures 122 W/m². Describe the shape of the source.

Solution

The intensity is proportional to the square of the peak electric field (Equation 34-20b), so the given data implies that the ratio of intensities is proportional to that of the inverse distances, $(150/122)^2 = 1.51 \approx (12/8)$. Thus, $\bar{S}r$ is roughly constant. The intensity near a long, cylindrically symmetric source, where r is the axial distance, has this space dependence (see Problem 53). ●

● **PART 4** CUMULATIVE PROBLEMS

Problem

1. An air-insulated parallel-plate capacitor has plate area 100 cm² and spacing 0.50 cm. The capacitor is charged to a certain voltage and then disconnected from the charging battery. A thin-walled, nonconducting box of the same dimensions as the capacitor is filled with water at 20.00°C. The box is released at the edge of the capacitor and moves without friction into the capacitor (Fig. 1). When it reaches equilibrium the water temperature is 21.50°C. What was the original voltage on the capacitor?

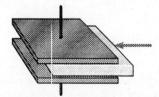

FIGURE 1   Cumulative Problem 1.

Solution

We assume that the entire difference between the initial and final values of the electrostatic energy stored in the capacitor is eventually dissipated as heat in the water (the dieletric medium), as mentioned on page 661. Thus, $U_0 - U = mc \, \Delta T$. From Equation 26-12, $U_0 - U = U_0 - U_0/\kappa = \frac{1}{2}C_0 V_0^2 (\kappa - 1)/\kappa$, where $C_0 = \varepsilon_0 A/d = (8.85 \text{ pF/m})(10^{-2} \text{ m}^2)/(5\times10^{-3} \text{ m}) = 17.7 \text{ pF}$, and $\kappa = 78$ for water, see Table 26-1. The amount of water is $m = (1 \text{ g/cm}^3)(100 \text{ cm}^2)(0.5 \text{ cm}) = 50 \text{ g}$, so putting everything together and solving for V_0, we find:

$$V_0 = \sqrt{\frac{2mc \, \Delta T}{C_0(\kappa - 1)/\kappa}} = \sqrt{\frac{2(50 \text{ g})(4.184 \text{ J/g.°C})(1.5°\text{C})}{(17.7 \text{ pF})(77/78)}}$$

$$= 5.99 \text{ MV.}$$

Problem

5. A coaxial cable consists of an inner conductor of radius a and an outer conductor of radius b; the space between the conductors is filled with insulation of dielectric constant κ (Fig. 4). The cable's axis is the z axis. The cable is used to carry electromagnetic energy from a radio transmitter to a broadcasting antenna. The electric field between the conductors points radially from the axis, and is given by $E = E_0\frac{a}{r}\cos(kz - \omega t)$. The magnetic field encircles the axis, and is given by $B = B_0\frac{a}{r}\cos(kz - \omega t)$. Here $E_0, B_0, k,$ and ω are constants. (a) Show, using appropriate closed surfaces and loops, that these fields satisfy Maxwell's equations. Your result shows that the cable acts as a "waveguide," confining an electromagnetic wave to the space between the conductors. (b) Find an expression for the speed at which the wave propagates along the cable.

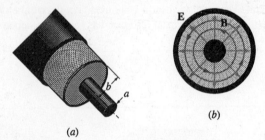

(a)

(b)

FIGURE 4 Cumulative Problem 5.

Solution

The method for (a) demonstrating that the given fields in a coaxial cable (so-called transverse electric and magnetic waves, or TEM-waves) satisfy Maxwell's equations, and (b) finding their speed, is the same as that used in Sections 34-4 and 5, except that cylindrical coordinates $r, \theta, z,$ and Maxwell's equations in the dielectric (see solution to Prob-

lem 34-69) must be used. Consider an infinitessimal volume of dielectric at a point (r, θ, z) bounded by mutually orthogonol coordinate displacements $dr, r \, d\theta, dz$. The dielectric contains no free charges or conduction currents, so the given radial electric field and circulating magnetic field satisfy Gauss's laws for electricity and magnetism in the dielectric.

In Faraday's law, take a CCW loop with sides dr and dz, at fixed θ, so that the normal to the area it bounds is parallel to $\mathbf{B}$, as shown. Then, since $\mathbf{E} \perp d\mathbf{z}$ and $\mathbf{B} \perp d\mathbf{r}$ and $d\mathbf{z}$,

$$\oint \mathbf{E} \cdot d\ell = -E \, dr + \left(E + \frac{\partial E}{\partial z} dz\right) dr = \frac{\partial E}{\partial z} dz \, dr$$

$$= -\frac{d\phi_B}{dt} = -\frac{d}{dt}(B \, dz \, dr) = -\frac{\partial B}{\partial t} dz \, dr,$$

or $\partial E/\partial z = -(E_0ak/r)\sin(kz - \omega t) = -\partial B/\partial t = -(B_0a\omega/r)\sin(kz - \omega t)$. Thus, $E_0k = B_0\omega$ (same as Equation 34-14).

In Ampere's law, take a CCW loop with sides $r \, d\theta$ and dz, at fixed r, so that the normal to its area is parallel to $\mathbf{E}$. Then, since $\mathbf{B} \perp d\mathbf{z}$ and $\mathbf{E} \perp r \, d\theta$ and $d\mathbf{z}$,

$$\oint \mathbf{B} \cdot d\ell = Br \, d\theta - \left(B + \frac{\partial B}{\partial z} dz\right) r \, d\theta = -\frac{\partial B}{\partial z} r \, d\theta \, dz$$

$$= \kappa\varepsilon_0\mu_0\frac{d\phi_E}{dt} = \kappa\varepsilon_0\mu_0\frac{d}{dt}(Er \, d\theta \, dz) = \kappa\varepsilon_0\mu_0\frac{\partial E}{\partial t}r \, d\theta \, dz,$$

or $-\partial B/\partial z = (B_0ak/r)\sin(kz - \omega t) = \kappa\varepsilon_0\mu_0(\partial E/\partial t) = \kappa\varepsilon_0\mu_0(E_0a\omega/r)\sin(kz - \omega t)$. Thus, $B_0k = \kappa\varepsilon_0\mu_0E_0\omega$ (same as Equation 34-15 except for κ).

Therefore, Maxwell's equations are satisfied provided E_0, B_0, k and ω are related by $B_0\omega = E_0k$ and $B_0k = E_0\omega\kappa\varepsilon_0\mu_0$. The wave speed follows by dividing these equations to eliminate the amplitudes. Then $\omega/k = k/\omega\kappa\varepsilon_0\mu_0$, or $(\omega/k)^2 = 1/\kappa\varepsilon_0\mu_0 = v^2$, and $v = 1/\sqrt{\kappa\varepsilon_0\mu_0} = c/\sqrt{\kappa}$.

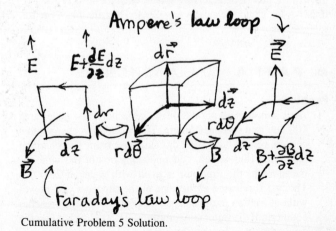

Cumulative Problem 5 Solution.

PART 5 OPTICS

● CHAPTER 35 REFLECTION AND REFRACTION

Section 35-2 Reflection

Problem

1. Through what angle should you rotate a mirror in order that a reflected ray rotate through 30°?

Solution

Since $\theta_1 = \theta_1'$ for specular reflection, (Equation 35-1) a reflected ray is deviated by $\phi = 180° - 2\theta_1$ from the incident direction. If rotating the mirror changes θ_1 by $\Delta\theta_1$, then the reflected ray is deviated by $\Delta\phi = |-2\Delta\theta_1|$ or twice this amount. Thus, if $\Delta\phi = 30°$, $|\Delta\theta_1| = 15°$.

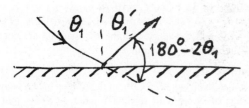

Problem 1 Solution.

Problem

5. Suppose the angle in Fig. 35-33 is changed to 75°. A ray enters the mirror system parallel to the axis. (a) How many reflections does it make? (b) Through what angle is it turned when it exits the system?

Solution

Now, after the first reflection, the ray leaves the top mirror at a grazing angle of $37\frac{1}{2}°$, and so makes a grazing angle of $180° - 75° - 37\frac{1}{2}° = 67\frac{1}{2}°$ with the bottom mirror. It is therefore deflected through an angle of $2(37\frac{1}{2}°) + 2(67\frac{1}{2})° = 210°$ CW, as it exits the system, after being reflected once from each mirror.

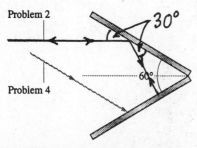

FIGURE 35-33 Problem 2 Solution.

Section 35-3 Refraction

Problem

9. Information in a compact disc is stored in "pits" whose depth is essentially one-fourth of the wavelength of the laser light used to "read" the information. That wavelength is 780 nm in air, but the wavelength on which the pit depth is based is measured in the $n = 1.55$ plastic that makes up most of the disc. Find the pit depth.

Solution

Equation 35-4 and the reasoning in Example 35-4 show that the wavelength in the plastic is $\lambda = \lambda_{air}/n = 780$ nm$/1.55 = 503$ nm. The pit depth is one quarter of this, or 126 nm.

Problem

13. A block of glass with $n = 1.52$ is submerged in one of the liquids listed in Table 35-1. For a ray striking the glass with incidence angle 31.5°, the angle of refraction is 27.9°. What is the liquid?

Solution

With the unknown liquid as medium 1, and the glass as medium 2, Snell's law gives $n_1 = n_2 \sin\theta_2/\sin\theta_1 = 1.52 \times \sin 27.9°/\sin 31.5° = 1.361$. This is the same as ethyl alcohol.

Problem

17. You're standing 2.3 m horizontally from the edge of a 4.5-m-deep lake, with your eyes 1.7 m above the water surface. A diver holding a flashlight at the lake bottom shines the light so you can see it. If the light in the water makes a 42° angle with the vertical, at what horizontal distance is the diver from the edge of the lake?

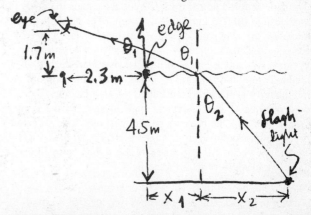

Problem 17 Solution.

Solution

Snell's law gives the angle of refraction (θ_1) in terms of the angle of incidence ($\theta_2 = 42°$) for the light path from the flashlight to your eye. These can be related to the other given distances by means of a carefully drawn diagram. Thus, $\theta_1 = \sin^{-1}(n_2 \sin \theta_2 / n_1) = \sin^{-1}(1.333 \sin 42°) = 63.1°$, where we used indices of refraction from Table 35-1, with $n_1 \approx 1$ for air. The geometry of the diagram makes the horizontal distances apparent: $\tan \theta_1 = (2.3 \text{ m} + x_1)/(1.7 \text{ m})$, or $x_1 = (1.7 \text{ m}) \tan 63.1° - 2.3 \text{ m} = 1.05 \text{ m}$, and $\tan \theta_2 = x_2/(4.5 \text{ m})$, or $x_2 = (4.5 \text{ m}) \tan 42° = 4.05 \text{ m}$. The total horizontal distance from the edge is $x_1 + x_2 = 5.11 \text{ m}$.

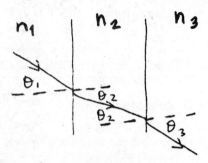

Problem 25 Solution.

Section 35-4   Total Internal Reflection

Problem

21. Find the critical angle for total internal refraction in (a) ice, (b) polystyrene, and (c) rutile. Assume the surrounding medium is air.

Solution

For $n_{air} \approx 1$, the critical angle for total internal reflection in a medium of refractive index n is $\theta_c = \sin^{-1}(1/n)$. (air is medium-2 in Equation 35-5). From Table 35-1, $n = 1.309$ (ice), 1.49 (polystyrene), and 2.62 (rutile), so $\theta_c = \sin^{-1}(1/1.309) = 49.8°$, $42.2°$ and $22.4°$, respectively, for these media.

Problem

25. Light propagating in a medium with refractive index n_1 encounters a parallel-sided slab with index n_2. On the other side is a third medium with index $n_3 < n_1$. Show that the condition for avoiding internal reflection at *both* interfaces is that the incidence angle at the n_1-n_2 interface be less than the critical angle for an n_1-n_3 interface. In other words, the index of the intermediate material doesn't matter.

Solution

Since medium-2 has parallel interfaces with media-1 and 3, the angle of refraction at the 1-2 interface equals the angle of incidence at the 2-3 interface, as shown. (The normals to the interfaces are also parallel, so the alternate angles, marked θ_2, are equal.) Thus Snell's law implies $n_1 \sin \theta_1 = n_2 \sin \theta_2 = n_3 \sin \theta_3$, so that the angles in media-1 and 3 are related as if media-2 were not present. Of course, there are conditions on the intensity of the light transmitted through medium-2 which do depend on n_2 and the critical angle, if any, at the 1-2 interface. If $n_2 < n_1$, the phenomenon of frustrated total reflection (i.e. transmission of light for angles greater than the critical angle) may occur if the thickness of the slab is on the order of a few wavelengths of the incident light.

Problem

29. What is the speed of light in a material for which the critical angle at an interface with air is 61°?

Solution

From Equations 35-5 and 2, $\sin \theta_c = n_{air}/n \approx \frac{1}{n} = v/c$, so $v = c \sin \theta_c = (3 \times 10^8 \text{ m/s}) \sin 61° = 2.62 \times 10^8 \text{ m/s}$. (The critical angle and the speed of light in a material are both related to the index of refraction.)

Problem

33. A scuba diver sets off a camera flash a distance h below the surface of water with refractive index n. Show that light emerges from the water surface through a circle of diameter $2h/\sqrt{n^2 - 1}$.

Solution

Light from the flash will strike the water surface at the critical angle for a distance $r = h \tan \theta_c$ from a point directly over the flash. Therefore, the diameter of the circle through which light emerges is $2r = 2h \tan \theta_c$. But $\sin \theta_c = 1/n$ (Equation 33-5 at the water-air interface), and $\tan^2 \theta_c = (\csc^2 \theta_c - 1)^{-1}$ (a trigonometric identity), so we can write $2r = 2h/\sqrt{n^2 - 1}$.

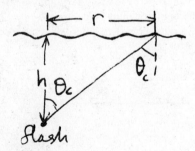

Problem 33 Solution.

Section 35-5 Dispersion

Problem

37. Two of the prominent spectral lines—discrete wavelengths of light—emitted by glowing hydrogen are hydrogen-α at 656.3 nm and hydrogen-β at 486.1 nm. Light from glowing hydrogen passes through a prism like that of Fig. 35-21, then falls on a screen 1.0 m from the prism. How far apart will these two spectral lines be? Use Fig. 35-19 for the refractive index.

Solution

The angular dispersion of H_α and H_β light in the prism of Fig. 35-21 can be found from the analysis in Example 35-6. For normal incidence on the prism, rays emerge with refraction angles of $\sin^{-1}(n \sin 40°)$. From Fig. 35-19, we estimate that $n_\alpha = 1.517$ and $n_\beta = 1.528$, so the angular dispersion is $\gamma = 79.2° - 77.2° = 1.98°$. We can assume that the size of the prism is small compared to the distance, r, to the screen, so the separation on the screen corresponding to γ is $\Delta x = \gamma \cdot r = (1.98°)(\pi/180°)(1 \text{ m}) = 3.45 \text{ cm}$.

Section 35-6 Reflection Coefficients and the Polarizing Angle

Problem

41. What is the refractive index of a material that transmits 92.4% of the light normally incident on it from air?

Solution

Equation 35-7, with $n_1 = 1$ for air, is $T = 4n/(1 + n)^2$. This is a quadratic equation for n, which has solution $n = (T^{-1/2} + \sqrt{T^{-1} - 1})^2$. (Note that only one root has $n \geq 1$, as required by Equation 35-2.) For $T = 0.924$, one finds $n = 1.76$. One can also obtain this result using Equation 35-6, which can be written as $n = (1 + \sqrt{R})/(1 - \sqrt{R}) = (1 + \sqrt{1 - T})/(1 - \sqrt{1 - T}) = (1 + \sqrt{1 - T})^2/T$, equivalent to the above solution for n.

Problem

45. The reflection coefficient for normally incident light is the same when a block of plastic is submerged in water and in diiodomethane. What is the refractive index of the plastic?

Solution

Equation 35-6 can be solved for the relative index of refraction, n_2/n_1 (medium-2 relative to medium-1) to get $n_2/n_1 = (1 + \sqrt{R})/(1 - \sqrt{R})$, where $n_2 > n_1$. Clearly, to get the same reflection coefficient in two different liquids, the index of refraction of the plastic must be greater than the index of one liquid (water) and less than that of the other (diiodomethane). Thus, $n_p/n_w = n_d/n_p$, or $n_p = \sqrt{n_w n_d} = \sqrt{(1.333)(1.738)} = 1.522$.

Paired Problems

Problem

49. Light propagating in air strikes a transparent crystal at incidence angle 35°. If the angle of refraction is 22°, what is the speed of light in the crystal?

Solution

Combining Equations 35-2 and 3 (or using the form of Snell's law preceding Equation 35-2), we find $v = c/n = c \sin \theta_2/\sin \theta_1 = (3 \times 10^8 \text{ m/s})\sin 22°/\sin 35° = 1.96 \times 10^8 \text{ m/s}$. (Note: $n_{\text{air}} \approx 1$.)

Problem

53. Light is incident from air on the flat wall of a polystyrene water tank. If the incidence angle is 40°, what angle does the light make with the tank normal in the water?

Solution

If the plastic wall of the tank has parallel faces, it does not effect the angle of refraction in the water (see solution to Problem 25). Then $n_1 \sin \theta_1 = n_3 \sin \theta_3$, or $\theta_3 = \sin^{-1}(n_1 \sin \theta_1/n_3) = \sin^{-1}(\sin 40°/1.333) = 28.8°$. (Media-1, 2, and 3 are air, polystyrene and water, respectively, in the solution to Problem 25.)

Problem

57. Repeat Problem 20 for the case $n = 1.75$, $\alpha = 40°$, and $\theta_1 = 25°$.

Solution

A general treatment of refraction through a prism of index of refraction $n_2 = n$, surrounded by air of index $n_1 = 1$, for the geometry of Fig. 35-36, is given in the solution to Problem 61. For $n = 1.75$, $\alpha = 40°$, and $\theta_1 = 25°$, the other angles defined there are:
$\theta_2 = \sin^{-1}(\sin \theta_1/n) = \sin^{-1}(\sin 25°/1.75) = 14.0°$,
$\phi_2 = \alpha - \theta_2 = 40° - 14.0° = 26.0°$,
$\phi_1 = \sin^{-1}(n \sin \phi_2) = \sin^{-1}(1.75 \sin 26.0°) = 50.2°$,
and $\delta = \theta_1 + \phi_1 - \alpha = 35.2°$.
(Note that ϕ_2 is less than the crictical angle for this prism, which is $\sin^{-1}(1/1.75) = 34.8°$.)

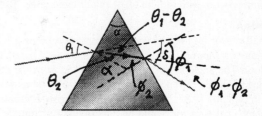

FIGURE 35-36 Problem 61 Solution.

Supplementary Problems

Problem

61. Light is incident with incidence angle θ_1 on a prism with apex angle α and refractive index n, as shown in Fig. 35-36. Show that the angle δ through which the outgoing beam deviates from the incident beam is given by

$$\delta = \theta_1 - \alpha + \sin^{-1}\left\{n \sin\left[\alpha - \sin^{-1}\left(\frac{\sin\theta_1}{n}\right)\right]\right\}.$$

Assume the surrounding medium has $n = 1$.

Solution

It is an exercise in ray tracing to determine the angles shown superposed on Fig. 35-36, from Snell's law and plane geometry: $\theta_2 = \sin^{-1}(\sin\theta_1/n)$ (Snell's law for the first refraction, with $n_1 = 1$ and $n_2 = n$), $\phi_2 = \alpha - \theta_2$ (α is the exterior angle to the triangle formed by the ray segment in the prism and the normals to the surfaces), $\phi_1 = \sin^{-1}(n \sin\phi_2)$ (Snell's law for the second refraction), and finally, $\delta = \theta_1 - \theta_2 + \phi_1 - \phi_2 = \theta_1 + \phi_1 - \alpha$ (the total deflection is the sum of the deflections at each refraction, clockwise deflection positive in Fig. 35-36). Writing down these steps in reverse order, substituting for each angle, one gets:

$$\delta = \theta_1 + \sin^{-1}\left[n \sin\left(\alpha - \sin^{-1}\left(\frac{\sin\theta_1}{n}\right)\right)\right] - \alpha.$$

(Note: Problems 20, 36, 55-58, and also Problems 27, 28, 30, 37 and 38, involve ray tracing through prisms, in which a sim-ilar, but not identical, analysis is useful. Of course, ϕ_1 must be a real angle, i.e., less than 90°, or total internal reflection occurs instead of the second refraction. This is determined by the given values of n, α and θ_1.)

Problem

65. (a) Differentiate the result of the preceding problem to show that the maximum value of ϕ occurs when the incidence angle θ is given by $\cos^2\theta = \frac{1}{3}(n^2 - 1)$. (b) Use this result and that of the preceding problem to find ϕ_{max} in water with $n = 1.333$.

Solution

One can differentiate ϕ, with respect to θ, directly, by using $d[\sin^{-1}(x/a)]/dx = 1/\sqrt{a^2 - x^2}$ (see the integral table in Appendix A.) Then

$$\frac{d\phi}{d\theta} = \frac{4}{\sqrt{n^2 - \sin^2\theta}}\frac{d(\sin\theta)}{d\theta} - 2 = \frac{4\cos\theta}{\sqrt{n^2 - \sin^2\theta}} - 2.$$

The condition for a maximum, $d\phi/d\theta = 0$, implies that $2\cos\theta_m = \sqrt{n^2 - \sin^2\theta_m}$, or $4\cos^2\theta_m = n^2 - (1 - \cos^2\theta)$, so $\cos^2\theta_m = \frac{1}{3}(n^2 - 1)$. If this value of θ is substituted into the expression for ϕ, after noting that $\sin^2\theta_m = \frac{1}{3}(4 - n^2)$, one gets $\phi_{max} = 4\sin^{-1}(\sqrt{(4 - n^2)/3n^2}) - 2\cos^{-1}\sqrt{(n^2 - 1)/3}$, which equals 42.1° for $n = 1.333$. (This is the average angle, above the anti-solar direction, that an observer sees a rainbow, because n is the average index of refraction for visible wavelengths.) ●

● CHAPTER 36   IMAGE FORMATION AND OPTICAL INSTRUMENTS

Sections 36-1 and 36-2   Plane and Curved Mirrors

Problem

1. A shoe store uses small floor-level mirrors to let customers view prospective purchases. At what angle should such a mirror be inclined so that a person standing 50 cm from the mirror with eyes 140 cm off the floor can see her feet?

Solution

A small mirror (M) on the floor intercepts rays coming from a customer's shoes (O), which are traveling nearly parallel to the floor. The angle to the customer's eye (E) from the mirror is twice the angle of reflection, so $\tan 2\alpha = h/d$, or $\alpha = \frac{1}{2}\tan^{-1}(140/50) = 35.2°$, for the given distances. Therefore, the plane of the mirror should be tilted by 35.2° from the vertical to provide the customer with a floor-level view of her shoes.

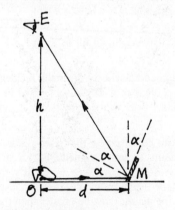

Problem 1 Solution.

Problem

5. An object is five focal lengths from a concave mirror. (a) How do the object and image heights compare? (b) Is the image upright or inverted?

Solution

(a) One can use Equation 36-2b directly to yield $h'/h = -f/(5f - f) = -\frac{1}{4}$. (b) A negative magnification applies to a real, inverted image.

Problem

9. A 12-mm-high object is 10 cm from a concave mirror with focal length 17 cm. (a) Where, (b) how high, and (c) what type is its image?

Solution

For an object on the mirror's axis, Equations 36-3 and 4 give: (a) $\ell' = f\ell/(\ell - f) = (17 \times 10 \text{ cm})/(10 - 17) = -24.3$ cm (i.e., behind the mirror). (c) A negative image distance indicates a virtual image. (b) $M = -\ell'/\ell = 24.3/10 = 2.43 = h'/h$, so the image is upright and 2.43×12 mm $= 29.1$ mm high.

Problem

13. At what two distances could you place an object from a 45-cm-focal-length concave mirror in order to get an image 1.5 times the object's size?

Solution

Substitute $M = \pm 1.5$ (for virtual and real images) into Equation 36-2b, and solve for the object distance. Then $\ell = f(1 - 1/M) = 45$ cm $(1 \mp 1/1.5) = 15$ cm (virtual image) or 75 cm (real image). (See Problems 11 and 15.)

Section 36-3 Lenses

Problem

17. A lens with 50-cm focal length produces a real image the same size as the object. How far from the lens are image and object?

Solution

For a real image the same size as the object, $h' = -h$, so $M = -1 = -\ell'/\ell$, or $\ell' = \ell$. The lens equation (Equation 36-7) then gives $(1/\ell) + (1/\ell') = 2/\ell = 1/f$, or $\ell = \ell' = 2f = 100$ cm.

Problem

21. A simple camera uses a single converging lens to focus an image on its film. If the focal length of the lens is 45 mm, what should be the lens-to-film distance for the camera to focus on an object 80 cm from the lens?

Solution

Set $\ell = 80$ cm and $f = 45$ mm in the lens equation and solve for ℓ'. The result is $\ell' = \ell f/(\ell - f) = (80 \times 45 \text{ cm})/(800 - 45) = 4.77$ cm.

Problem

25. A lens has focal length $f = 35$ cm. Find the type and height of the image produced when a 2.2-cm-high object is placed at distances (a) $f + 10$ cm and (b) $f - 10$ cm.

Solution

The lens equation and magnification for a thin (converging, i.e., positive f) lens, Equations 36-6 and 7, give $M = -\ell'/\ell = -f/(\ell - f)$, so $h' = Mh = -fh/(\ell - f)$. (a) If $f = 35$ cm and $\ell = f + 10$ cm, then $h' = -(35 \text{ cm})(2.2/10) = -7.7$ cm. A negative image height signifies a real, inverted image. (b) If $\ell = f - 10$ cm, then $h' = -(35 \text{ cm})(2.2)/(-10) = +7.7$ cm, which represents a virtual, erect image of the same size.

Section 36-4 The Lensmaker's Formula

Problem

29. You're standing in a wading pool and your feet appear to be 30 cm below the surface. How deep is the pool?

Solution

The image formed by a single refracting interface between two media, for paraxial rays, is described by Equation 36-8, with sign conventions for distances defined in the paragraph following. For the flat surface of the wading pool, $R = \infty$, $n_1 = 1.333$ for water, ℓ is the depth of the pool (your feet, the object, are on the bottom), $n_2 = 1$ for air, and $\ell' = -30$ cm (for a virtual image at the apparent depth). Thus, $(1.333/\ell) + (1/(-30 \text{ cm})) = (n_2 - n_1)/\infty = 0$, or $\ell = 40$ cm. (This problem could also be solved directly from Snell's law, as in the previous chapter, without the paraxial ray approximation.)

Problem

33. Rework Example 36-6 for a fish 15 cm from the *far* wall of the tank.

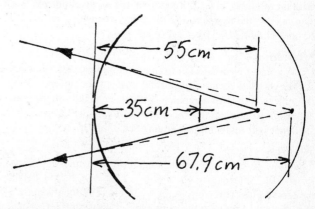

Problem 33 Solution.

Solution

As in Example 36-6, use Equation 36-8 with $R = -35$ cm, $n_1 = 1.333$, $n_2 = 1$, but with $\ell = 70$ cm $- 15$ cm $= 55$ cm (as distance from the near wall). Then $\ell' = [(1 - 1.333)/(-35 \text{ cm}) - 1.333/55 \text{ cm}]^{-1} = -67.9$ cm. In this case, the object is closer to the refracting surface than its image (see sketch and compare with Fig. 36-29b).

Problem

37. Two lenses made of the same material have the same focal length. One is plano-convex, the other double convex with both curvatures the same. How do the curvature radii of the two lenses compare?

Solution

The plano-convex lens, $R_1 = R$ and $R_2 = \infty$ (or $R_1 = \infty$ and $R_2 = -R$), has focal length $R/(n - 1)$, while the double convex lens, $R_1 = -R_2 = R'$, has focal length $R'/2(n - 1)$. (See Equation 36-10.) These are equal if $R = \frac{1}{2}R'$ (i.e., the plano-convex lens is more strongly curved, but both lenses are assumed thin).

Problem

41. A plano-convex lens has curvature radius 20 cm and is made from glass with $n = 1.5$. Use the generalized lens-maker's formula given in Problem 71 to find the focal length when the lens is (a) in air, (b) submerged in water ($n = 1.333$) and (c) embedded in glass with $n = 1.7$. Comment on the sign of your answer to (c).

Solution

In the generalized lens maker's formula, $n_r = n_{\text{lens}}/n_{\text{ext}}$ is the index of refraction of the lens relative to the external medium. For a plano-convex lens, $R_1 = 20$ cm and $R_2 = \infty$ (or $R_1 = \infty$ and $R_2 = -20$ cm) so $f = 20$ cm$/(n_r - 1)$. (a) In air, the relative index of refraction of the lens is $1.5/1$, so $f = 40$ cm. (b) In water, $n_r = 1.5/1.333$ and $f = 160$ cm. (c) In a medium of higher index of refraction, the relative index is less than one, so the lens acts as a diverging lens with $f = 20$ cm$/[(1.5/1.7) - 1] = -170$ cm.

Problem

45. An object placed 15 cm from a plano-convex lens made of crown glass focuses to a virtual image twice the size of the object. If the lens is replaced with an identically shaped one made from diamond, what type of image will appear and what will be its magnification? See Table 35-1.

Solution

Equations 36-6 and 2b for the magnification (which is positive for a virtual image) give $M_g = 2 = -\ell'/\ell = f_g/(f_g - \ell)$ for the crown glass lens, so $f_g = 2\ell = 30$ cm. The focal length of a diamond lens with the same radii of curvature is $f_d = (n_g - 1)f_g/(n_d - 1) = 30$ cm $(1.520 - 1)/(2.419 - 1) = 11.0$ cm (use Equation 36-8 and Table 35-1). An object

15 cm from the diamond lens produces a real, inverted image (negative M) magnified by $M_d = 11.0/(11.0 - 15) = -2.74$.

Section 36-5   Optical Instruments

Problem

49. A camera's zoom lens covers the focal length range from 38 mm to 110 mm. (a) You point the camera at a distant object and photograph it first at 38 mm and then with the camera zoomed out to 110 mm. Compare the sizes of its images on the two photos. (b) If the camera's lowest f-ratio is 3.8 at 38 mm, what is it at 110 mm? Assume the effective lens area doesn't change.

Solution

(a) The image size can be determined from the lens equation, $h'/h = -\ell'/\ell = -f/(\ell - f)$. The ratio of the image sizes, for two different focal lengths, is $h'_1/h'_2 = [-f_1/(\ell - f_1)] \div [-f_2/(\ell - f_2)] \approx f_1/f_2$, if $\ell \gg f_1$ or f_2 (as for a distant object). Then the image size at 110 mm is approximately $110/38 = 2.89$ times larger than at 38 mm. (b) The f-ratio is the focal length divided by the diameter of the lens, f/d. If d remains constant while the lens is zoomed, then $f_1/d = (f_1/f_2)(f_2/d)$, so the f-ratio at 110 mm is also 2.89 times the f-ratio at 38 mm. This lens is $f/3.8$ at 38 mm and $f/(2.89 \times 3.8)$ or $f/11$ at 110 mm. (Note: $f/3.8$ at 38 mm is just a numerical coincidence in this problem.)

Problem

53. A 300-power compound microscope has a 4.5-mm-focal-length objective lens. If the distance from objective to eyepiece is 10 cm, what should be the focal length of the eyepiece?

Solution

For a 300× microscope, with $f_o = 4.5$ mm and $L = 100$ mm we can solve Equation 36-12 for $f_e = (100/4.5)(25 \text{ cm}/300) = 1.85$ cm.

Paired Problems

Problem

57. (a) How far from a 1.2-m-focal length concave mirror should you place an object in order to get an inverted image 1.5 times the size of the object? (b) Where will the image be?

Solution

(a) Equation 36-2b for the mirror's magnification can be rewritten as $-1/M = (\ell/f) - 1$, so $\ell = f(1 - 1/M)$. For a concave mirror with $f = 1.2$ m > 0, and a real image with $M = -1.5$, one finds $\ell = (1.2 \text{ m})(1 + 1/1.5) = 2.0$ m. (b) From Equation 36-4, $\ell' = -M\ell = 1.5(2.0 \text{ m}) = 3.0$ m. (A real image is, of course, in front of the mirror.)

Problem

61. How far from a 1.6-m focal length concave mirror should you place an object to get an upright image magnified by a factor of 2.5?

Solution

The analysis in the solutions to Problems 57 and 59 shows that $\ell = f(1 - 1/M) = (1.6 \text{ m})(1 - 1/2.5) = 96$ cm.

Problem

65. An object is 68 cm from a plano-convex lens whose curved side has curvature radius 26 cm. The refractive index of the lens is 1.62. Where and of what type is the image?

Solution

The focal length of the lens is given by Equation 36-10, with $R_1 = 26$ cm and $R_2 = \infty$ (or $R_1 = \infty$ and $R_2 = -26$ cm), so $f^{-1} = (n - 1)/R_1 = (1.62 - 1)/26 \text{ cm} = (41.9 \text{ cm})^{-1}$. An object at $\ell = 68$ cm is imaged at $\ell'^{-1} = f^{-1} - \ell^{-1} = (41.9 \text{ cm})^{-1} - (68 \text{ cm})^{-1} = (109 \text{ cm})^{-1}$. This is a real, inverted image, on the opposite side of the lens from the object.

Supplementary Problems

Problem

69. Show that identical objects placed equal distances on either side of the focal point of a concave mirror or converging lens produce images of equal size. Are the images of the same type?

Solution

A concave mirror or a converging lens are both represented by positive focal lengths in the lens or mirror equations. For either, the object and image sizes are related by $h'/h = -\ell'/\ell = -f/(\ell - f)$. One can easily see that for $\ell - f = \pm x$ (where $0 < x < f$ is implicit in the statement of this problem), the image size is the same. However, for $\ell = f + x$, the image is real ($h' < 0$) and for $\ell = f - x$, it is virtual ($h' > 0$). (A lens has two symmetrically placed focal points; this problem makes sense for the one on the same side of the lens as the object.)

Problem

73. A Newtonian telescope like that of Fig. 36-48c has a primary mirror with 20-cm diameter and 1.2-m focal length. (a) Where should the flat diagonal mirror be placed to put the focus at the edge of the telescope tube? (b) What shape should the flat mirror have to minimize blockage of light to the primary?

Solution

(a) The focal point of the primary mirror is 1.2 m from the mirror apex, and the radius of the telescope tube (presumably the same diameter as the mirror) is 0.1 m. Since the total distance of the light path from the primary mirror to the focus is not changed by the insertion of the secondary mirror, the secondary mirror should be (at least) 0.1 m closer than the focal point, or 1.1 m from the primary mirror apex. (b) The cone of rays reflected by the circular primary mirror is sliced by the plane of the secondary mirror into an ellipse (recall the definition of conic sections), which is the shape which blocks the least amount of incoming light.

● CHAPTER 37 INTERFERENCE AND DIFFRACTION

Section 37-1 Interference

Problem

1. Find the minimum thickness of a soap film ($n = 1.33$) in which 550-nm light will undergo constructive interference.

Solution

The condition for constructive interference from a soap film is Equation 37-1a, in which the minimum thickness corresponds to the integer $m = 0$. Thus, $2nd_{min} = \frac{1}{2}\lambda$, or $d_{min} = \lambda/4n = 550 \text{ nm}/4(1.33) = 103$ nm. (Recall that Equation 37-1a applies to normal incidence on a thin film in air.)

Problem

5. As a soap bubble ($n = 1.33$) evaporates and thins, the reflected colors gradually disappear. (a) What is its thickness just as the last vestige of color vanishes? (b) What is the last color seen?

Solution

The minimum thickness of the bubble, which produces interference colors, is $d_{min} = \lambda_{min}/4n$, where λ_{min} is the shortest visible wavelength, normally 400 nm violet light. (See the solution to Problem 1.) Thus, $d_{min} = 400 \text{ nm}/4(1.33) = 75.2$ nm.

Problem

9. An oil film with refractive index 1.25 floats on water. The film thickness varies from 0.80 μm to 2.1 μm. If 630-nm light is incident normally on the film, at how many locations will it undergo enchanced reflection?

Solution

In a thin film of oil between air and water ($n_{air} < n_{oil} < n_{water}$), there are 180° phase changes for reflection at both boundaries. Therefore, for normally incident light, Equation 37-1b gives the condition for constructive interference, $d = m\lambda/2n = m(630 \text{ nm})/2(1.25) = (0.252 \text{ }\mu\text{m})m$. Varying thickness of 0.80 μm $\leq d \leq$ 2.1 μm implies that $3.17 \leq m \leq 8.33$. Since m is an integer, $4 \leq m \leq 8$, or 5 bright maxima occur at locations corresponding to the allowed integers from 4 to 8.

Problem

13. You apply a slight pressure with your finger to the upper of a pair of glass plates forming an air wedge as in Fig. 37-46. The wedge is illuminated from above with 500-nm light, and you place your finger where, initially, there is a dark band. If you push gently so the band becomes light, then dark, then light again, by how much have you deflected the plate?

Solution

The difference in the thickness of air in the wedge between a bright and an adjacent dark band is $\frac{1}{4}\lambda$ (one quarter wavelength in air), so the upper plate was depressed by $\frac{3}{4}\lambda = 375$ nm.

Problem

17. The evacuated box of the previous problem is filled with chlorine gas, whose refractive index is 1.000772. How many bright fringes pass a fixed point as the tube fills?

Solution

Since the wavelength of the light is different in a gas (e.g. chlorine or air) and in vacuum ($\lambda_{gas} = \lambda_{vac}/n_{gas}$) there is a difference in the number of wave cycles in the enclosed interferometer arm, when the box is evacuated or filled with gas. The light travels the length of the arm twice, out and back, and each cycle of difference results in one fringe shift. Thus, the number of fringes in the shift is

$$\frac{2 \times 42.5 \text{ cm}}{\lambda_{gas}} - \frac{2 \times 42.5 \text{ cm}}{\lambda_{vac}} = \frac{(n_{gas} - 1)(85 \text{ cm})}{(641.6 \text{ nm})} \approx 1022,$$

where $n_{gas} - 1 = 7.72 \times 10^{-4}$ for chlorine gas, and we dropped approximately three quarters of a fringe.

Section 37-3 Double-Slit Interference

Problem

21. A double-slit experiment has slit spacing 0.12 mm. (a) What should be the slit-to-screen distance L if the bright fringes are to be 5.0 mm apart when the slits are illuminated with 633-nm laser light? (b) What will be the fringe spacing with 480-nm light?

Solution

The particular geometry of this type of double-slit experiment is described in the paragraphs preceding Equations 37-3a and b. (a) The spacing of bright fringes on the screen is $\Delta y =$

$\lambda L/d$, so $L = (0.12 \text{ mm})(5 \text{ mm})/(633 \text{ nm}) = 94.8$ cm. (b) For two different wavelengths, the ratio of the spacings is $\Delta y'/\Delta y = \lambda'/\lambda$, therefore $\Delta y' = (5 \text{ mm})(480/633) = 3.79$ mm.

Problem

25. Light shines on a pair of slits whose spacing is three times the wavelength. Find the locations of the first- and second-order bright fringes on a screen 50 cm from the slits. *Hint:* Do Equations 37-3 apply?

Solution

Since $d = 3\lambda$, the angles are not small and Equations 37-3 do not apply. The interference maxima occur at angles given by Equation 37-2a, $\theta = \sin^{-1}(m\lambda/d) = \sin^{-1}(m/3)$, so only two orders are present, for values of $m = 1$ and 2 ($\theta < 90°$). If we assume that the slit/screen geometry is as shown in Fig. 37-15, then $y = L\tan\theta = L\tan(\sin^{-1}(m/3)) = Lm/\sqrt{9 - m^2}$. (Consider a right triangle with hypotenuse of 3 and opposite side m, or use $\tan\theta = \sin\theta \div \sqrt{1 - \sin^2\theta}$.) For $m = 1$ and 2, and $L = 50$ cm, this gives $y_1 = (50 \text{ cm})(1/\sqrt{8}) = 17.7$ cm, and $y_2 = (50 \text{ cm})(2/\sqrt{5}) = 44.7$ cm.

Problem

29. Laser light at 633 nm falls on a double-slit apparatus with slit separation 6.5 μm. Find the separation between (a) the first and second and (b) the third and fourth bright fringes, as seen on a screen 1.7 m from the slits.

Solution

Since $d \sim 10\lambda$ for this interference process, the small angle approximation is not particularly accurate, especially for higher orders. The angular position and position on the screen (for the usual slit/screen configuration) for bright fringes are $\theta_m = \sin^{-1}(m\lambda/d)$ and $y_m = L\tan\theta_m$, so the separation of two bright fringes on the screen is $\Delta y_{m_1 m_2} = y_{m_2} - y_{m_1} = L[\tan(\sin^{-1}(m_2\lambda/d)) - \tan(\sin^{-1}(m, \lambda/d))]$. (a) For $L = 1.7$ m, $d = 6.5$ μm, $\lambda = 633$ nm, $m_1 = 1$ and $m_2 = 2$, one finds $\Delta y_{12} = 17.1$ cm. (b) For $m_1 = 3$ and $m_2 = 4$, and the same other data, one finds $\Delta y_{34} = 20.0$ cm. (The approximate separation implied by Equation 37-3a is $\Delta y = \lambda L/d = 16.6$ cm.)

Section 37-4 Multiple-Slit Interference and Diffraction Gratings

Problem

33. A 5-slit system with 7.5-μm slit spacing is illuminated with 633-nm light. Find the angular positions of (a) the first 2 maxima and (b) the third and sixth minima.

Solution

(a) Primary maxima occur at angles $\theta = \sin^{-1}(m\lambda/d)$. The first two (after the central peak, $m = 0$) are for $m = 1$ and 2 at $\theta_1 = \sin^{-1}(633 \text{ nm}/7.5 \ \mu\text{m}) = 4.84°$ and $\theta_2 = \sin^{-1}(2 \times 633 \text{ nm}/7.5 \ \mu\text{m}) = 9.72°$. (b) Minima occur at angles $\theta' = \sin^{-1}(m'\lambda/Nd)$, where $m' = \pm 1, \pm 2, \ldots$, but excluding multiples of $N = 5$, in this case. The third minimum is for $m' = 3$ and the sixth for $m' = 7$ (because $m' = 5$ doesn't count). Then $\theta'_3 = \sin^{-1}(3\lambda/5d) = 2.90°$ and $\theta'_7 = \sin^{-1}(7\lambda/5d) = 6.79°$. (These minima would be difficult to observe because the secondary maxima between them are faint.)

Problem

37. Light is incident normally on a grating with 10,000 lines per cm. What is the maximum order in which (a) 450-nm and (b) 650-nm light will be visible?

Solution

The grating condition is $\sin\theta = m\lambda/d$, and, of course, for the diffracted light to be visible, $\theta < 90°$, or $m\lambda/d < 1$. Therefore, the highest order visible is the greatest integer m less than d/λ. For this grating, $d = 1 \text{ cm}/10^4 = 10^3 \text{ nm}$, so for $\lambda = 450 \text{ nm}$ and 650 nm, the highest visible orders are less than $10^3/450 = 2.22$ and $10^3/650 = 1.54$, or second and first, respectively.

Problem

41. Estimate the number of lines per cm in the grating used to produce Fig. 37-24. (See page 972 of text.)

Solution

The number of lines per cm ($1/d$ in cm^{-1}) is easily estimated from the angular position of the central 550 nm line in a particular order, as shown in the figure, i.e. $1/d = \sin\theta/m\lambda$. For example, in fifth order, this line is at $\theta = 61°$ (average of right and left values) so $1/d = \sin 61°/5(550 \text{ nm}) = 3.18 \times 10^3/\text{cm}$ or about 3200 lines/cm.

Problem

45. Echelle spectroscopy uses relatively course gratings in high order. Compare the resolving power of an 80 line/mm echelle grating used in twelfth order with a 600 line/mm grating used in first order, assuming the two have the same width.

Solution

If the echelle and grating have the same width, then the number of lines in each is proportional to the given spacings, $N/N' = 80/600$. The ratio of the resolving powers (Equation 37-7) is then $mN/m'N' = 12 \times 80/1 \times 600 = 1.6$, so the echelle has about 60% greater resolving power than the grating.

Sections 37-5 and 37-6   Single-Slit Diffraction and the Diffraction Limit

Problem

49. For what ratio of slit width to wavelength will the first minima of a single-slit diffraction pattern occur at $\pm 90°$?

Solution

When $\theta = 90°$, in Equation 37-9, and $m = 1$ for the first minimum then $a/\lambda = 1$.

Problem

53. Find the intensity as a fraction of the central peak intensity for the second secondary maximum in single-slit diffraction, assuming the peak lies midway between the second and third minima.

Solution

The second and third minima lie at angles $\sin\theta_2 = 2\lambda/a$ and $\sin\theta_3 = 3\lambda/a$. If we take the mid-value (as in Example 37-7) to be at $\sin\theta = 5\lambda/2a$, then the intensity at this angle, relative to the central intensity, is $\bar{S}/\bar{S}_0 = [\sin(5\pi/2)/(5\pi/2)]^2 = 4/25\pi^2 = 1.62 \times 10^{-2}$.

Problem

57. What is the minimum spot diameter to which a camera set at f-ratio of 16 can focus parallel light with 650-nm wavelength? *Hint:* Equation 37-13b gives the minimum angular spacing between the central maximum and first minimum; here you want the angular spread of the circle that marks the minimum.

Solution

The diffraction limit for a lens opening of diameter D, focusing light of wavelength λ is $\theta_{min} = 1.22 \ \lambda/D$. The radius of a spot, at the focal length of the lens, with this angular spread, is $r = f\theta_{min}$ (see Hint). The minimum spot diameter is $2f\theta_{min} = 2(1.22)\lambda f/D = 2(1.22)(650 \text{ nm}) \times 16 = 25.4 \ \mu\text{m}$ (since f/D is the f-ratio).

Problem

61. Under the best conditions, atmospheric turbulence limits the resolution of ground-based telescopes to about 1 arc second (1/3600 of a degree) as shown in Fig. 37-50. For what aperture sizes is this limitation more severe than that of diffraction at 550 nm? Your answer shows why large ground-based telescopes do not produce better images than small ones, although they do gather more light.

Solution

The aperture satisfying the Rayleigh criterion (Equation 37-13b) for $\theta_{min} = 1'' = \pi/(180 \times 3600) = 4.85 \times 10^{-6}$ (radians), at the given wavelength, is $D = 1.22(550 \text{ nm}) \div 4.85 \times 10^{-6} = 13.8 \text{ cm}$, or about $5\frac{1}{2}$ in. The resolution of all larger diameter ground based telescopes is limited by atmospheric conditions, at this wavelength.

Paired Problems

Problem

65. Find the total number of lines in a 2.5-cm-wide diffraction grating whose third-order spectrum has the 656-nm hydrogen-α spectral line at an angular position of 37°.

Solution

The grating condition for normally incident light (same as Equation 37-2a) and the given data imply a grating spacing of $d = m\lambda/\sin\theta = 3(656\text{ nm})/\sin 37° = 3.27\ \mu\text{m}$, or a grating constant of $d^{-1} = 3.06\times10^3$ lines/cm. On a grating 2.5 cm wide, the total number of lines is $(3.06\times10^3/\text{cm})(2.5\text{ cm}) = 7.65\times10^3$ ($= 7645$, to within a hundreth of a line).

Problem

69. What diameter optical telescope would be needed to resolve a Sun-sized star 10 light-years from Earth? Take $\lambda = 550$ nm. Your answer shows why stars appear as point sources in optical astronomy.

Solution

The angular size of the sun, at a distance of 10 ly, is only $\theta = 2(6.96\times10^5\text{ km})/10(9.46\times10^{12}\text{ km}) = 1.47\times10^{-8}$ (see Appendix E), so even a diffraction-limited space telescope would need an aperture of $D = 1.22\lambda/\theta = 1.22(550\text{ nm}) \div (1.47\times10^{-8}) = 45.6$ m to resolve it in visible light (see Equation 37-13b). (However, ground based optical interferometers are currently being developed.)

Supplementary Problems

Problem

73. In Fig. 37-6 the mth Newton's ring appears a distance r from the center of the lens. Show that the curvature radius of the lens is given approximately by $R = r^2/(m + \frac{1}{2})\lambda$, where the approximation holds when the thickness of the air space is much less than the curvature radius.

Solution

As explained in the solution to Problem 11, there is constructive interference between rays reflected (normally) from the upper and lower surfaces of the film of air which separates the lens and the glass plate, producing a bright Newton's ring when $2d = (m + \frac{1}{2})\lambda$ (Equation 37-1a with $n_{\text{air}} = 1$). The thickness of the air film at the position of the ring (a distance r from the central axis of the lens) is $d = R - R\cos\theta$, where $\sin\theta = r/R$. If $d \ll R$, then $\theta \ll 1$ and the small angle approximation gives $\theta \approx \sin\theta = r/R$, and $\cos\theta = \sqrt{1 - \sin^2\theta} \approx 1 - \frac{1}{2}\theta^2$. Then $d \approx R(1 - \cos\theta) \approx R\cdot\frac{1}{2}(r/R)^2 = r^2/2R$. Substituting this into the interference condition, one gets $R = r^2/(m + \frac{1}{2})\lambda$.

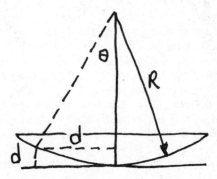

Problem 73 Solution.

Problem

77. The signal from a 103.9-MHz FM radio station reflects off a building 400 m away, effectively producing two sources of the same signal. You're driving at 60 km/h along a road parallel to a line between the station's antenna and the building and located a perpendicular distance of 6.5 km from them. How often does the signal appear to fade when you're driving roughly opposite the transmitter and building?

Solution

If we assume a constant phase difference between the direct and reflected waves, essentially a two-slit interference pattern is produced along the road traveled by the car, with minima (or maxima) spaced approximately $\Delta y = \lambda L/d$ apart, along the road and near the perpendicular bisector of the line between the sources. (A constant phase difference in Equations 37-3a and b cancels out in calculating the spacing.) The time between dead spots on this section of road, for a passing car with speed v, is $\Delta t = \Delta y/v = \lambda L/vd$. Using $\lambda = c/f$ and the numbers given, we find $\Delta t = (3\times10^8\text{ m/s})(6.5\text{ km}) \times (3.6\text{ s}/60\text{ m})/(400\text{ m})(103.9\text{ MHz}) = 2.82$ s. (The actual positions of the maxima and minima would, of course, depend on the phase difference between the direct and reflected waves.)

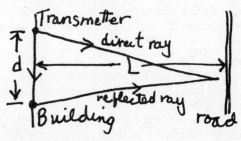

Problem 77 Solution.

Problem

81. In a double-slit experiment, a thin glass plate with refractive index 1.56 is placed over one of the slits. The fifth bright fringe now appears where the second dark fringe previously appeared. How thick is the plate if the incident light has wavelength 480 nm?

Solution

Without the glass plate, the second dark fringe appears at an angular position given by $d \sin \theta = 3\lambda/2$ (Equation 37-2b with $m = 1$). The glass plate, with thickness Δ, introduces an additional optical path difference of $(\Delta/\lambda_{glass}) - (\Delta/\lambda) = (n - 1)(\Delta/\lambda)$, where $n = \lambda/\lambda_{glass}$ is the refractive index of the plate. The fifth bright fringe occurs at an angular position for which the total path difference is five wavelengths, or $(d \sin \theta/\lambda) + (n - 1)(\Delta/\lambda) = 5$ (this is a modified Equation 37-2a with $m = 5$). Since the angle is the same in both cases, we can substitute $d \sin \theta/\lambda = \frac{3}{2}$ to obtain $5 = (\frac{3}{2}) + (n - 1)(\Delta/\lambda)$, or $\Delta = 7\lambda/2(n - 1) = 7(480 \text{ nm})/2(0.56) = 3.00 \ \mu\text{m}$.

● PART 5 CUMULATIVE PROBLEMS

Problem

1. A *grism* is a grating ruled onto a prism, as shown in Fig. 1. The grism is designed to transmit undeviated one wavelength of the spectrum in a given order, as refraction in the prism compensates for the deviation at the grating. Find an equation relating the separation d of the grooves that constitute the grating, the wedge angle α of the prism, the refractive index n, the undeviated wavelength λ_0, and order m_0.

Solution

Consider the interference of rays of wavelength λ, from the same plane wavefront incident normally on the upper surface of the grism in Fig. 1, which emerge from adjacent slits of the grating (grooves). The two rays will be in phase (producing an intensity maximum) when their path lengths, in wavelengths, differ by an integer, m. From the sketch, $|PQ/\lambda_n - P'Q'/\lambda| = m$, where $\lambda_n = \lambda/n$ is the wavelength in the grism and n is its refractive index. Since $PQ = d \sin \alpha$, $P'Q' = d \sin \beta$, where $d = P'Q$ is the groove separation, and $\delta = \alpha - \beta$ is the deviation, this condition can be written as $|n \sin \alpha - \sin (\alpha - \delta)| = m\lambda/d$ (the grism equation). That a given wavelength (λ_0) be transmitted undeviated ($\delta = 0$) requires $(n - 1) \sin \alpha = m_0 \lambda_0/d$, which is the sought for equation.

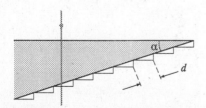

FIGURE 1 Cumulative Problem 1 Solution.

Problem

5. In one type of optical fiber, called a *graded-index fiber,* the refractive index varies in a way that results in light rays being guided along the fiber on curved trajectories, rather than undergoing abrupt reflections. Figure 3 shows a simple model that demonstrates this effect; it also describes the basic optical effect in mirages. A slab of transparent material has refractive index $n(y)$ that varies with position y perpendicular to the slab face. A light ray enters the slab at $x = 0$, $y = 0$, making an angle θ_0 with the normal just inside the slab. The refractive index at this point is $n(y = 0) = n_0$. (a) By writing $\sin \theta$ in Snell's law in terms of the components dx and dy of the ray path, show that that path (written in the form of x as a function of y) is given by

$$x = \int_0^y \frac{n_0 \sin \theta_0}{\sqrt{[n(y)]^2 - n_0^2 \sin^2 \theta_0}} \, dy.$$

(b) Suppose $n(y) = n_0(1 - ay)$, where $n_0 = 1.5$ and $a = 1.0 \text{ mm}^{-1}$. If $\theta_0 = 60°$, find an explicit expression for x as a function of y, and plot your result to give the actual ray path. Explain the shape of your curve in terms of what happens when the ray reaches a point where $n(y) = n_0 \sin \theta_0$. What happens beyond this point?

Solution

(a) Snell's law (Equation 35-3) relates the refractive index and the direction of the ray path, at any y, to the values at $y = 0$, that is $n \sin \theta = n_0 \sin \theta_0$ (n and θ are both functions of y). The slope of the ray path (with the y axis) is $\tan \theta = dx/dy = \sin \theta/\sqrt{1 - \sin^2 \theta}$, so substitution and separation of variables yields $dx = n_0 \sin \theta_0 [n^2 - n_0^2 \sin^2 \theta_0]^{-1/2} \, dy$. The given expression for x as a function of y follows immediately

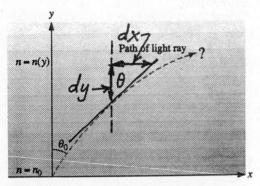

FIGURE 3 Cumulative Problem 5 Solution.

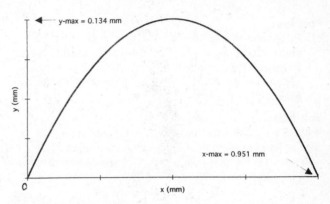

Cumulative Problem 5 Solution.

by integration from the origin ($x = 0 = y$). (b) If $n(y) = n_0(1 - ay)$, the integral in part (a) can be evaluated by using the last entry in the first column of the integral table on page A-10. An intermediate substitution of $z = (1 - ay)/\sin\theta_0$ facilitates this, with the result:

$$x = \int_0^y \frac{\sin\theta_0 \, dy}{\sqrt{(1 - ay)^2 - \sin^2\theta_0}} = \left(\frac{\sin\theta_0}{a}\right) \int_z^{1/\sin\theta_0} \frac{dz}{\sqrt{z^2 - 1}}$$

$$= \left(\frac{\sin\theta_0}{a}\right) \left[\ln\left(\frac{1 + \cos\theta_0}{\sin\theta_0}\right) - \ln(z + \sqrt{z^2 - 1}) \right].$$

The given value $\theta_0 = 60°$ (or $\sin\theta_0 = \frac{1}{2}\sqrt{3}$) reduces this further to $x = (\sqrt{3}/2a)[\ln\sqrt{3} - \ln(z + \sqrt{z^2 - 1})]$, where $z = 2(1 - ay)/\sqrt{3}$. (Note that if the ln's are combined and the identity $1/(z + \sqrt{z^2 - 1}) = z - \sqrt{z^2 - 1}$ is used, the expression on p. A-35 of the Answers to Odd-Numbered Problems is obtained, since

$$\frac{\sqrt{3}}{z + \sqrt{z^2 - 1}} = \sqrt{3}\left[\frac{2(1 - ay)}{\sqrt{3}} - \sqrt{\frac{4(1 - a^2y^2)}{3} - 1}\right]$$

$$= 2\left[1 - ay - \sqrt{\frac{1}{4} - 2ay + a^2y^2}\right].$$

However, this expression is not well suited for plotting.) At this point, it is convenient to exploit the relation between the natural logarithm and inverse hyperbolic functions, namely, $\ln(z + \sqrt{z^2 - 1}) = \cosh^{-1}z$, which allows one to express y (or z) as a function of x. The result is $z = \cosh(\ln\sqrt{3} - 2ax/\sqrt{3})$, or $ay = 1 - (\sqrt{3}/2)\cosh(\ln\sqrt{3} - 2ax/\sqrt{3})$, which is plotted below, for $a = 1 \text{ mm}^{-1}$. Since $\sin\theta = (n/n_0)\sin\theta_0 \le 1$, the above expressions are valid for $0 \le y \le (1 - \sin\theta_0)/a = 0.134$ mm (or $1 \le z \le 1/\sin\theta_0 = 2/\sqrt{3}$). The ray path is bent back towards the x axis after reaching its maximum penetration, where $n = n_0\sin\theta_0$.

PART 6 MODERN PHYSICS

● **CHAPTER 38** THE THEORY OF RELATIVITY

Section 38-2 Matter, Motion, and the Ether

Problem

1. Consider an airplane flying at 800 km/h airspeed between two points 1800 km apart. What is the round-trip travel time for the plane (a) if there is no wind? (b) if there is a wind blowing at 130 km/h perpendicular to a line joining the two points? (c) if there is a wind blowing at 130 km/h along a line joining the two points? Ignore relativistic effects. (Why are you justified in doing so?)

Solution

Since the velocities are small compared to c, we can use the non-relativistic Galilean transformation of velocities in Equation 3-10, $\mathbf{u} = \mathbf{u'} + \mathbf{v}$, where $\mathbf{u}$ is the velocity relative to the ground (S), $\mathbf{u'}$ is that relative to the air (S'), and $\mathbf{v}$ is that of S' relative to S (in this case, the wind velocity). We used a notation consistent with that in Equations 38-11 and 12. (a) If $\mathbf{v} = 0$ (no wind), $\mathbf{u} = \mathbf{u'}$ (ground speed equals air speed) and the round-trip travel time is $t_a = 2d/u = 2(1800 \text{ km})/(800 \text{ km/h}) = 4.5$ h. (b) If $\mathbf{v}$ is perpendicular to $\mathbf{u}$, then $u'^2 = u^2 + v^2$ or $u = \sqrt{800^2 - 130^2}$ km/h = 789 km/h, and the round-trip travel time is $t_b = 2d/u = 4.56$ h. (c) If $\mathbf{v}$ is parallel or antiparallel to $\mathbf{u}$ on alternate legs of the round-trip, then $u = u' \pm v$ and the travel time is

$$t_c = \frac{d}{u'+v} + \frac{d}{u'-v} = \frac{1800 \text{ h}}{800+130} + \frac{1800 \text{ h}}{800-130}$$
$$= 4.62 \text{ h}.$$

Note that $t_a < t_b < t_c$, as mentioned following Equation 38-2.

Problem

5. Suppose the speed of light differed by 100 m/s in two perpendicular directions. How long should the arms be in a Michelson interferometer if this difference is to cause the interference pattern to shift one-half cycle (i.e., a light fringe shifts to where a dark one was) relative to the pattern if there were no speed difference? Assume 550-nm light.

Solution

For equal length arms, different velocities of light, say $c_\parallel \neq c_\perp$, introduce a phase difference of half a cycle when $\pm\pi = \Delta\phi = k_\parallel 2L - k_\perp 2L = 2\omega L(c_\parallel^{-1} - c_\perp^{-1})$ (recall that $k = \omega/c$). Since $\Delta c = |c_\parallel - c_\perp| \ll c$, this can be written as $2\omega L \Delta c/c^2 = \pi$. If we use $\omega = 2\pi f = 2\pi c/\lambda$ (think of λ and c as average values for the two directions), then $2(2\pi c/\lambda)L \Delta c/c^2 = \pi$, or $L = \lambda c/(4 \Delta c) = (550 \text{ nm} \times 3\times10^8 \text{ m/s})/(4 \times 100 \text{ m/s}) = 41.3$ cm.

Section 38-4   Space and Time in Relativity

Problem

9. Earth and Sun are 8.3 light-minutes apart, as measured in their rest frame. (a) What is the speed of a spacecraft that makes the trip in 5.0 min according to its on-board clocks? (b) What is the trip time as measured by clocks in the Earth-Sun frame?

Solution

Note that the distance is given in the system S, where the Earth and the Sun are practically at rest (the orbital speed of the earth is very small compared to the speed of light or the speed of the spacecraft), but the time interval is given in system S', where the spacecraft is at rest, i.e., $\Delta x = 8.3$ c·min and $\Delta t' = 5$ min. (One light-minute, the distance light travels in one minute, equals c multiplied by a minute, so c·min is a convenient notation for this light-unit.) (a) Equations 38-4 and

5, for time dilation and Lorentz contraction, relate the given quantities to $\Delta x'$ and Δt, so that the spacecraft's speed, $v = \Delta x/\Delta t = \Delta x'/\Delta t'$, can be determined. For example, if we use Equation 38-8 for shorthand notation, Equations 38-4 and 5 become $\Delta x = \gamma \Delta x'$ and $\Delta t = \gamma \Delta t'$. Then $v = \Delta x/(\gamma \Delta t') = (\Delta x/\gamma)/\Delta t' = (8.3 \text{ c·min}) \div (\gamma 5 \text{ min})$, or $\gamma v/c = 8.3/5 = 1.66$, which is easily solved for $v/c = 0.857$. (Square the equation and substitute for γ: $\gamma^2(v/c)^2 = (v/c)^2/(1 - v^2/c^2) = (1.66)^2$. Then $(v/c)^2 = (1.66)^2/[1 + (1.66)^2]$.) (b) The equation $\gamma v/c = 1.66$ is also easily solved for $\gamma = \sqrt{(1.66)^2 - 1} = 1.32$. (Use the fact that $(v/c)^2 = 1 - 1/\gamma^2$.) Then $\Delta t = (1.32)(5 \text{ min}) = 9.69$ min.

Problem

13. You wish to travel to a star N light-years from Earth. How fast must you go if the one-way journey is to occupy N years of your life?

Solution

This problem is similar to Problem 9(a), with $\Delta x = N$ ly and $\Delta t' = N$ y. Thus, $\gamma v/c = \Delta x/c\Delta t' = N$ ly/cN y $= 1$, and $v/c = 1/\sqrt{2}$.

Problem

17. Two distant galaxies are receding from Earth at $0.75c$, in opposite directions. How fast does an observer in one galaxy measure the other to be moving?

Solution

Our galaxy (S') is moving with speed $v = 0.75c$ relative to one of the galaxies mentioned in the question (S), and the other galaxy is moving with speed $u' = 0.75c$ relative to us. (All velocities are assumed to be along the common x-x' axes.) Then, the relativistic velocity addition formula (Equation 38-11) gives $u = (u' + v)/(1 + u'v/c^2) = (0.75c + 0.75c)/[1 + (0.75)^2] = 0.960c$.

Problem

21. Earth and Sun are 8.33 light-minutes apart. Event A occurs on Earth at time $t = 0$, and event B on the Sun at time $t = 2.45$ min, as measurd in the Earth-Sun frame. Find the time order and time difference between A and B for observers (a) moving on a line from Earth to Sun at $0.750c$, (b) moving on a line from Sun to Earth at $0.750c$, and (c) moving on a line from Earth to Sun at $0.294c$.

Solution

The relative speed of the Earth and the Sun is small compared to c ($v/c \approx 10^{-4}$), so we may consider the Sun to be approximately at rest, a distance 8.33 c·min from Earth (system S). Events A and B have coordinates $x_A = 0$, $t_A = 0$, $x_B = 8.33$ c·min, and $t_B = 2.45$ min in S, where we chose the Earth as the origin and the x axis in the direction of the Sun. An observer in system S', moving along the x axis with speed v, sees these events separated by a time interval $\Delta t' = t'_B - t'_A = \gamma[t_B - x_B(v/c^2) - t_A + x_A(v/c^2)] = \gamma[2.45 \text{ min} - 8.33 \text{ min } (v/c)]$. (Coordinates in S and S' are related by the Lorentz transformation, see Table 38-1.) (a) If $v = 0.75c$ and $\gamma = 1.51$, then $\Delta t' = -5.74$ min, or event B occurs before A in S'. (b) If $v = -0.75c$, then $\Delta t' = 13.1$ min, and B occurs after A in S'. (c) Since $2.45/8.33 \approx 0.294$, $\Delta t' = 0$ if $v = 0.294c$, and A and B are essentially simultaneous in S'.

Problem

25. Derive the Lorentz transformations for time, Equations 38-9 and 38-10, from the transformations for space.

Solution

If we solve Equation 38-7 for t', and substitute for x' from Equation 38-6, we obtain Equation 38-9:

$$t' = \frac{x}{\gamma v} - \frac{x'}{v} = \frac{x}{\gamma v} - \frac{\gamma(x - vt)}{v}$$

$$= \gamma\left[\left(1 - \frac{v^2}{c^2}\right)\frac{x}{v} - \frac{x}{v} + t\right] = \gamma\left(t - \frac{vx}{c^2}\right).$$

By a similar procedure, one may solve for t and obtain Equation 38-10.

Section 38-5 Energy and Momentum in Relativity

Problem

29. At what speed will the momentum of a proton (mass 1 u) equal that of an alpha particle (mass 4 u) moving at 0.5c?

Solution

The momenta of the proton and alpha particle are equal when $m_p u_p/\sqrt{1 - u_p^2/c^2} = m_\alpha u_\alpha/\sqrt{1 - u_\alpha^2/c^2}$, see Equation 38-12. Square and solve for u_p to obtain:

$$\frac{u_p}{c} = \left[1 + \left(\frac{m_p}{m_\alpha}\right)^2\left(\frac{c^2}{u_\alpha^2} - 1\right)\right]^{-1/2}$$

$$= \left[1 + \left(\frac{1}{4}\right)^2\left(\frac{1}{0.5^2} - 1\right)\right]^{-1/2}$$

$$= \frac{1}{\sqrt{1 - 3/16}} = 0.918.$$

Problem

33. Find (a) the total energy and (b) the kinetic energy of an electron moving at 0.97c.

Solution

For this electron, $v/c = 0.97$, $\gamma = 4.11$, and $m_e c^2 = 0.511$ MeV. (a) From Equation 38-14, $E = \gamma mc^2 = (4.11)(0.511 \text{ MeV}) = 2.10$ MeV. (b) From Equation 38-13, $K = (\gamma - 1)mc^2 = (3.11)(0.511 \text{ MeV}) = 1.59$ MeV.

Problem

37. Among the most energetic cosmic rays ever detected are protons with energies around 10^{20} eV. Find the momentum of such a proton, and compare with that of a 25-mg insect crawling at 2 mm/s (Fig. 38-36).

Solution

This energy is so much greater than the proton's rest energy ($mc^2 = 938$ MeV) that $p \approx E/c = 10^{20}$ eV/c. (See Equation 38-1). The atomic unit of momentum in conventional SI units is 1 eV/c = $(1.602\times10^{-19} \text{ J})/(2.998\times10^8 \text{ m/s}) = 5.34\times10^{-28}$ kg·m/s, so $p \approx 5.34\times10^{-8}$ kg·m/s. This is about equal to the momentum of the insect, $mv = (25\times10^{-6} \text{ kg})(2\times10^{-3} \text{ m/s}) = 5\times10^{-8}$ kg·m/s.

Problem

41. Find the speed of an electron with kinetic energy (a) 100 eV. (b) 100 keV, (c) 1 MeV, (d) 1 GeV. Use suitable approximations where possible.

Solution

For the electron, $m_e c^2 = 511$ keV, so γ corresponding to kinetic energy $K = (\gamma - 1)m_e c^2$ is $\gamma = 1 + K/511$ keV, from which $v/c = \sqrt{1 - 1/\gamma^2}$ can be found (see Equations 38-13 and 8). (a) If $K \ll 511$ keV, the non-relativistic expression, $v/c = \sqrt{2K/mc^2}$, can be used (see solution to Problem 43), so $v/c = \sqrt{2(100 \text{ eV})/511 \text{ keV}} = 1.98\times10^{-2}$. (b) If $K = 100$ keV, then $\gamma = 1.20$ and $v/c = 0.548$, while (c) if $K = 1$ MeV, $\gamma = 2.96$ and $v/c = 0.941$. (d) If $K \gg 511$ keV, γ is large and $v/c \approx 1$. If one expands the square root in $v/c = \sqrt{1 + (1 + K/m_e c^2)^{-2}}$, in powers of $m_e c^2/K \ll 1$, one obtains $v/c \approx 1 - \frac{1}{2}(m_e c^2/K)^2 = 1 - \frac{1}{2}(0.511\times10^{-3})^2 = 1 - 1.31\times10^{-7}$.

Section 38-6 What Is Not Relative

Problem

45. Show from the Lorentz transformations that the spacetime interval of Equation 38-17 has the same value in all frames of reference.

Solution

Consider two frames, S and S', related by the Lorentz transformations in Table 38-1. (Since the equations are linear, they apply to differences of coordinates also.) We have
$$c^2 \Delta t'^2 - \Delta x'^2 - \Delta y'^2 - \Delta z'^2 =$$
$$c^2\gamma^2(\Delta t - v \Delta x/c^2)^2 - \gamma^2(\Delta x - v \Delta t)^2 - \Delta y^2 - \Delta z^2 =$$
$$\gamma^2(c^2 \Delta t^2 - 2v \Delta x \Delta t + v^2 \Delta x^2/c^2 - \Delta x^2 + 2v \Delta x \Delta t - v^2 \Delta t^2) - \Delta y^2 - \Delta z^2 =$$
$$\gamma^2(1 - v^2/c^2)(c^2 \Delta t^2 - \Delta x^2) - \Delta y^2 - \Delta z^2 =$$
$$c^2 \Delta t^2 - \Delta x^2 - \Delta y^2 - \Delta z^2.$$
Therefore, $\Delta s'^2 = \Delta s^2$ and the spacetime interval is invariant.

Paired Problems

Problem

49. An extraterrestrial spacecraft passes Earth and 4.5 s later, according to its clocks, it passes the Moon. Find its speed.

Solution

(Note: This and the next problem are essentially the same as Problem 9a, whose result, $v/c = (\Delta x/c \, \Delta t') \div \sqrt{1 + (\Delta x/c \, \Delta t')^2}$, can be used directly. However, a repetition of the reasoning may be useful.) In the spacecraft's frame (S') the distance from the Earth to the Moon appears Lorentz contracted from the proper distance (3.85×10^5 km in the Earth-Moon frame, S), $\Delta x' = \Delta x \sqrt{1 - v^2/c^2} = \Delta x/\gamma$, so its speed is $v = \Delta x'/\Delta t' = (\Delta x/\gamma)/(4.5 \text{ s})$. Thus $\gamma v/c = (3.85 \times 10^5 \text{ km}) \div (3 \times 10^5 \text{ km/s})(4.5 \text{ s}) = 0.285$ and $v/c = 0.285/\sqrt{1 + (0.285)^2} = 0.274$ (square and solve for v/c). (Alternatively, time dilation gives $\Delta t = \gamma \Delta t'$ in the Earth-Moon frame and $v = \Delta x/\Delta t$ leads to the same result.)

Problem

53. Event A occurs at $x = 0$ and $t = 0$ in a frame of reference S. Event B occurs at $x = 3.8$ light-years, $t = 1.6$ years in S. Find (a) the distance and (b) the time between A and B in a frame S' moving at $0.80c$ along the x axis of S.

Solution

The coordinates of the events in S and S', are related by the Lorentz transformation in Table 38-1, with $v/c = 0.8$ and $\gamma = 5/3$. (a) $x'_B - x'_A = \gamma[x_B - x_A - v(t_B - t_A)] = (5/3)[3.8 \text{ ly} - (0.8 \, c)(1.6 \text{ y})] = (5/3)(3.8 - 1.28) \text{ ly} = 4.20$ ly. (b) $t'_B - t'_A = \gamma[t_B - t_A - (v/c^2)(x_B - x_A)] = (5/3)[1.6 \text{ y} - (0.8/c)(3.8 \text{ ly})] = -2.40$ y, i.e., B occurs before A in S'. (Since the light travel time from the position of A to that of B is greater than the magnitude of the time difference (3.8 y versus 1.6 y in S, or 4.2 y versus 2.4 y is S', the events are not causily connected.)

Supplementary Problems

Problem

57. How fast would you have to travel to reach the Crab Nebula, 6500 light-years from Earth, in 20 years? Give your answer to 7 significant figures.

Solution

As in Problems 9 and 49, the distance to the Crab Nebula, $\Delta x = 6500$ ly, is specified in the Earth's system (S), while the time interval, $\Delta t' = 20$ y, is in the traveler's system (S'), so either Lorentz contraction ($\Delta x = \gamma \, \Delta x'$) or time dilation ($\Delta t = \gamma \, \Delta t'$) give $v = \Delta x/\Delta t = \Delta x'/\Delta t' = \Delta x/\gamma \, \Delta t'$, or $\gamma v/c = \Delta x/c \, \Delta t' = 6500 \text{ ly}/20c \cdot \text{y} = 325$. Then $v/c = 325/\sqrt{1 + (325)^2}$. A good calculator like the HP-28S gives $v/c = 0.99999527$, but one can also use the binomial expansion to find $v/c = [1 + (1/325)^2]^{-1/2} \approx 1 - \frac{1}{2}(1/325)^2 = 1 - 4.73 \times 10^{-6}$.

Problem

61. When the speed of an object increases by 5%, its momentum goes up by a factor of 5. What was the original speed?

Solution

As in Problem 55, we write $v_2 = 1.05v_1$ and $p_2 = 5p_1$. Following the same procedure, we find $1.05/\gamma_1 = 5/\gamma_2$, or $(1.05)^2(1 - v_1^2/c^2) = 5^2(1 - v_2^2/c^2) = 5^2(1 - (1.05)^2v_1^2/c^2)$, or $v/c = \sqrt{[25 - (1.05)^2]/24(1.05)^2} = 0.950$.

Problem

65. Use Equation 9-7 to estimate the size to which you would have to squeeze each of the following before escape speed at its surface approximated the speed of light: (a) Earth; (b) the Sun; (c) the Milky Way galaxy, containing about 10^{11} solar masses. Your answers show why general relativity is not needed for most astronomical calculations.

Solution

Equation 9-7 for the escape speed from the surface of a spherical object of mass M and radius R is $v_{esc} = \sqrt{2GM/R}$. If this is set equal to c, then $R_S = 2GM/c^2$. (R_S is called the Schwarzschild radius. An object of mass M and radius less than R_S would collapse under its own gravity to form a black hole.) (b) The Schwarzschild radius for a solar mass object is $2GM_\odot/c^2 = 2(6.67 \times 10^{-11} \text{ N·m}^2/\text{kg}^2) \times (1.99 \times 10^{30} \text{ kg})/(3 \times 10^8 \text{ m/s})^2 = 2.92$ km, or roughly 3 km. Since R_s is proportional to M, the Schwarzschild radius for any massive object is $R_s = (M/M_\odot)3$ km. (a) For the Earth, this is $(5.97/1.99 \times 10^6)(3 \text{ km}) \approx 9$ mm, while (c) for the Milky Way galaxy, $R_s \approx 3 \times 10^{11}$ km ≈ 0.03 ly.

● CHAPTER 39   INSIDE ATOMS AND NUCLEI

Useful numerical values of Planck's constant, in SI and atomic units, are:

$h = 6.626 \times 10^{-34}$ J·s $= 4.136 \times 10^{-15}$ eV·s $= 1240$ eV·nm/c,

and

$\hbar = h/2\pi = 1.055 \times 10^{-34}$ J·s $= 6.582 \times 10^{-16}$ eV·s
$= 197.3$ MeV·fm/c.

Section 39-1   Toward the Quantum Theory

Problem

1. Find the energy in electron-volts of (a) a 1.0-MHz radio photon, (b) a 5.0×10^{14}-Hz optical photon, and (c) a 3.0×10^{18}-Hz x-ray photon.

Solution

The energy of a photon (Equation 39-1) is $E_\gamma = hf$, where h is Planck's constant. For the frequencies given, the photon energies in atomic units are (a) 4.14 neV (b) 2.07 eV, and (c) 12.4 keV.

Problem

5. Find the rate of photon production by (a) a radio antenna broadcasting 1.0 kW at 89.5 MHz, (b) a laser producing 1.0 mW of 633-nm light, and (c) and x-ray machine producing 0.10-nm x rays with a total power of 2.5 kW.

Solution

The rate of photon emission is the power output (into photons) divided by the photon energy, $\mathcal{P}/E_\gamma = \mathcal{P}/hf = \mathcal{P}\lambda/hc$. For the devices specified, rates are
(a) 1 kW/$(6.626 \times 10^{-34}$ J·s $\times 89.5$ MHz) $= 1.69 \times 10^{28}$ s^{-1},
(b) (1 mW $\times$ 633 nm)/$(6.626 \times 10^{-34}$ J·s $\times 3 \times 10^8$ m/s) $= 3.18 \times 10^{15}$ s^{-1}, and (c) 1.26×10^{18} s^{-1}.

Problem

9. A Rydberg hydrogen atom makes a downward transition to the $n = 225$ state. If the photon emitted has energy 9.32 μeV, what was the original state?

Solution

Combining Equations 39-1 and 3, one can write the energy of a photon emitted in a hydrogen atom transition between states $n_1 \rightarrow n_2$ as $E_\gamma = hc/\lambda = hcR_H(n_2^{-2} - n_1^{-2})$. (In a Rydberg atom, n_1 and n_2 are both very large, see Example 39-1.) Solving for n_1, we find $n_1 = [n_2^{-2} - (E_\gamma/hcR_H)]^{-1/2} = [(225)^{-2} - (9.32 \ \mu\text{eV}/13.6 \text{ eV})]^{-1/2} = 229$. (The constant $hcR_H = (1240 \text{ eV·nm})(0.01097 \text{ (nm)}^{-1}) = 13.6$ eV, called the ionization energy, is the energy difference between an ionized hydrogen atom, $n_1 = \infty$, and the ground state, $n_2 = 1$.)

Problem

13. Electron microscopes can usually resolve smaller objects than optical microscopes because they illuminate their subjects with electrons whose de Broglie wavelengths are much smaller than that of light (Fig. 39-37). What is the minimum electron speed that would make an electron microscope superior to an optical microscope using 450-nm light?

Solution

$\lambda = h/p < 450$ nm implies $p = \gamma m v > h/450$ nm, or $\gamma v/c > hc/(mc^2)(450 \text{ nm}) = (1240 \text{ eV·nm})(511 \text{ keV})^{-1} \times (450 \text{ nm })^{-1} = 5.39 \times 10^{-6}$. Then $v/c > (5.39 \times 10^{-6}) \div \sqrt{1 + (5.39 \times 10^{-6})^2} \approx 5.39 \times 10^{-6}$, or $v > (5.39 \times 10^{-6}) \times (3 \times 10^5$ km/s) $= 1.62$ km/s (see Chapter 38). (Since $\lambda \gg h/mc = 0.00243$ nm, for an electron, if you used the non-relativistic expression for momentum, $p = mv$, then you would get the same result, $v > h/m\lambda = (6.63 \times 10^{-34}$ J·s)$(9.11 \times 10^{-31}$ kg)$^{-1}(450$ nm)$^{-1} = 1.62$ km/s, see Example 39-2.)

Section 39-2   Quantum Mechanics

Problem

17. An electron is moving in the $+x$ direction with speed measured at 5.0×10^7 m/s, to an accuracy of $\pm 10\%$. What is the minimum uncertainty in its position?

Solution

The non-relativistic expression, $p = mv$, gives a rough estimate of Δx in Equation 39-5 (since $v/c = 1/6$ and $\gamma = 1.014$). It is $\Delta x \gtrsim \hbar/m \, \Delta v = (1.055 \times 10^{-34}$ J·s) $\times (9.11 \times 10^{-31}$ kg)$^{-1}(10^7$ m/s)$^{-1} = 11.6$ pm, where $\Delta v = 20\%$ of 5×10^7 m/s is the full range of uncertainty in v. (If the relativistic mometum $p = \gamma m v$ is used, then $\Delta p = m \, \Delta(\gamma v) = 9.50 \times 10^{-27}$ kg·m/s, and $\Delta x \gtrsim \hbar/\Delta p = 11.1$ pm.)

Section 39-3   Nuclear Physics

Problem

21. What is the half-life of a radioactive material if 75% of it decays in 5.0 hours?

Solution

If 75% decays in 5 h, then 25% remains after 5 h. Since 25% = 50% × 50%, 5 h is two half-lives, or $t_{1/2} = 2.5$ h. [In general, the amount remaining after time t is $N/N_0 = (1/2)^{t/t_{1/2}}$, where N_0 is the orginal amount at $t = 0$. In this problem, $t_{1/2} = t \ln(1/2)/\ln(1/4) = \frac{1}{2}(5 \text{ h}).]$

Problem

25. The energy released in the fission of a single uranium-235 nucleus is about 200 MeV. Estimate the mass of uranium that fissioned in the Hiroshima bomb, shown in Fig. 39-39, whose explosive yield was 12.5 kilotons (see Appendix C for a useful conversion factor).

Solution

The explosive yield of 1 megaton is about 4.18×10^{15} J (from Appendix C), so the Hiroshima bomb released the energy in about $(12.5)(4.18 \times 10^{12}$ J$)/(200$ MeV$)(1.6 \times 10^{-13}$ J/MeV$) = 1.63 \times 10^{24}$ fissions of U^{235}. This number of atoms has a mass of about $(1.63 \times 10^{24})(235$ u$)(1.66 \times 10^{-27}$ kg/u$) = 0.637$ kg. (Of course, the uranium in the bomb wasn't enriched to 100% U^{235}, so the actual mass would have been somewhat greater.)

Section 39-4   Elementary Particles

Problem

29. What is the quark composition of the antiproton?

Solution

Since a proton is formed from the combination uud of quarks, an antiproton is $\bar{u}\bar{u}\bar{d}$.

Paired Problems

Problem

33. (a) Find the energy of the highest energy photon that can be emitted as the electron jumps between two adjacent energy levels in the Bohr hydrogen atom. (b) Which energy levels are involved?

Solution

The energy of the emitted photon in a hydrogen atom transition between adjacent states ($n_1 \to n_2 = n_1 - 1$) is $E_\gamma = hc/\lambda = hcR_H[(n_1 - 1)^{-2} - n_1^{-2}] = hcR_H(2n_1 - 1)n_1^{-2}(n_1 - 1)^{-2}$ (see Equations 39-1 and 3). This is a maximum for the smallest allowed $n_1 > n_2 \geq 1$, which is $n_1 = 2$. Then (a) $E_{\gamma,\max}(\frac{3}{4}) hcR_H = (\frac{3}{4}) \times (13.6$ eV$) = 10.2$ eV (see solution to Problem 9 for the constant), and (b) $n_2 = n_1 - 1 = 1$. (This is the Lyman alpha photon at 122 nm.)

Problem

37. An electron is known to be within ± 0.05 nm of the nucleus of an atom. What is the uncertainty in its velocity?

Solution

The minimum uncertainty in momentum (Equation 39-5) is $\Delta p = \hbar/\Delta x = 1.055 \times 10^{-34}$ J$\cdot$s$/10^{-10}$ m $= 1.055 \times 10^{-24}$ kg$\cdot$m/s $= (197.3$ eV$\cdot$nm/c$)/(0.1$ nm$) = 1.973$ keV/c (if $\Delta x = 2(0.05$ nm$)$). This is small compared to $mc = 511$ keV/c for an electron, so non-relativistic expressions can be used. Then $\Delta v = \Delta p/m = (1.055 \times 10^{-24}$ kg$\cdot$m/s$)/(9.11 \times 10^{-31}$ kg$) = 1.16 \times 10^6$ m/s $= (1.973$ keV/c$)/(511$ keV/c$^2) = 3.86 \times 10^{-3}$ c. (We used both SI and atomic units for pedagogical reasons.)

Supplementary Problems

Problem

41. Use the result of the preceding problem to estimate the minimum possible kinetic energy for (a) an electron confined to a region of atomic dimensions, about 0.1 nm and (b) a proton confined to a region of nuclear dimensions, about 1 fm. Your answers show the order of magnitude of the energies to be expected in atomic and nuclear reactions, respectively.

Solution

For a particle confined to a one-dimensional region of size Δx, the minimum momentum (the uncertainty in which is given in the previous problem) is $\Delta p = 2p \geq \hbar/\Delta x$ (see Equation 39-5). If $pc \ll mc^2$ (the particle's rest energy), then the non-relativistic kinetic energy must satisfy $K = p^2/2m \geq (\hbar/2 \Delta x)^2/2m = \hbar^2/8m \Delta x^2 = (\hbar c)^2/8(mc^2)\Delta x^2$. (a) For an electron ($mc^2 = 511$ keV) confined to atomic dimensions ($\Delta x = 0.1$ nm), $K \geq (197.3$ eV$\cdot$nm$)^2/8(511$ keV$)(0.1$ nm$)^2 = 0.952$ eV $\simeq 1$ eV (typical of atomic transitions). (b) For a proton ($mc^2 = 938$ MeV) confined to nuclear dimensions ($\Delta x \simeq 1$ fm), $K \geq (197.3$ MeV$\cdot$fm$)^2/8(938$ MeV$)(1$ fm$)^2 = 5.19$ MeV $\simeq 5$ MeV (typical of nuclear transitions).

PHYSICS WITH MODERN PHYSICS

● **CHAPTER 39X**   LIGHT AND MATTER: WAVES OR PARTICLES?

Useful constants and combinations in SI and atomic units, for the problems in Chapters 39X–45 are:
$h = 6.626 \times 10^{-34}$ J$\cdot$s $= 4.136 \times 10^{-15}$ eV$\cdot$s $= 1240$ eV$\cdot$nm/c

$\hbar = h/2\pi = 1.055 \times 10^{-34}$ J$\cdot$s $= 6.582 \times 10^{-16}$ eV$\cdot$s
$= 197.3$ MeV$\cdot$fm/c
$c = 2.998 \times 10^8$ m/s, $ke^2 = 1.440$ eV$\cdot$nm,

$1u = 1.661 \times 10^{-27}$ kg $= 931.5$ MeV/c^2
$ke^2 = 1.440$ eV·nm,
$k_B = 1.381 \times 10^{-23}$ J/K $= 8.617 \times 10^{-5}$ eV/K

Section 39X-2 Blackbody Radiation

Note: Most answers in this section are very sensitive to the exact values used for the constants h and c.

Problem

1. If you increase the temperature of a blackbody by a factor of 2, by what factor does its radiated power increase?

Solution

The Stefan-Boltzmann law (Equation 39X-1) says that the total radiated power, or luminosity, of a blackbody is proportional to T^4, so doubling the absolute temperature increases the luminosity by a factor of $2^4 = 16$.

Problem

5. According to Planck's theory, what is the minimum non-zero energy of a molecule with vibration frequency 3.4×10^{14}-Hz?

Solution

In Planck's treatment, the energy of the lowest nonzero oscillator state is $hf = (4.136 \times 10^{-15}$ eV·s$)(3.4 \times 10^{14}$ Hz$) = 1.41$ eV.

Problem

9. Find the temperature range over which a blackbody will radiate the most energy at a wavelength in the visible range (400 nm–700 nm).

Solution

The Wien displacement law, for 400 nm $< \lambda_{max} <$ 700 nm, gives 2.898 mm·K/700 nm $= 4.14 \times 10^3$ K $< T <$ 7.25×10^3 K $= 2.898$mm·K/400nm. (See Equation 39X-2, i.e.)

Problem

13. Use the series expansion for e^x (Appendix A) to show that Planck's law (Equation 39X-3) reduces to the Rayleigh-Jeans law (Equation 39X-5) when $\lambda \gg hc/kT$.

Solution

For $\lambda \gg hc/kT$, the exponent in Planck's law (Equation 39X-3) is small and $e^x - 1 \approx x = hc/\lambda kT$. Then the radiance becomes $R(\lambda, T) \approx 2\pi hc^2/\lambda^5(hc/\lambda kT) = 2\pi ckT/\lambda^4$, which is the Rayleigh-Jeans law.

Section 39x-3 Photons

Problem

17. A red laser at 650 nm and a blue laser at 450 nm emit photons at the same rate. How do their total power outputs compare?

Solution

The ratio of the photon energies (Equation 39X-6) is $E_{blue}/E_{red} = f_{blue}/f_{red} = \lambda_{red}/\lambda_{blue} = 650/450 = 1.44$. Since the lasers emit photons at the same rate, this is also the ratio of their power outputs.

Problem

21. Electrons in a photoelectric experiment emerge from an aluminum surface with maximum kinetic energy of 1.3 eV. What is the wavelength of the illuminating radiation?

Solution

Einstein's equation for the photoelectric effect (Equation 39X-7) gives $K_{max} = (hc/\lambda) - \phi$, or $\lambda = hc/(K_{max} + \phi) = 1240$ eV·nm/$(1.3 + 4.28)$ eV $= 222$ nm. (We used the work function listed in Table 39X-1.)

Problem

25. Which materials in Table 39X-1 exhibit the photoelectric effect *only* for wavelengths shorter than 275 nm?

Solution

The energy of a photon of wavelength 275 nm is $hc/\lambda = 4.51$ eV. Such photons cannot eject photoelectrons from materials whose work functions are greater than this energy, which includes copper, silicon, and nickel in Table 39X-1. (Higher energy, or shorter wavelength, is necessary.)

Problem

29. The maximum electron energy in a photoelectric experiment is 2.8 eV. When the wavelength of the illuminating radiation is increased by 50%, the maximum electron energy drops to 1.1 eV. Find (a) the work function of the emitting surface and (b) the original wavelength.

Solution

The photoelectric effect equations (Equation 39X-7) for the two experimental runs are $K_{max} = (hc/\lambda) - \phi = 2.8$ eV, and $K'_{max} = (hc/\lambda') - \phi = (hc/1.5\,\lambda) - \phi = 1.1$ eV. (b) Subtracting there, one gets $hc/3\lambda = 1.7$ eV, or $\lambda = 1240$ eV·nm/5.1 eV $= 243$ nm. (a) Substituting this wavelength into either photoelectric effect equation, one finds $\phi = (5.1 - 2.8)$ eV $= (3.4 - 1.1)$ eV $= 2.3$ eV.

Problem

33. What is the minimum photon energy for which it is possible for the photon to lose half its energy undergoing Compton scattering with an electron?

Solution

The solution to Problem 31 shows that if the photon loses half its energy, then $E - E' = \frac{1}{2}E = E\,\Delta\lambda/(\lambda + \Delta\lambda)$, or $\Delta\lambda = \lambda = hc/E$. Thus, $E = hc/\Delta\lambda = hc/\lambda_c(1 - \cos\theta)$. Since $\cos\theta \geq -1$, and $\lambda_c = h/m_e c$, the minimum energy is $hc \div (h/m_e c)(1 - (-1)) = \frac{1}{2}m_e c^2 = \frac{1}{2}(511$ keV$) = 255$ keV.

Section 39X-4 Atomic Spectra and the Bohr Atom

Problem

37. What is the maximum wavelength of light that can ionize hydrogen in its ground state? In what spectral region is this?

Solution

The energy of the ground state of hydrogen is -13.6 eV (Equation 39X-12b), which has the same magnitude as the ionization energy. A photon with at least this energy has wavelength $\lambda \leq hc/|E| = 1240$ eV·nm/13.6 eV $= 91.2$ nm. This is the same as the Lyman series limit (Equation 39X-9 with $n_2 = 1$ and $n_1 = \infty$) $R_H^{-1} = hc/13.6\text{eV}$, and lies in the ultraviolet.

Problem

41. At what energy level does the Bohr hydrogen atom have diameter 5.18 nm?

Solution

The diameter of a hydrogen atom in the Bohr model is (Equation 39X-13) $2r_n = 2n^2a_0$, so $n^2 = 5.18$ nm$/2(0.0529$ nm$) = 49.0$, or $n = 7$. This is the sixth excited state.

Problem

45. Helium with one of its two electrons removed acts very much like hydrogen, and the Bohr model successfully describes it. Find (a) the radius of the ground-state electron orbit and (b) the photon energy emitted in a transition from the $n = 2$ to the $n = 1$ state in this singly ionized helium. *Hint:* The nuclear charge is $2e$.

Solution

Modifying the treatment of the Bohr atom in the text for singly ionized helium (He II), by replacing the nuclear charge with $2e$, one gets $r_n = -2ke^2/2E_n$ and $E_n = -(2ke^2)^2m/2\hbar^2n^2$. Thus, $E_n = -4(ke^2/2n^2a_0) = -2^2(13.6$ eV$)/n^2$, and $r_n = \frac{1}{2}n^2a_0$. (In general, replacing the nuclear charge with Ze gives results for any one-electron Bohr atom.) (a) The radius of the ground state of He II is $\frac{1}{2}a_0 = 0.0265$ nm. (b) The energy released in the transition $n = 2$ to $n = 1$ is $\Delta E = 4(13.6$ eV$)(1 - \frac{1}{4}) = 40.8$ eV. (There is also a small change in m for helium, different from the correction in hydrogen, for the motion of the nucleus.)

Section 39X-5 Matter Waves

Problem

49. A proton and an electron have the same de Broglie wavelength. How do their speeds compare, assuming both are much less than that of light?

Solution

The same de Broglie wavelength means the same momentum, so at non-relativistic speeds, $m_pv_p = m_ev_e$, or $v_e/v_p = m_p/m_e = 1836$ (use the "best known values" of physical constants given on the inside front cover).

Problem

53. A Davisson-Germer type experiment using nickel gives peak intensity in the reflected beam when the incident and reflected beams have an angular separation of 100°. What is the electron energy?

Solution

The Bragg angle in the Davisson-Germer experiment, for a deflection of 100°, is $\theta = (180° - \phi)/2 = (180° - 100°)/2 = 40°$ (see Fig. 39X-18). The Bragg condition (for the first order intensity peak) gives $\lambda = 2d \sin \theta$, while the kinetic energy (for a non-relativistic electron) is $K = p^2/2m = (h/\lambda)^2/2m = (hc/\lambda)^2/(2mc^2)$. For scattering from a nickel crystal, $d = 91$ pm, so $K = (hc/2d \sin \theta)^2/(2mc^2) = (1.24$ keV·nm$/91$ pm $\sin 40°)^2/8(511$ keV$) = 110$ eV.

Section 39X-6 The Uncertainty Principle

Problem

57. A proton has velocity $\mathbf{v} = (1500 \pm 0.25)\hat{\imath}$ m/s. What is the uncertainty in its position?

Solution

Take the uncertainty in velocity to be the full range of variation given, i.e., $\Delta v = 0.25$ m/s $- (-0.25$ m/s$) = 0.5$ m/s, and use $\Delta p = m \Delta v$ in Equation 39X-15 (since $v/c \ll 1$). Then $\Delta x \gtrsim \hbar/m \Delta v = (1.055 \times 10^{-34}$ J·s$)(1.67 \times 10^{-27}$ kg$)^{-1}(0.5$ m/s$)^{-1} = 126$ nm.

Problem

61. A proton is moving along the x axis with speed $v = (1500 \pm 0.25)$ m/s, but its direction $(+$ or $-x)$ is unknown. Find the uncertainty in its position. *Hint:* What is the difference between the two extreme possibilities for the *velocity?*

Solution

The uncertainty in the proton's momentum is $\Delta p_x = 2mv = 2(1.67 \times 10^{-27}$ kg$)(1500$ m/s$) = 5.01 \times 10^{-24}$ kg·m/s, since p_x could vary between $+mv$ and $-mv$ if the direction is unknown. The Equation 39X-15 gives $\Delta x \gtrsim \hbar/\Delta p_x = (1.055 \times 10^{-34}$ J·s$) \div (5.01 \times 10^{-24}$ kg·m/s$) = 21.1$ pm.

Problem

65. The lifetimes of unstable particles set energy-time uncertainty limits on the accuracy with which the rest energies— and hence the masses—of those particles can be known.

The particle known as the neutral pion has rest energy 135 MeV and lifetime 8×10^{-17} s. Find the uncertainty in its rest energy (a) in eV and (b) as a fraction of the rest energy.

Solution

(a) The maximum time for a measurement of the pion's rest mass is about one lifetime, $\Delta t = 8\times10^{-17}$ s, so Equation 39X-16 gives $\Delta E \gtrsim \hbar/\Delta t = 6.582\times10^{-16}$ eV·s ÷ 8×10^{-17} s = 8.23 eV (this is the width of the neutral pion). (b) As a fraction of the rest energy, this is $\Delta E/E = 8.23$ eV ÷ 135 MeV = 6.10×10^{-8}.

Paired Problems

Problem

69. A photocathode ejects electrons with maximum energy 0.85 eV when illuminated with 430-nm blue light. Will it eject electrons when illuminated with 633-nm red light, and if so what will be the maximum electron energy?

Solution

Einstein's photoelectric effect equation, and the data for the blue light, give the work function of the photocathode material, $\phi = E_\gamma - K_{max} = hc/\lambda - K_{max} =$ (1240 eV·nm/430 nm) − 0.85 eV = 2.03 eV. The energy of a photon of the red light is only $hc/\lambda = $ (1240 eV·nm) ÷ (633 nm) = 1.96 eV, and is insufficient to eject photoelectrons. (633 nm is greater than the cutoff wavelength of $\lambda_{cutoff} = hc/\phi = 610$ nm for the photocathode material.)

Problem

73. (a) Find the energy of the highest energy photon that can be emitted as the electron jumps between two adjacent allowed energy levels in the Bohr hydrogen atom. (b) Which energy levels are involved?

Solution

The energy of the photon emitted in a hydrogen atom transition between adjacent states $(n_1 \to n_2 = n_1 - 1)$ is

$$E_\gamma = \frac{hc}{\lambda} = hcR_H \left[\frac{1}{(n_1 - 1)^{-2}} - \frac{1}{n_1^{-2}} \right] = \frac{hcR_H(2n_1 - 1)}{n_1^{-2}(n_1 - 1)^{-2}}$$

(see Equations 39X-6 and 9 and the discussion of the Bohr Atom in the text). This is a maximum for the smallest allowed $n_1 > n_2 \geq 1$, which is $n_1 = 2$. Then (a) $E_{\gamma,max} = (\frac{3}{4}) hcR_H = (\frac{3}{4})(13.6$ eV) = 10.2 eV, and (b) $n_2 = n_1 - 1 = 1$. (This is the Lyman alpha photon at 122 nm.)

Problem

77. An electron is known to be within about 0.1 nm of the nucleus of an atom. What is the uncertainty in its velocity?

Solution

We suppose that "within 0.1 nm" means ±0.1 nm, so we may take the uncertainty in the electron's position to be $\Delta x = $ 0.2 nm (as in Example 39X-7). Then the minimum uncertainty in momentum is $\Delta p \gtrsim \hbar/\Delta x = $ 1.055×10^{-34} J·s/2×10^{-10} m = (197.3 eV·nm/c)/0.2 nm $\simeq$ 5×10^{-25} kg·m/s $\simeq 1$ keV/c. This is small compared to $mc = 511$ keV/c, so non-relativistic expressions can be used. Then $\Delta v = \Delta p/m \approx (5\times10^{-25}$ kg·m/s)/(9.11×10^{-31} kg) $\simeq$ 6×10^5 m/s $\simeq (1$ keV/c)/(511 keV/c^2) $\simeq 2\times10^{-3}$ c. (We included both SI and atomic units for pedagogical reasons.)

Supplementary Problems

Problem

81. A photon's wavelength is equal to the Compton wavelength of a particle with mass m. Show that the photon's energy is equal to the particle's rest energy.

Solution

The Compton wavelength of a particle is $\lambda_c = h/mc$. The energy of a photon with this wavelength is $E_\gamma = hc/\lambda_c = $ $hc/(h/mc) = mc^2$, the same as the particle's rest energy.

Problem

85. A photon undergoes a 90° Compton scattering off a stationary electron, and the electron emerges with *total* energy $\gamma m_e c^2$, where γ is the relativistic factor introduced in Chapter 38. Find an expression for the initial photon energy.

Solution

For Compton scattering at 90°, Equation 39-8 gives $\lambda' = $ $\lambda + \lambda_c$. In terms of the photon energies ($\lambda = hc/E_0$, $\lambda' = hc/E'$) and the electron's Compton wavelength ($\lambda_c = hc/mc^2$), this can be written as $1/E' = 1/E_0 + 1/mc^2$, or $E' = E_0 mc^2/(E_0 + mc^2)$. The recoil electron's kinetic energy is $K_e = (\gamma - 1)mc^2 = E_0 - E' = $ $E_0 - E_0 mc^2/(E_0 + mc^2) = E_0^2/(E_0 + mc^2)$. This is a quadratic equation in E_0, namely $E_0^2 - (\gamma - 1) mc^2(E_0 + mc^2) = 0$. The positive solution for E_0 is $\frac{1}{2}\left[(\gamma - 1)mc^2 + \sqrt{(\gamma - 1)^2 m^2 c^4 + 4(\gamma - 1)m^2 c^4}\right]$ $= \frac{1}{2}mc^2 \times \left[(\gamma - 1) + \sqrt{(\gamma - 1)(\gamma + 3)}\right]$.

● **CHAPTER 40** QUANTUM MECHANICS

Section 40-2 The Schrödinger Equation

Problem

1. What are the units of the wave function $\psi(x)$ in a one-dimensional situation?

Solution

The one-dimensional wave function is related to the probability by Equation 40-2, $dP = \psi^2(x)\,dx$. Since probability (a pure number) is dimensionless, the units of ψ must be the square root of the inverse of the units of x (a length), or (meters)$^{-1/2}$.

Problem

5. Use a table of definite integrals or symbolic math software to help evaluate the normalization constant A in the wave function of Problem 2.

Solution

The normalization condition for $\psi = Ae^{-x^2/a^2}$ is $1 = \int_{-\infty}^{\infty} A^2 e^{-2x^2/a^2}\,dx = A^2\sqrt{\pi/(2/a^2)} = A^2 a\sqrt{\pi/2}$. (See, for example, the CRC table of definite integrals.) Then $A = (2/\pi)^{1/4}a^{-1/2}$.

Section 40-3 The Infinite Square Well

Problem

9. What is the width of an infinite square well in which a proton cannot have an energy less than 100 eV?

Solution

The lowest possible energy of the proton (in a one-dimensional infinite square well) is its ground-state energy, which we take to be 100 eV. We can solve for the well-width from Equation 40-6, with $n = 1$: $L = h/\sqrt{8\,mE_1} = $ (1240 eV·nm/c) $\div \sqrt{8(938 \text{ MeV}/c^2)(100 \text{ eV})} = 1.43$ pm.

Problem

13. An electron drops from the $n = 7$ to the $n = 6$ level of an infinite square well 1.5 nm wide. Find (a) the energy and (b) the wavelength of the photon emitted.

Solution

(a) $\Delta E = (n_i^2 - n_f^2)h^2/8\,mL^2 = (7^2 - 6^2)(1240 \text{ eV·nm/c})^2 \div 8(511 \text{ keV}/c^2)(1.5 \text{ nm})^2 = 2.17$ eV (from Equation 40-6).
(b) $\lambda = hc/\Delta E = 8mc^2 L^2/(n_i^2 - n_f^2)hc = 571$ nm.

Problem

17. Repeat Example 40-1 for a proton trapped in a nuclear-size square well of width 1 fm. Comparison with the result of Example 40-1 gives a rough estimate of the energy difference between nuclear and chemical reactions.

Solution

The ground-state energy of a proton in an infinite square well of width 1 fm can be estimated from Equation 40-6:

$$E_1 = \frac{h^2}{8mL^2} = \frac{(6.63 \times 10^{-34} \text{ J·s})^2}{8(1.67 \times 10^{-27} \text{ kg})(10^{-15} \text{ m})^2} = 3.29 \times 10^{-11} \text{ J}.$$

In atomic units:

$$E_1 = \frac{(hc)^2}{8mc^2 L^2} = \frac{(1240 \text{ MeV-fm})^2}{8(938 \text{ MeV})(1 \text{ fm})^2} = 205 \text{ MeV}.$$

Comparison with Example 40-1 shows that the energy scale for nuclear reactions is millions of times greater than for chemical reactions.

Problem

21. Sketch the probability density for the $n = 2$ state of an infinite square well extending from $x = 0$ to $x = L$, and determine where the particle is most likely to be found.

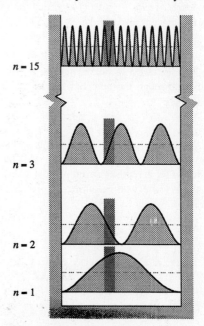

FIGURE 40-12 For reference.

Solution

The probability density for the $n = 2$ state of a one-dimensional infinite square well is $\psi_2(x)^2 = (2/L)\sin^2(2\pi x/L)$ (see Equation 40-7), a graph of which is shown in Fig. 40-12. The probability density has maxima when $2\pi x/L = \pi/2$ and $3\pi/2$, or at $x = L/4$ and $3L/4$, where the particle is most likely to be found (in a measurement with given Δx).

Sections 40-4, 40-5, and 40-6 The Harmonic Oscillator, Quantum Tunneling, and the Finite Potential Wells

Problem

25. What is the ground-state energy for a particle in a harmonic oscillator potential whose classical angular frequency is $\omega = 1.0\times10^{17}$ s^{-1}?

Solution

The ground-state energy of a one-dimensional harmonic oscillator is $E_0 = \frac{1}{2}\hbar\omega = \frac{1}{2}(6.582\times10^{-16}$ eV·s$)(1.0\times10^{17}$ s$^{-1}) = 32.9$ eV (see Equation 40-11 with $n = 0$).

Problem

29. For what quantum numbers is the spacing between adjacent energy levels in a harmonic oscillator less than 1% of the actual energy?

Solution

The energy levels are $E_n = (n + \frac{1}{2})\hbar\omega$ (Equation 40-11) so the spacing of adjacent levels is $\Delta E = \hbar\omega$. Thus, $\Delta E/E_n = 1/(n + \frac{1}{2}) < 0.01$ for $n + \frac{1}{2} > 100$, or $n > 99$.

Problem

33. The probability that a particle of mass m and energy $E < U$ will tunnel through the potential barrier of Fig. 40-17 is approximately

$$P = e^{-2\sqrt{2m(U-E)}L/\hbar},$$

where L is the barrier width. Evaluate this probability (a) for a 2.8-eV electron incident on a 1.0-nm wide barrier 4.0 eV high and (b) for a 1200-kg car moving at 15 m/s striking a 1.0-m-thick stone wall requiring, classically, 150 kJ to breach it.

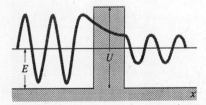

FIGURE 40-17 For reference.

Solution

(a) For the electron, we use <u>atomic units</u> to calculate the probability. $P = \exp\{-2\sqrt{2mc^2(U - E)}L/hc\} = \exp\{-2\sqrt{2}(511$ keV$)(4.0 - 2.8)$ eV $(1.0$ nm$)/1240$ eV·nm$\} = e^{-1.79} = 0.168$. (b) For the car, SI units are more appropriate, and $E = \frac{1}{2}mv^2 = 135$ kJ. The probability of penetrating a 150 kJ high barrier, 1 m thick, is $P = \exp\{-2\sqrt{2}(1200$ kg$)(150 - 135)$kJ $(1$ m$) \div (6.626\times10^{-34}$ J·s$)\} = \exp\{-1.81\times10^{37}\} = 10^{-7.87\times10^{36}}$, which is vanishingly small. (We used $e^x = 10^{x/\ln 10}$ to express the exponential as a power of ten.)

Section 40-7 The Schrödinger Equation in Three Dimensions

Problem

37. The generalization of the Schrödinger equation to three dimensions is

$$-\frac{\hbar^2}{2m}\left(\frac{\partial^2\psi}{\partial x^2} + \frac{\partial^2\psi}{\partial y^2} + \frac{\partial^2\psi}{\partial z^2}\right) + U(x, y, z)\psi = E\psi.$$

(a) For a particle confined to the cubical region $0 \le x \le L$, $0 \le y \le L$, $0 \le z \le L$, show by direct substitution that the equation is satisfied by wave functions of the form $\psi(x, y, z) = A\sin(n_x\pi x/L)\sin(n_y\pi y)L)\sin(n_z\pi z/L)$, where the n's are integers and A is a constant. (b) In the process of working part (a), verify that the energies E are given by Equation 40-12.

Solution

For the given wave function $\psi(x, y, z)$, the second partial derivatives are $\partial^2\psi/\partial x^2 = -(n_x\pi/L)^2\psi$, $\partial^2\psi/\partial y^2 = -(n_y\pi/L)^2\psi$, and $\partial^2\psi/\partial z^2 = -(n_z\pi/L)^2\psi$. Substituting into the Schrödinger equation, with a potential for a cubical box ($U = 0$ inside and $U = \infty$ outside), we find

$$-\frac{\hbar^2}{2m}\left(\frac{\partial^2\psi}{\partial x^2} + \frac{\partial^2\psi}{\partial y^2} + \frac{\partial^2\psi}{\partial z^2}\right)$$

$$= -\frac{1}{2m}\left(\frac{h}{2\pi}\right)^2\left[-\left(\frac{n_x\pi}{L}\right)^2 - \left(\frac{n_y\pi}{L}\right)^2 - \left(\frac{n_z\pi}{L}\right)^2\right]\psi$$

$$= (h^2/8mL^2)(n_x^2 + n_y^2 + n_z^2)\psi = E\psi,$$

thus (a) demonstrating that $\psi(x, y, z)$ is a solution, and (b) displaying the energy. This wave function also satisfies the boundary conditions appropriate for confinement ($\psi = 0$ at $x = 0$ or L, $y = 0$ or L, and $z = 0$ or L, points where $U = \infty$), since n_x, n_y and n_z are integers.

Paired Problems

Problem

41. An alpha particle (mass 4 u) is trapped in a uranium nucleus of diameter 15 fm. Treating the system as a one-dimensional square well, what would be the minimum energy for the alpha particle?

Solution

Equation 40-6, with $n = 1$, gives the lowest alpha-particle energy, in a one-dimensional infinite square well of width 15 fm, as $E = h^2/8mL^2 = (hc/L)^2/8(mc^2) = (1240 \text{ MeV·fm}/15 \text{ fm})^2 \div 8(4 \times 931.5 \text{ MeV}) = 0.229 \text{ MeV}$ (use 1 u = 931.5 MeV/c^2).

Problem

45. A harmonic oscillator emits a 1.1-eV photon as it undergoes a transition between adjacent states. What is its classical oscillation frequency?

Solution

For a one-dimensional harmonic oscillator, $\Delta E = \hbar\omega = hf$ for transitions between adjacent states (see Equation 40-11.), so $f = \Delta E/h = 1.1 \text{ eV}/4.136\times10^{-15} \text{ eV·s} = 2.66\times10^{14} \text{ Hz}$.

Supplementary Problems

Problem

49. For what quantum state is the probability of finding a particle in the left-hand quarter of an infinite square well equal to 0.303?

Solution

A straightforward generalization of Example 40-2 shows that the probability of finding a particle, in the quantum state n, in the left-hand quarter of a one-dimensional infinite square well is just

$$P = \int_0^{L/4} x_n^2(x)\, dx = \frac{2}{L}\left| \frac{x}{2} - \frac{\sin(2n\pi x/L)}{(4n\pi/L)} \right|_0^{L/4}$$
$$= \frac{1}{4} - \frac{\sin(n\pi/2)}{2\pi n}.$$

The probability is equal to $\frac{1}{4}$ for any even n, and is greater than $\frac{1}{4}$ for $n = 3 + 4n'$ (or less than $\frac{1}{4}$ for $n = 1 + 4n'$) where $n' = 0, 1, 2, \ldots$ (This follows from the fact that $\sin(n\pi/2)$ alternates between ± 1 for integer n.) In this problem, the probability is greater than $\frac{1}{4}$, so $0.303 - 0.25 = \frac{1}{2}\pi n$, or $n = \frac{1}{2}\pi(0.053) = 3.00$, for the third quantum state.

Problem

53. Consider an infinite square well with a steplike potential in the bottom, as shown in Fig. 40-33. Without solving any equations, sketch what you think the wave function should look like for a particle whose energy is (a) less than the step height and (b) greater than the step height.

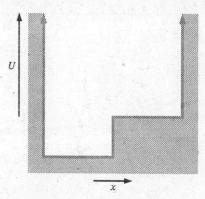

FIGURE 40-33 Problem 53.

Solution

The solutions of the Schrödinger equation ($\psi'' = -[2m(E - U)/\hbar^2]\psi$) are oscillatory in regions where the kinetic energy of the particle is positive (ψ'' has the opposite sign to ψ if $E - U > 0$, so ψ always curves towards the x axis), while the solutions are exponential in classically forbidden regions (ψ'' has the same sign as ψ if $E - U < 0$, so ψ always curves away from the x axis). In regions where $E > U$, the wave function oscillates more rapidly when $E - U$ is greater (i.e. the deBroglie wavelength, $\lambda = h/\sqrt{2m(E - U)}$, is smaller). Except where the potential is infinite (in which case $\psi \equiv 0$), ψ & ψ' are continuous. In general, where oscillatory wave functions join, as at the step

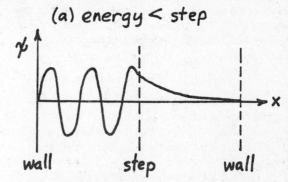

Problem 53 Solution (a).

(b) energy > step

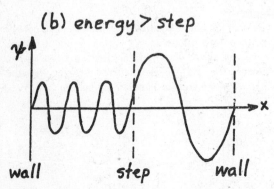

Problem 53 Solution (b).

in part (b), the amplitude of the oscillations is greater where the wavelength is greater (classically, the probability density is greater where the velocity is smaller). (If the functions join at a point where $\psi' = 0$, the amplitudes are the same, indicative of a resonance condition.) Typical wave functions are sketched above for the potential in Fig. 37-37.

Problem

57. The equation derived in the preceding problem cannot be solved algebraically since the unknown ε appears in a trig function and under the square root. It can be solved graphically, by plotting both sides on the same graph and determining where they intersect. It can also be solved by trial and error on a calculator, by using a calculator with a root-finding routine, or by computer. Use one of these methods to find all possible values of ε for (a) $\mu = 2$, (b) $\mu = 20$, and (c) $\mu = 50$. *Note:* The number of solutions varies with μ; there may be no solutions, meaning no bound states are possible, or there may be one or more bound states.

Solution

(The solutions of Problems 55–57 have been consolidated below.) In the interior of the well, $U(x) = 0$, while in the exterior, $U(x) = U_0$. After multiplication by $2m/\hbar$ and a rearrangement of terms, Equation 37-1 becomes:

$$\frac{d^2\psi}{dx^2} = \begin{cases} -(2mE/\hbar^2)\psi, & \text{for } 0 \leq x \leq L \\ (2m(U_0 - E)/\hbar^2)\psi, & \text{for } L \leq x \leq \infty. \end{cases}$$

Since $E > 0$, ψ is oscillatory inside the well. The solution which satisfies the condition $\psi = 0$ at $x = 0$ (where $U(x) = \infty$) is $\psi_{\text{in}} = A\sin(\sqrt{2mE}\,x/\hbar)$. Outside the well, the type of solution (exponential or oscillatory) depends on whether $U_0 - E$ is positive or negative. The former corresponds to a bound state and the latter to an unbound or scattering state. Here, we have assumed that $E < U_0$ (exponential ψ), so the normalizable solution outside is $\psi_{\text{out}} = Be^{-\sqrt{2m(U_0-E)}\,x/\hbar}$. Direct substitution of $\varepsilon/L^2 = 2mE/\hbar^2$

and $\mu/L^2 = 2mU_0/\hbar^2$ yields $\psi_{\text{in}} = A\sin(\sqrt{\varepsilon}\,A/L)$ and $\psi_{\text{out}} = Be^{-\sqrt{\mu-\varepsilon}\,x/L}$.

At $x = L$, $\psi_{\text{in}} = \psi_{\text{out}}$, therefore $A\sin\sqrt{\varepsilon} = Be^{-\sqrt{\mu-\varepsilon}}$, which is the first equation: $|d\psi_{\text{in}}/dx|_{x=L} = (\sqrt{\varepsilon}\,A/L)\cos\sqrt{\varepsilon} = |d\psi_{\text{out}}/dx|_{x=L} = -(\sqrt{\mu-\varepsilon}\,B/L)e^{-\sqrt{\mu-\varepsilon}}$, which, after multiplication by L, yields the second equation: $\sqrt{\varepsilon}\,A\cos\sqrt{\varepsilon} = -\sqrt{\mu-\varepsilon}\,Be^{-\sqrt{\mu-\varepsilon}}$. These equations can be combined, as indicated in the problem, to yield:

$$A\sin\sqrt{\varepsilon} = Be^{-\sqrt{\mu-\varepsilon}} = -\frac{\sqrt{\varepsilon}\,A\cos\sqrt{\varepsilon}}{\sqrt{\mu-\varepsilon}},$$

or

$$\tan\sqrt{\varepsilon} = -\sqrt{\frac{\varepsilon}{\mu-\varepsilon}}.$$

Before a graphical solution is attempted, it is convenient to rewrite the above condition for the bound state energies $(0 < \varepsilon < \mu)$ as $-\sqrt{\varepsilon}\cot\sqrt{\varepsilon} = \sqrt{\mu - \varepsilon}$. The right-hand side, as a function of $\sqrt{\varepsilon}$, represents a quarter of a circle of radius $\sqrt{\mu}$. The left-hand side is a function which ranges from 0 to ∞ in the intervals $\frac{1}{2}\pi \leq \sqrt{\varepsilon} \leq \pi$, $\frac{3}{2}\pi \leq \sqrt{\varepsilon} \leq 2\pi$, $\frac{5}{2}\pi \leq \sqrt{\varepsilon} \leq 3\pi$, etc., for which physical solutions are possible. From the sketch below, it can be seen that there are no bound states for $\sqrt{\mu} < \frac{1}{2}\pi$ (or $\mu < 2.47$), there is one bound state for $\frac{1}{2}\pi < \sqrt{\mu} < \frac{3}{2}\pi$ (or $2.47 < \mu < 22.2$), there are two bound states for $\frac{3}{2}\pi < \sqrt{\mu} < \frac{5}{2}\pi$ (or $22.2 < \mu < 61.7$), etc. Values of ε can be found from the graph, or by numerical computation. For the specified cases, they are: (a) none for $\mu = 2$, (b) $\varepsilon = 6.44$ for $\mu = 20$, and (c) $\epsilon = 7.52$ and 29.3 for $\mu = 50$.

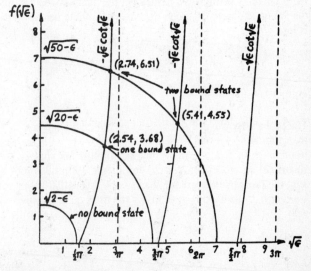

Problem 57 Solution.

CHAPTER 41 ATOMIC PHYSICS

Section 41-1 The Hydrogen Atom

Problem

1. Using physical constants accurate to four significant figures (see the inside front cover), verify the numerical values of the Bohr radius a_0 and the hydrogen ground-state energy E_1.

Solution

$$a_0 = \frac{4\pi \varepsilon_0 \hbar^2}{me^2} = \frac{4\pi(8.854 \text{ pF/m})(6.626\times10^{-34} \text{ J·S}/2\pi)^2}{(9.109\times10^{-31} \text{ kg})(1.602\times10^{-19} \text{ C})^2}$$
$$= 5.293\times10^{-11} \text{ m},$$

while

$$E_1 = -\frac{\hbar^2}{2ma_0^2} = \frac{-(6.626\times10^{-34} \text{ J·s}/2\pi)^2}{2(9.109\times10^{-31} \text{ kg})(5.292\times10^{-11} \text{ m})^2}$$
$$= -2.180\times10^{-18} \text{ J} = -13.61 \text{ eV}.$$

Problem

5. What is the maximum possible magnitude for the orbital angular momentum of an electron in the $n = 7$ state of hydrogen?

Solution

The quantum number ℓ can take integer values from 0 to $n - 1$, so its maximum value is 6. From Equation 41-10, $L = \sqrt{6 \times 7}\hbar = \sqrt{42}\hbar$.

Problem

9. Determine the principal and orbital quantum numbers for a hydrogen atom whose electron has energy -0.850 eV and orbital angular momentum of magnitude $\sqrt{12}\hbar$.

Solution

Equations 41-9 and 10 can be written as $n^2 = -13.6 \text{ eV}/E$, and $\ell(\ell + 1) = (L/\hbar)^2$. For the given state, $n^2 = 13.6/0.850 = 16 = (4)^2$, and $\ell(\ell + 1) = 12 = 3(4)$, so this is a 4f-state ($n = 4$, $\ell = 3$).

Problem

13. Give a symbolic description for the state of the electron in a hydrogen atom when the total energy is -1.51 eV and the orbital angular momentum is $\sqrt{6}\hbar$.

Solution

From Equation 41-9, $n = \sqrt{13.6/1.51} = 3.00$, and from Equation 41-10, $\ell(\ell + 1) = 6 = 2(3)$. Thus, $n = 3$ and $\ell = 2$, and this is a 3d-state.

Problem

17. A hydrogen atom has energy $E = -0.850$ eV. What are the maximum possible values for (a) the magnitude of its orbital angular momentum and (b) the component of that angular momentum on a chosen axis?

Solution

Equation 41-9 shows that $n = \sqrt{-13.6 \text{ eV}/E} = \sqrt{13.6/0.850} = 4$ for this hydrogen atom state. (a) The fact that $\ell_{max} = n - 1$ and Equation 41-10 imply $L_{max} = \sqrt{12}\,\hbar = 3.65\times10^{-34}$ J·s. (b) From Equation 41-11, with $m_{\ell, max} = \ell_{max}$, one finds $L_{z, max} = 3\hbar = 3.16\times10^{-34}$ J·s.

Section 41-2 Electron Spin

Problem

21. Verify the value of the Bohr magneton in Equation 41-15.

Solution

Values from the inside front cover substituted into Equation 41-15 yield:

$$\mu_B = \frac{e\hbar}{2m} = \frac{(1.602\times10^{-19} \text{ C})(6.626\times10^{-34} \text{ J·s})}{4\pi(9.109\times10^{-31} \text{ kg})}$$
$$= 9.273\times10^{-24} \text{ A·m}^2.$$

Problem

25. What are the possible j values for a hydrogen atom in the $3D$ state?

Solution

For the $3D$-state, $\ell = 2$ and $s = \frac{1}{2}$ (hydrogen has one electron) so the possible j-values are $j = \ell - \frac{1}{2} = \frac{3}{2}$ and $j = \ell + \frac{1}{2} = \frac{5}{2}$.

Section 41-3 The Pauli Exclusion Principle

Problem

29. Suppose you put five electrons into an infinite square well. (a) How do the electrons arrange themselves to achieve the lowest total energy? (b) Give an expression for this energy in terms of the electron mass m, the well width L, and Planck's constant.

Solution

(a) The energy levels for a one-dimensional infinite square well are spacially non-degenerate (see Fig. 40-10), so the Pauli principle allows, at most, two electrons per level (one with spin up, one with spin down). Thus, two electrons may occupy the ground state ($n = 1$), two the first excited state ($n = 2$), and one the second excited state ($n = 3$). (b) The energy of this configuration is $2E_1 + 2E_2 + E_3 = 2E_1 + 2(2)^2E_1 + (3)^2E_1 = 19E_1 = 19h^2/8mL^2$, where we used Equation 40-6 for the energies with E_1 denoting the ground state energy. (For a higher dimensional potential, degeneracy alters these conclusions accordingly.)

Section 41-4 Multielectron Atoms and the Periodic Table

Problem

33. Write the full electronic structure of scandium.

Solution

Scandium ($Z = 21$), the first of the transition metals, has one $3d$-electron in addition to the configuration of the preceding element, calcium ($Z = 20$). (See explanation in the text just before Example 41-7.) Thus $1s^2 2s^2 2p^6 3s^2 3p^6 4s^2 3d^1$ is the full electronic configuration for scandium.

Section 41-5 Transitions and Atomic Spectra

Problem

37. An electron in a highly excited state of hydrogen ($n_1 \gg 1$) drops into the state $n = n_2$. What is the lowest value of n_2 for which the emitted photon will be in the infrared ($\lambda > 700$ nm)?

Solution

The wavelength emitted in a hydrogen atom transition from states n_1 to n_2 is given by $\Delta E = hc/\lambda = 13.6 \text{ eV}(n_2^{-2} - n_1^{-2})$ (subtract energy levels in Equation 41-9 or see Chapter 39). When $n_1 \gg 1$, $n_1^{-2} \approx 0$ and $hc/\lambda \approx 13.6 \text{ eV}/n_2^2$. Therefore, for photons in the infrared ($\lambda > 700$ nm), $n_2 = \sqrt{\lambda(13.6 \text{ eV})/hc} > \sqrt{(700 \text{ nm})(13.6 \text{ eV})/(1240 \text{ eV·nm})} = 2.77$, or $n_2 \geq 3$. (Under similar conditions, i.e., $n_1 \approx \infty$, $n_2 \geq 105$ for microwave photons, $\lambda > 1$ mm.)

Problem

41. The $4s \rightarrow 3p$ transition in sodium produces a doublet spectral line at 1138.1 nm and 1140.4 nm. Combine this fact with the discussion in Example 41-8 to find an accurate value for the energy difference between the 3s and 4s states in sodium.

Solution

Because the s-levels have no fine structure splitting, the transitions and photon energies given in this problem and in Example 41-8 can be combined to give the $4s - 3s$ energy difference. The transitions are $4s \xrightarrow{1140.4 \text{ nm}} 3P_{3/2} \xrightarrow{589.0 \text{ nm}} 3s$, and $4s \xrightarrow{1138.1 \text{ nm}} 3P_{1/2} \xrightarrow{589.6 \text{ nm}} 3s$, and the energy difference is $\Delta E = hc(\lambda_{4s-3p}^{-1} + \lambda_{3p-3s}^{-1})$. From either set of transitions:

$$1240 \text{ eV·nm}\left(\frac{1}{1140.4 \text{ nm}} + \frac{1}{589.0 \text{ nm}}\right)$$
$$= 1240 \text{ eV·nm}\left(\frac{1}{1138.1 \text{ nm}} + \frac{1}{589.6 \text{ nm}}\right) = 3.193 \text{ eV}.$$

Problem

45. What is the approximate minimum accelerating voltage for an x-ray tube with an iron target to produce the $L\alpha$ line? *Hint:* See Problem 44.

Solution

In the operation of an x-ray tube, the bombarding electrons (with kinetic energy equal to the electronic charge times the accelerating voltage, $K_e = e\Delta V$) effectively eject an inner shell electron from the target atoms (as hinted at in the previous problem). The ionization energy of an L-shell electron in iron can be estimated (as in Example 41-9) from Equation 41-9, with $Z_{\text{eff}} = 26 - 9 = 17$, or $(13.6 \text{ eV})(17)^2/2^2 = 983$ eV. Thus, about 980 V accelerating voltage is necessary.

Problem

49. An ensemble of square-well systems of width 1.17 nm all contain electrons in highly excited states. They undergo all possible transitions in dropping toward the ground state, obeying the selection rule that Δn must be odd. (a) What wavelengths of visible light are emitted? (b) Is there any infrared emission? If so, how many spectral lines fall in the infrared?

Solution

The energy levels for an electron in an infinite square well are given by Equation 40-6. In atomic units, $E_n = n^2(hc)^2 \div 8mc^2L^2 = n^2(1240 \text{ eV·nm})^2/8(0.511 \text{ MeV})(1.17 \text{ nm})^2 = n^2(0.275 \text{ eV})$. The transition energies are proportional to the differences of squares of integers, $\Delta E_{fi} = (n_i^2 - n_f^2) \times (0.275 \text{ eV})$, where $\Delta n = n_i - n_f$ must be odd. (a) Photons in the visible region have energies between $hc/700$ nm $= 1.77$ eV and $hc/400$ nm $= 3.10$ eV, so $n_i^2 - n_f^2 = (n_i + n_f)(n_i - n_f) = (2n_i - \Delta n)\Delta n$ should lie between $1.77/0.275 = 6.45$ and $3.10/0.275 = 11.3$. For $\Delta n = 1$ transitions, this implies $6.45 < 2n_i - 1 < 11.3$, or $3.72 < n_i < 6.14$. Thus, $n_i = 4$, 5, or 6, and the transitions $4 \rightarrow 3$, $5 \rightarrow 4$, or $6 \rightarrow 5$ lead to visible photons with wavelengths $\lambda = hc/\Delta E_{fi} = 1240 \text{ eV·nm}/(2n_i - 1)(0.275 \text{ eV}) = 645$ nm, 501 nm, and 410 nm, respectively. The lowest energy $\Delta n = 3$ transition ($n_i^2 - n_f^2 = 4^2 - 1^2 = 15 > 11.3$) lies outside the visible region, in the ultraviolet. (b) The two $\Delta n = 1$ transitions $3 \rightarrow 2$ and $2 \rightarrow 1$ ($n_i^2 - n_f^2 = 5$ or $3 < 6.45$) both lie in the infrared at wavelengths 0.903 μm and 1.50 μm, respectively.

Paired Problems

Problem

53. Find the spacing between adjacent orbital angular momentum values for hydrogen in the $n = 2$ state.

Solution

For the $n = 2$ states in hydrogen, the ℓ-values can be 0, or 1. The orbital angular momentum values are $L = \sqrt{\ell(\ell + 1)}\hbar = 0$, or $\sqrt{2}\hbar$, and their spacing is $\sqrt{2}\hbar$.

Problem

57. Estimate the energy of the $L\alpha$ x-ray transition in arsenic.

Solution

For the $L\alpha$ line in As ($Z = 33$), we may use $Z_{eff} = Z - 9$ (as in Example 41-9) to estimate the transition energy $\Delta E(L\alpha) = (13.6 \text{ eV})(33 - 9)^2(2^{-2} - 3^{-2}) = 1.09 \text{ keV}$.

Supplementary Problems

Problem

61. Substitute the ψ_{2s} wave function (Equation 41-8) into Equation 41-4 to verify that the equation is satisfied and that the energy is given by Equation 41-7 within $n = 2$.

Solution

First, we find the derivatives of $\psi_{2s} = Ae^{-r/2a_0}(2 - r/a_0)$, where $A = 1/4\sqrt{2\pi a_0^3}$, using the product rule and collecting terms:

$$d\psi_{2s}/dr = -Ae^{-r/2a_0}(2 - r/2a_0)/a_0,$$
$$d^2\psi_{2s}/dr^2 = Ae^{-r/2a_0}(3 - r/2a_0)/2a_0^2.$$

Next, we expand the derivatives on the left-hand side of Equation 41-4, substitute and factorize:

$$-\frac{\hbar^2}{2m}\left(\frac{d^2\psi_{2s}}{dr^2} + \frac{2}{r}\frac{d\psi_{2s}}{dr}\right)$$
$$= -\frac{\hbar^2}{2m}Ae^{-r/2a_0}\left[\frac{1}{2a_0^2}\left(3 - \frac{r}{2a_0}\right) - \frac{2}{ra_0}\left(2 - \frac{r}{2a_0}\right)\right]$$
$$= \frac{\hbar^2}{2m}Ae^{-r/2a_0}\frac{1}{ra_0}\left(4 - \frac{r}{2a_0}\right)\left(1 - \frac{r}{2a_0}\right)$$
$$= \frac{\hbar^2}{4mra_0}\left(4 - \frac{r}{2a_0}\right)\psi_{2s}.$$

When this is substituted into the full Equation 41-4, we can cancel a common factor of ψ_{2s} and use $\hbar^2/ma_0 = ke^2$:

$$\frac{\hbar^2}{4mra_0}\left(4 - \frac{r}{2a_0}\right) - \frac{ke^2}{r} = \frac{\hbar^2}{mra_0} - \frac{\hbar^2}{8ma_0^2} - \left(\frac{\hbar^2}{ma_0}\right)\frac{1}{r}$$
$$= \frac{-\hbar^2}{8ma_0^2} = E.$$

This is just the left-hand part of Equation 41-7 with $n = 2$.

CHAPTER 42 MOLECULAR AND SOLID-STATE PHYSICS

Section 42-2 Molecular Energy Levels

Problem

1. Find the energies of the first four rotational states of the HCl molecules described in Example 42-1.

Solution

The energies of rotational states (above the $j = 0$ state) are given by Equation 42-2, where for the HCl molecule, $\hbar^2/I = 2.63$ meV (from Example 42-1). Thus, $E_{rot} = j(j + 1)\hbar^2/2I = \frac{1}{2}j(j + 1)2.63$ meV. For $j = 0, 1, 2,$ and 3, $E_{rot} = 0, 2.63$ meV, 7.89 meV, and 15.78 meV.

Problem

5. Photons of wavelength 1.68 cm excite transitions from the rotational ground state to the first rotational excited state in a gas. What is the rotational inertia of the gas molecules?

Solution

The energy of the absorbed photon equals the difference in energy between the $j = 1$ and $j = 0$ rotational levels, which is (see Example 42-1) $\hbar^2/I = \Delta E_{1\to0} = hc/\lambda = 1240$ eV·nm/1.68 cm $= 7.38\times10^{-5}$ eV $= 1.18\times10^{-23}$ J. Therefore, $I = (1.055\times10^{-34}$ J·s$)^2/(1.18\times10^{-23}$ J$) = 9.41\times10^{-46}$ kg·m².

Problem

9. The rotational spectrum of diatomic oxygen (O_2) shows spectral lines spaced 0.356 meV apart in energy. Find the atomic separation in this molecule. *Hint:* See Example 42-1, but remember that the oxygen atoms have equal mass.

Solution

The separation of the rotational spectral lines in energy is $\Delta(\Delta E) = \hbar^2/I$ (see Example 42-1), or $\hbar^2/I = 0.356$ meV for O_2. In a diatomic molecule, with equal mass atoms and atomic separation R, each atom rotates about the center of mass at a distance of $R/2$, so $I = 2(m_0)(R/2)^2 = (8 \text{ u})R^2$, where the mass of an oxygen atom is about $m_0 = 16$ u. Then $R^2 = I/8$ u $= \hbar^2/(8$ u$)(0.356$ meV$)$, or

$R = \hbar c/\sqrt{(8 \text{ u}c^2)(0.356 \text{ meV})} = (197.3 \text{ eV·nm}) \div \sqrt{(8 \times 931.5 \text{ meV}) \times (0.356 \text{ meV})} = 0.121$ nm.

Problem

13. The energy between adjacent vibrational levels in diatomic nitrogen is 0.293 eV. What is the classical vibration frequency of this molecule?

Solution

From $\Delta E_{vib} = \hbar\omega = hf$, for adjacent levels, we find $f = 0.293$ eV$/4.136\times10^{-15}$ eV·s $= 7.08\times10^{13}$ Hz for the N_2-molecule.

Section 42-3   Solids

Problem

17. Express the 7.84-eV ionic cohesive energy of NaCl in kilocalories per mole of ions.

Solution

Using conversion factors from Appendix C, we find:
$(7.84 \text{ eV})(1.602\times10^{-19} \text{ J/eV})(1 \text{ kcal}/4184 \text{ J}) \times (6.022\times10^{23}/\text{mol}) = 181$ kcal/mol.

Problem

21. (a) Differentiate Equation 42-4 to obtain an expression for the force on an ion in an ionic crystal. (b) Use your result to find the force on an ion in NaCl if the crystal could be compressed to half its equilibrium spacing (see Example 42-3 for relevant parameters). Compare with the electrostatic attraction between the ions at a separation of $\frac{1}{2}r_0$. Your result shows how very "stiff" this ionic crystal is.

Solution

(a) The force in the r direction (positive is repulsive) is:

$$F_r = -\frac{dU}{dr} = \alpha\frac{ke^2}{r_0}\frac{d}{dr}\left[\frac{r_0}{r} - \frac{1}{n}\left(\frac{r_0}{r}\right)^n\right]$$
$$= \alpha\frac{ke^2}{r_0}\left[-\frac{r_0}{r^2} - \frac{n}{n}\left(\frac{r_0}{r}\right)^{n-1}\left(-\frac{r_0}{r^2}\right)\right]$$
$$= \alpha\frac{ke^2}{r_0^2}\left[\left(\frac{r_0}{r}\right)^{n+1} - \left(\frac{r_0}{r}\right)^2\right].$$

(b) If $r = \frac{1}{2}r_0$, $F_r = (\alpha ke^2/r_0^2)(2^{n+1} - 2^2) = \alpha(2^{n-1} - 1)F_{el}$, where $F_{el} = 4ke^2/r_0^2$ is the magnitude of the electrostatic attraction between a Na$^+$ and Cl$^-$ ion at a distance of $\frac{1}{2}r_0$. If we use values from Example 42-3 ($\alpha = 1.748$, $r_0 = 0.282$ nm, and $n = 8.22$), the repulsive force is $(1.748)(2^{7.22} - 1) = 259$ times the electrostatic attractive force. Since $F_{el} = 4(9\times10^9 \text{ N·m}^2/\text{C}^2)(1.6\times10^{-19} \text{ C})^2 \div (0.282 \text{ nm})^2 = 1.16\times10^{-8}$ N, $F_r = 259(1.16\times10^{-8} \text{ N}) = 3.00\times10^{-6}$ N.

Problem

25. Suppose the charge carriers in a material were protons, with density 10^{28} m^{-3}—comparable to that of electrons in a metal. What would be the order of magnitude of the Fermi energy?

Solution

Since the Fermi energy is proportional to the reciprocal of the charge carrier's mass (see solution to Problem 23), we expect the magnitude for protons to be $1/1836$ times the Fermi energy for electrons (with comparable density), or about three orders of magnitude smaller (i.e., $\sim$meV). In fact, if one uses $n = 10^{28}$ m^{-3} and $mc^2 = 938$ MeV (for protons) in the expression for E_F in Example 42-4, one finds $E_F = (3n/\pi)^{2/3}(hc)^2/8(mc^2) = (3\times10^{28} \text{ m}^{-3}/\pi)^{2/3} \times (1.24\times10^{-6} \text{ eV·m})^2/8(938\times10^6 \text{ eV}) = 9.22\times10^{-4}$ eV $\approx$ 1 meV.

Problem

29. What is the shortest wavelength of light that could be produced by electrons jumping the band gap in a material from Table 42-1? What is the material?

Solution

The wavelength emitted depends on the energy gap, since $\lambda = hc/E_{gap}$. The maximum wavelength for the materials in Table 42-1 (corresponding to the smallest gap) is for InAs, $\lambda = 1240$ eV·nm$/0.35$ eV $= 3.54$ μm (in the infrared). The minimum is for ZnS, $\lambda = 1240$ eV·nm$/3.6$ eV $= 344$ nm (in the ultraviolet).

Problem

33. The Sun radiates most strongly at about 500 nm, the peak of its Planck curve. The semiconductor zinc sulfide has a band gap of 3.6 eV. (a) What is the maximum wavelength absorbed by ZnS? (b) Would ZnS make a good photovoltaic cell? Why or why not?

Solution

(a) The photon wavelength corresponding to the band gap in ZnS is 344 nm (see solution to Problem 29) so photons with greater wavelengths (less energy) would not be readily absorbed (see statement of Problem 32). (b) ZnS would be a poor photovoltaic material since most of the energy in the solar spectrum is near the peak wavelength at 500 nm, and would not be absorbed.

Section 42-4   Superconductivity

Problem

37. The critical magnetic field in niobiom-titanium superconductor is 15 T. What current is required in a 5000-turn solenoid 75 cm long to produce a field of this strength?

Solution

The magnetic field inside a long thin solenoid is $B = \mu_0 nI$ (see Equation 30-12), so $I = 15 \text{ T}/(5000/0.75 \text{ m})(4\pi \times 10^{-7} \text{ T·m/A}) = 1.79 \text{ kA}$.

Paired Problems

Problem

41. What wavelength of infrared radiation is needed to excite a transition between the $n = 0$, $j = 3$ state and the $n = 1$, $j = 2$ state in KCl, for which the rotational inertia is 2.43×10^{-45} kg·m^2 and the classical vibration frequency is 8.40×10^{12} Hz?

Solution

The difference in energy between these vibrational-rotational levels is $\Delta E = \Delta E_{\text{vib}} + \Delta E_{\text{rot}} = hf - 3\hbar^2/I = (4.136 \times 10^{-15} \text{ eV·s})(8.40 \times 10^{12} \text{ Hz}) - 3(6.582 \times 10^{-16} \text{ eV·s})^2 \times (1.602 \times 10^{-19} \text{ J/eV})/(2.43 \times 10^{-45} \text{ J·s}^2) = 34.7$ meV, corresponding to a photon wavelength of $\lambda = hc/\Delta E = 35.8 \ \mu$m. (See Example 42-2 for ΔE_{rot} between the $j = 2$ and $j = 3$ levels, and Equation 42-3 for ΔE_{vib} between the $n = 1$ and $n = 0$ levels. Note that $\Delta n = -\Delta j$ in this transition.)

Supplementary Problems

Problem

45. What would be the Fermi energy in a one-dimensional infinite square well 10 nm wide and holding 100 electrons? Assume two electrons (with opposite spins) per energy level.

Solution

With two electrons of opposite spin in each level (which is non-degenerate for the one-dimensional infinite square well), the highest filled level for 100 electrons is $n = 50$. Thus, the Fermi energy is $E_F = n^2(hc/L)^2/8mc^2 = (50 \times 1240 \text{ eV·nm}/10 \text{ nm})^2/8(511 \text{ keV}) = 9.40$ eV (see Equation 40-6).

Problem

49. The transition from the ground state to the first rotational excited state in diatomic oxygen (O_2) requires about 356 μeV. At what temperature would the thermal energy kT be sufficient to set diatomic oxygen into rotation? Would you ever find diatomic oxygen exhibiting the specific heat of a monatomic gas at normal pressure?

Solution

With reference to the solution of Problem 47, O_2 would behave like a monatomic gas if there were insufficient thermal energy to excite rotational states. This implies that $k_B T \leq 356 \ \mu$eV, or $T \leq 356 \ \mu\text{eV}/(86.17 \ \mu\text{eV/K}) \simeq 4.1$ K. Such behavior is not observed, however, since the normal boiling point of oxygen (below which O_2 liquifies) is 90.2 K (see Table 20-1).

Problem

53. The Madelung constant (Section 42-3) is notoriously difficult to calculate because it is the sum of an alternating series of nearly equal terms. But it can be calculated for a hypothetical one-dimensional crystal consisting of a line of alternating positive and negative ions, evenly spaced (Fig. 42-42). Show that the potential energy of an ion in this "crystal" can be written

$$U = -\alpha \frac{ke^2}{r_0},$$

where the Madelung constant α has the value $2 \ln 2$. *Hint:* Study the series expansions listed in Appendix A.

FIGURE 42-42 Problem 53.

Solution

For any ion in Fig. 42-42, there are two oppositely charged ions at distances of r_0, two similarly charged ions at distances of $2r_0$, two opposite ions at $3r_0$, etc. Thus, the electrostatic potential energy of any ion is therefore

$$U = -\frac{2ke^2}{r_0} + \frac{2ke^2}{2r_0} - \frac{2ke^2}{3r_0} + \cdots$$

$$= -\frac{2ke^2}{r_0}\left(1 - \frac{1}{2} + \frac{1}{3} - \cdots\right) = -\alpha\frac{ke^2}{r_0}.$$

Comparison of this with the series expansion of $\ln(1 + x)$ in Appendix A, shows that $\alpha = 2 \ln 2$ for this "crystal". (Note: The convergence of this series, which needs special consideration for $x = 1$, is discussed in many first-year calculus textbooks.)

● CHAPTER 43 NUCLEAR PHYSICS

Section 43-1 Discovery of the Nucleus

Problem

1. In a head-on collision of a 9.0-MeV α particle and a nucleus in a gold foil, what is the minimum distance before electrical repulsion reverses the α particle's direction? Assume the gold nucleus remains at rest. Your answer shows how Rutherford set upper limits on the size of nuclei.

Solution

In Rutherford scattering of α-particles (charge $2e$) from approximately infinitely massive gold nuclei (charge Ze, with $Z = 79$ and no recoil assumed), we may equate the Coulomb potential energy at the distance of closest approach, $k(2e)(Ze)/r_{min}$, to the initial kinetic energy, 9 MeV in this case (see Equation 26-1 and use the conservation of energy). Therefore, 9 MeV = $2Zke^2/r_{min}$, or r_{min} = $2(79) \times (1.44 \text{ MeV·fm})/(9 \text{ MeV}) = 25.3$ fm. This is about 3.6 times the radius of a gold nucleus (Equation 43-1 gives $R = 1.2$ fm $(197)^{1/3} \approx 7.0$ fm). However, for lighter nuclei (where r_{min} is comparable to R), deviations from the Rutherford scattering law, due to nuclear forces, allow an upper limit to be set on the nuclear size.

Section 43-2 Building Nuclei: Elements, Isotopes, and Stability

Problem

5. How do (a) the number of nucleons and (b) the nuclear charge compare in the two nuclei $^{35}_{17}$Cl and $^{35}_{19}$K?

Solution

(a) The mass number (number of nucleons) is $A = 35$ for both, but (b) the charge, Ze, of a potassium nucleus, $Z = 19$, is two electronic charge units greater than that for a chlorine nucleus, $Z = 17$.

Section 43-3 Properties of the Nucleus

Problem

9. Write the symbol for a boron nucleus with twice the radius of ordinary hydrogen ($^1_1$H).

Solution

Since $R \sim A^{1/3}$ in Equation 43-1, a nucleus with twice the radius has eight times the mass number of $^1_1$H, or $A = 8$. Reference to the periodic table shows $Z = 5$ for boron, so the symbol for this nucleus is $^8_5$B.

Problem

13. Find the energy needed to flip the spin state of a proton in Earth's magnetic field, whose magnitude is about 30 μT.

Solution

As explained in Example 43-2, $\Delta U = 2\mu_p B = 2(30 \ \mu T) \times (1.41 \times 10^{-26} \text{J/T}) = 8.46 \times 10^{-31}$ J $= 5.28 \times 10^{-12}$ eV. (The frequency of a photon with this energy, 1.28 kHz, is in the audible range!)

Section 43-4 Binding Energy

Problem

17. Determine the atomic mass of nickel-60, given that its binding energy is very nearly 8.8 MeV/nucleon.

Solution

The total nuclear binding energy of $^{60}_{28}$Ni is the number of nucleons ($A = 60$) times the given binding energy per nucleon, or $E_b = 60(8.8 \text{ MeV})(1 \text{ u·c}^2/931.5 \text{ MeV}) = 0.567$ u·c². If we express Equation 43-3 in terms of atomic masses, by adding $Z = 28$ electron rest energies ($m_e c^2$) to both sides, and neglect atomic binding energies (as mentioned in the text following Example 43-3), we obtain:

$$M(^{60}_{28}\text{Ni}) = 28M(^1_1\text{H}) + (60 - 28)m_n - E_b/c^2$$
$$= 28(1.00783 \text{ u}) + 32(1.00867 \text{ u}) - 0.567 \text{ u}$$
$$= 59.930 \text{ u} .$$

(The actual binding energy of ^{60}Ni is so close to 8.8 MeV/nucleon that the accuracy of the atomic mass just calculated is better than one might expect from data given to two figures.)

Problem

21. The mass of a lithium-7 nucleus is 7.01435 u. Find the binding energy per nucleon.

Solution

The binding energy per nucleon can be found from Equation 43-3 as written, since the nuclear mass of $^7_3$Li is given. Then $E_b/A = [3(1.00728 \text{ u}) + (7 - 3)(1.00867 \text{ u}) - 7.01435 \text{ u}] \times (931.5 \text{ MeV/u})/7 = 5.61$ MeV/nucleon. (See note in solution of Problem 19 on accuracy.) The binding energy per nucleon, for very light nuclides, is low, because the nuclear force is not yet saturated for so few nucleons.

Section 43-5 Radioactivity

Problem

25. The decay constant for argon-46 is 0.0835 s^{-1}. What is its half-life?

Solution

Equation 43-6 gives $t_{1/2} = \ln 2/0.0835 \text{ s}^{-1} = 8.30$ s.

Problem

29. A milk sample shows an iodine-131 activity level of 450 pCi/L. What is its activity in Bq/L?

Solution

Since 1 Ci = 3.7×10^{10} Bq, the activity of the milk sample in SI units is $(450 \text{ pCi/L})(3.7 \times 10^{10} \text{ Bq/Ci}) = 16.7$ Bq/L.

Problem

33. How many atoms in a 1-g sample of U-238 decay in 1 minute? *Hint:* This time is so short compared with the half-life that you can consider the activity essentially constant.

Solution

Since N is essentially constant over time intervals much shorter than the half-life, the number decaying in the sample is just the activity, λN. (Consider Equation 43-4 for $dt = 1$ min and $\lambda = \ln 2/(4.46 \times 10^9 \text{ y})$; dN/N is so small that N is essentially constant). In 1 g of ^{238}U, the number of atoms is approximately Avogadro's number divided by the atomic weight, $N \approx 6.02 \times 10^{23}/238 = 2.53 \times 10^{21}$, and λ from Table 43-2 and Equation 43-6 is $\lambda = \ln 2/(4.46 \times 10^9 \text{ y}) = 2.95 \times 10^{-16} \text{ min}^{-1}$. Thus, the number decaying is $\lambda N = 7.47 \times 10^5$ decays/min.

Problem

37. Nitrogen-13 has a 10-min half life. A sample of ^{13}N contains initially 10^5 atoms. Plot the number of atoms as a function of time from $t = 0$ to $t = 1$ hour. Make your horizontal axis (time) linear, but your vertical axis logarithmic. Why is the curve a straight line? What is the significance of its slope?

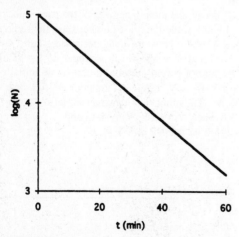

Problem 37 Solution.

Solution

The number of nuclei remaining after time t is, from Equation 43-5b, $N(t) = N_0 2^{-t/t_{1/2}}$, so $\log N = \log N_0 - t(\log 2/t_{1/2}) = 5 - t(0.0301 \text{ min}^{-1})$, where $N_0 = 10^5$ and $t_{1/2} = 10$ min as given. (Note that logarithmic graphs are normally base ten.) This is a straight line with slope $-\log 2/t_{1/2}$ and intercept $\log N_0$, i.e., the smaller the half-life, the steeper the slope. The time range specified in this case covers six half-lives.

Problem

41. Marie Curie and Pierre Curie won the 1903 Nobel Prize for isolating 0.1 g of radium-226 chloride (^{226}RaCl$_2$). What was the activity of their sample, in Bq and in curies?

Solution

The activity, or the number of decays per unit time, of a sample of radioactive material is the magnitude of Equation 43-4, or λN. The decay constant for ^{226}Ra is given by Equation 43-6 and its half-life in Table 43-2: $\lambda = \ln 2/t_{1/2} = 0.693/1600$ y $= 1.37 \times 10^{-11} \text{ s}^{-1}$. The number of ^{226}Ra nuclei in 10^{-1} g of RaCl$_2$ is one-tenth of Avogadro's number divided by the molecular weight, or about $N = 10^{-1} \times 6.022 \times 10^{23}/(226 + 2 \times 35.45) \approx 2.03 \times 10^{20}$. Thus, the activity of this sample was $\lambda N = (1.37 \times 10^{-11} \text{ s}^{-1}) \times (2.03 \times 10^{20}) = 2.78$ GBq $= (2.78 \times 10^9 \text{ Bq}) \times (1 \text{ Ci}/3.7 \times 10^{10} \text{ Bq}) = 75.2$ mCi. (Note that the activity of 1 g of ^{226}Ra is approximately $1.37 \times 10^{-11} \text{ Bq} \times 6.02 \times 10^{23}/226 = 3.65 \times 10^{10} \text{ Bq} \approx 1$ Ci.)

Problem

45. Analysis of a moon rock shows that 82% of its initial K-40 has decayed to Ar-40, a process with a half-life of 1.2×10^9 years. How old is the rock?

Solution

If 82% of the original ^{40}K decayed, then 18% remains in a rock of age t. From Equation 43-5b and the given half-life, $t = t_{1/2} \ln (N_0/N)/\ln 2 = (1.2 \text{ Gy}) \ln (1/0.18)/\ln 2 = 2.97$ Gy. (A type of lunar highlands rock rich in potassium (K), rare earth elements (REE), and phosphorus (P), is called KREEP norite.)

Problem

49. Today, uranium-235 comprises only 0.72% of natural uranium; essentially all the rest is U-238. Use the half-lives given in Table 43-2 to determine the percentage of uranium-235 in natural uranium when Earth formed, about 4.5 billion years ago.

Solution

Suppose that when the Earth formed ($t = 0$), natural uranium consisted of just the two longest-lived isotopes in Table 43-2 (^{234}U has an abundance of 0.0057%). Then the percentage of

^{235}U today (t = 4.5 Gy) is 0.0072 = $N^{235}/(N^{235} + N^{238})$, or N^{238} = 138 N^{235}. The original amounts of the two isotopes are given by Equation 43-5b, with half-lives from Table 43-2, as $N_0^{235} = N^{235}\,2^{4.5/0.704}$ and $N_0^{238} = N^{238}\,2^{4.5/4.46}$, so that the original percentage must have been

$$\frac{N_0^{235}}{N_0^{235} + N_0^{238}} = \frac{1}{1 + (N_0^{238}/N_0^{235})}$$

$$= \frac{1}{1 + (N^{238}/N^{235})2^{(4.5/4.46) - (4.5/0.704)}}$$

$$= \frac{1}{1 + (138)2^{-5.38}} = \frac{1}{1 + 3.30} = 23.2\%$$

(Actually, from current data on nuclear reactions and models of nucleosynthesis in supernova explosions, one can predict the isotopic abundances of U^{235} and U^{238} when they were produced. By reversing the above argument, one can then estimate the age of the elements in the nebula from which the solar system formed.)

Paired Problems

Problem

53. The nuclear mass of $^{48}_{22}$Ti is 47.9359 u. Find the binding energy per nucleon.

Solution

As in the solution to Problem 21, E_b/A = [22(1.00728) + 26(1.00867) − 47.9359](931.5 MeV)/48 = 8.72 MeV/nucleon.

Problem

57. A sample of oxygen-15 ($t_{1/2}$ = 2.0 min) is produced in a hospital's cyclotron. What should be the initial activity concentration if it takes 3.5 min to get the O-15 to a patient undergoing a PET scan for which an activity of 0.5 mCi/L is necessary?

Solution

Multiplying Equation 43-5b by the decay rate, λ, one finds that the activity produced must be $(\lambda N_0) = (\lambda N)2^{t/t_{1/2}}$ = (0.5 mCi/L)$2^{3.5/2.0}$ = 1.68 mCi/L. (see solution to Problem 39.)

Supplementary Problems

Problem

61. How cool would you have to get a material before the thermal energy kT was insufficient to excite protons in a 35-T magnetic field from their lower to upper spin state?

Solution

The thermal energy is less than the spin-flip energy when $k_B T < \Delta U = 2\mu_p B$ (see Example 43-2), or $T < 2 \times$ (1.41×10^{-26} J/T)(35 T)/(1.38×10^{-23} J/K) = 71.5 mK.

Problem

65. Some human lung cancers in smokers may be caused by polonium-210, which arises in the decay series of uranium-238 that occurs naturally in fertilizers used on tobacco plants. Write equations for (a) the production and (b) the decay of ^{210}Po. (c) How might the health effects of Po-210 differ if its half-life were 1 day or 10 years instead of the actual 138 days?

Solution

From Fig. 43-19, one sees that polonium-210 is (a) produced by the β-decay of bismuth-210, $^{210}_{83}$Bi → $^{210}_{83}$Po + e^- + $\bar{\nu}$, and (b) decays by α-decay into lead-206, $^{210}_{84}$Po → $^{206}_{82}$Pb + α. (c) Biological effects depend on the activity and energy release of the decay, as well as on the type of radiation and the properties of the absorbing tissue. In a radioactive decay series, where the half-life of the progenitor is much greater than that of any decay product, the activity of each product in the chain is approximately the same (secular equilibrium). Since a smoker replenishes the supply of ^{210}Po to his or her lungs on a daily basis, it is unlikely that a change in half-life, within the given range, will make much difference.

Problem

69. Nickel-65 beta decays by electron emission with decay constant λ = 0.275 h^{-1}. (a) What is the daughter nucleus? (b) In a sample that is initially pure Ni-65, how long will it be before there are twice as many daughter nuclei as parent nuclei?

Solution

(a) In a β^- decay, the atomic number of the parent nucleus increases by one while the mass number stays the same ($Z \to Z + 1$ and $A \to A$). Thus, the daughter nucleus has Z = 28 + 1 = 29 and A = 65, which is copper-65 (or $^{65}_{29}$Cu). (b) The number of parent nuclei at time t (in an initially pure sample at t = 0) is $N = N_0 e^{-\lambda t}$ (Equation 43-5a), or $t = \lambda^{-1} \ln(N_0/N)$. The number of daughter nuclei is $N_d = N_0 - N$, so when $N_d = 2N$, N_0/N = 3 and $t \ln 3/(0.275$ h$^{-1})$ = 3.99 h.

● **CHAPTER 44** NUCLEAR ENERGY: FISSION AND FUSION

Section 44-1 Energy from the Nucleus

Problem

1. The masses of the neutron, the deuterium nucleus, and the ^{3}He nucleus are 1.008665 u, 2.013553 u, and 3.014932 u, respectively. Use the Einstein mass-energy relation to verify the 3.27-MeV energy release in the D-D fusion reaction of Equation 44-4a.

Solution

The energy release (called the Q-value for the reaction) is the difference in the mass-energy of the initial and final reactants, $Q = [2M(^2_1\text{H}) - M(^3_2\text{He}) - M(^1_0n)]c^2 =$ $[2 \times 2.013553 - 3.014932 - 1.008665]\mu c^2 =$ $(0.003509)(931.5 \text{ MeV}) = 3.27 \text{ MeV}.$

Sections 44-2 and 44-3 Nuclear Fission and its Applications

Problem

5. Neutron-induced fission of ^{235}U results in the fission fragments iodine-139 and yttrium-95. How many neutrons are released?

Solution

In this reaction, the conservation of charge (atomic number Z) is the same as the conservation of the number of protons, so the conservation of the number of nucleons (mass number A) is equivalent to the conservation of the number of neutrons. The numbers of neutrons in the nuclei ^{235}U, ^{139}I, and ^{95}Y are $A-Z = 235 - 92 = 143$, $139 - 53 = 86$, and $95 - 39 = 56$, respectively, so the number of neutrons released in the reaction is $1 + 143 - 86 - 56 = 2$.

Problem

9. Find the explosive yield in equivalent tonnage of TNT for the chain reaction analyzed in Example 44-3.

Solution

The supercritical chain reaction (or explosion) of Example 44-3 released 1.1×10^{15} J of energy, equivalent to $(1.1 \times 10^{15} \text{ J}) \times (1 \text{ kt}/4.18 \times 10^{12} \text{ J}) = 263 \text{ kt}$ of TNT. (We used the energy equivalent for kilotons of TNT given in Section 44-3 and Appendix C.)

Problem

13. The 1974 Threshold Test Ban Treaty limits underground nuclear tests to a maximum yield of 150 kt. (a) How much ^{235}U would be needed for a device with this yield, assuming 30% efficiency? (b) What would be the diameter of this mass, assembled into a sphere? The density of uranium is 18.7 g/cm^3.

Solution

(a) Use of the conversion factor in the solution to Problem 11 gives the amount of ^{235}U as $(150 \text{ kt})(50.9 \text{ g/kt})/30\% = 25.5 \text{ kg}.$ (b) A uniform sphere of mass M and density ρ has diameter $D = (6M/\pi\rho)^{1/3} = (6 \times 25.5 \text{ kg}/\pi \times 18.7 \times 10^3 \text{ kg/m}^3)^{1/3} = 13.7 \text{ cm}.$ (The volume of a sphere is $\frac{1}{6}\pi D^3$.)

Problem

17. The temperature in a typical reactor core is 600 K. What is the thermal speed of a neutron at this temperature? *Hint:* See Section 20-1.

Solution

The thermal speed can be calculated from Equation 20-3:

$$v_{\text{rms}} = \sqrt{\frac{3k_B T}{m}} = \sqrt{\frac{3(1.38 \times 10^{-23} \text{ J/K})(600 \text{ K})}{1.67 \times 10^{-27} \text{ kg}}}$$

$$= 3.86 \text{ km/s}.$$

Problem

21. If a reactor is shut down abruptly, neutron absorption by ^{135}Xe prevents rapid start-up. The xenon has a 9.2-h half-life. How long must reactor operators wait until the ^{135}Xe level drops to one-tenth of its peak value? (Your answer neglects formation of additional ^{135}Xe from the decay of ^{135}I.)

Solution

From Equation 43-5b, the time for decay to one tenth the initial amount of ^{135}Xe (with no production assumed) is $t = t_{1/2} \ln(N_0/N)/\ln 2 = (9.2 \text{ h}) \ln 10/\ln 2 = 30.6 \text{ h}.$

Problem

25. In the dangerous situation of prompt criticality in a fission reactor, the generation time drops to 100 μs as prompt neutrons alone sustain the chain reaction. If a reactor goes prompt critical with $k = 1.001$, how long does it take for a 100-fold increase in reactor power?

Solution

Following the same reasoning as in Example 44-5 (or in the solution to the previous problem), we have $k^n = 10^2$, or $n \log k = 2$. The time required for this increase in power is the number of generations times the generation time, or $n\tau = (2/\log 1.001) \times 100 \text{ }\mu\text{s} = 0.461 \text{ s}.$

Section 44-4 Nuclear Fusion

Problem

29. Fusion researchers often express temperature in energy units, giving the value of kT rather than T. What is the temperature in kelvins of a 2-keV plasma?

Solution

A thermal energy of $kT = 2$ keV corresponds to a temperature of $T = 2$ keV$(1.6\times10^{-16}$ J/keV$)/1.38\times10^{-23}$ J/K$) = 2.3\times10^7$ K.

Problem

33. How much heavy water (deuterium oxide, ^2H_2O or D_2O) would be needed to power a 1000-MW D–D fusion power plant for 1 year?

Solution

In one year, a 1 GW power plant produces 1 GW·y $= (10^9$ J/s$)$ $(3.156\times10^7$ s$)(1$ MeV$/1.602\times10^{-13}$ J$) = 1.97\times10^{29}$ MeV of energy. If we use 7.2 MeV/deuteron as the average energy release in a D–D reactor (see Example 44-6), then 2.74×10^{28} deuterons are required. The molecular weight of D_2O is about 20 u, so about $\frac{1}{2}(20$ u$)(1.66\times10^{-27}$ kg/u$) \times (2.74\times10^{28}) = 454$ kg of heavy water (each molecule of which contains two deuterons) would be needed.

Problem

37. The proton-proton cycle consumes four protons while producing about 27 MeV of energy. (a) At what rate must the Sun consume protons to produce its power output of about 4×10^{26} W? (b) The present phase of the Sun's life will end when it has consumed about 10% of its original protons. Estimate how long this phase will last, assuming the Sun's 2×10^{30} kg mass was initally 71% hydrogen.

Solution

(a) The number of protons consumed per second is the power output divided by the energy release per proton. This is about

$$\left(\frac{4\times10^{26} \text{ J/s}}{27 \text{ MeV/4 protons}}\right)\left(\frac{1 \text{ MeV}}{1.6\times10^{-13} \text{ J}}\right) = 3.7\times10^{38} \text{ protons/s}.$$

(Note: four protons are consumed in releasing 27 MeV.)
(b) 10% of the Sun's original protons is about $(0.1)(0.71) \times (2\times10^{30}$ kg$/1.67\times10^{-27}$ kg/proton$) = 8.5\times10^{55}$ protons. The consumption of this many protons at the rate found in part (a) would take about $(8.5\times10^{55}$ protons$)/(3.7\times10^{38}$ protons/s$) = 2.3\times10^{17}$ s $= 7.3$ billion years. The present age of the Sun is about 4.5 billion years.

Paired Problems

Problem

41. The total power generated in a nuclear power reactor is 1500 MW. How much ^{235}U does it consume in a year?

Solution

The number of ^{235}U-atoms which undergo fission to produce the energy generated in one year is $(1.5\times10^9$ W$) \times (3.156\times10^7$ s$)/(200$ MeV/fission$)(1.602\times10^{-13}$ J/MeV$) = 1.48\times10^{27}$. These atoms have a total mass of about $(235$ u$) \times (1.48\times10^{27})(1.66\times10^{-27}$ kg/u$) = 576$ kg. (See Example 44-2).

Problem

45. This problem and the next one explore the differences between magnetic and inertial-confinement fusion. (a) What is the mean thermal speed of a deuteron in the PLT device of Problem 40? (b) At this speed, how long would it take a deuteron to cross the plasma column? Using this value as a "confinement time" in the absence of magnetic confinement and the density given in Problem 40, calculate the Lawson parameter and show that it falls far short of the value needed for D–T fusion.

Solution

(a) We can use Equation 20-3 to find the thermal speed of the deuterons ($m \approx 2$ u): $v_{rms} = \sqrt{3k_BT/m} = \sqrt{3(1.38\times10^{-23} \text{ J/K})(6\times10^7 \text{ K})/(2 \times 1.66\times10^{-27} \text{ kg})} = 865$ km/s. (b) It would take only $(90$ cm$)/(865$ km/s$) = 1.04$ μs to cross the plasma column at this speed, if there were no collisions or other confinement mechanism. Without confinement, the Lawson parameter would be $n\tau = (4\times10^{19}$ m$^{-3})(1.04$ μs$) = 4\times10^{13}$ s/m^2, about 4×10^{-7} times smaller than the desired criterion for D–T fusion (Equation 44-5).

Supplementary Problems

Problem

49. Use a graphical or numerical method to find the multiplication factor necessary to fission 10^{22} uranium-235 nuclei in 2.0 μs, if the generation time is 10 ns.

Solution

The number of generations is $n = 2$ μs$/10$ ns $= 200$, so Equation 44-2 gives $10^{22}(k - 1) = k^{201} - 1 \approx k^{201}$. A numerical solution can be obtained from Newton's method, applied to the function $f(k) = k^{201} - 10^{22}(k - 1)$, with a first guess of $k_0 = 10^{22/201} = 1.2866$. Then $f(k_0) = 10^{22}(2 - k_0)$ and $f'(k_0) = 201 k_0^{200} - 10^{22} = 10^{22}(201 - k_0)/k_0$, so the next approximation is $k_1 = k_0 - f(k_0)/f'(k_0) = 199 k_0/(201 - k_0) = 1.2820$. Another iteration gives $k_2 = k_1 - f(k_1)/f'(k_1) = 1.2793$, so the root seems to be converging toward 1.279.

A more accurate value of k can be obtained by graphical methods, if a suitable PC and software are handy. Plotting $10^{22}(k - 1)$ and $k^{201} - 1$ versus k on the same axes, as shown below, one obtains $k = 1.27847$.

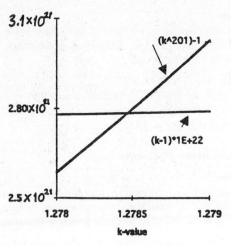

Problem 49 Solution.

Problem

53. The volume of the fireball produced in a nuclear explosion is roughly proportional to the weapon's explosive energy yield. For weapons exploded at ground level, how would the land area subject to a given level of damage scale with the weapon's yield? Your result shows one reason military strategists favor multiple smaller warheads.

Solution

If the volume of the fireball is proportional to the yield, then the radius of the fireball goes like the cube root of the yield, i.e., $R \sim (\text{yield})^{1/3}$. The land area effected by the fireball goes like the square of its radius, i.e., $R^2 \sim (\text{yield})^{2/3}$. Thus, doubling the yield increases the effected land area by $2^{2/3} = 1.59$, which is less area than would be effected by exploding two bombs on non-overlapping targets.

● **CHAPTER 45** FROM QUARKS TO THE COSMOS

Section 45-1 Particles and Forces

Problem

1. How long could a virtual photon of 633-nm red laser light exist without violating conservation of energy?

Solution

In order to test the conservation of energy in a process involving one virtual photon, a measurement of energy, with uncertainty less than the photon's energy ($\Delta E < hc/\lambda$), must be performed in a time interval less than the virtual photon's lifetime ($\Delta t < \tau$). Thus, $\Delta E \, \Delta t < hc\tau/\lambda$. But Heisenberg's principle limits the product of these uncertainties to $\Delta E \, \Delta t \gtrsim \hbar$, so $hc\tau/\lambda > \hbar$ or $\tau > \lambda/2\pi c = (633\times10^{-9}$ m$)/(2\pi \times 3\times10^8$ m/s$) \simeq 3\times10^{-16}$ s, for the given wavelength. In other words, if the lifetime of a virtual photon of wavelength 633 nm were less than 3×10^{-16} s, no measurement showing a violation of conservation of energy would be possible.

Section 45-2 Particles and More Particles

Problem

5. Use Table 45-1 to find the total strangeness before and after the decay $\Lambda^0 \to \pi^- + p$, and use your answer to determine the force involved in this reaction.

Solution

The Λ^0 has strangeness -1, while the p and π^- both have zero strangeness. Thus $\Delta S = 1$ for this decay (final minus initial strangeness). Since strangeness is conserved in strong and electromagnetic interactions, the decay must be a weak interaction.

Problem

9. Which of the following reactions (a) $\Lambda^0 \to \pi^+ + \pi^-$ and (b) $K^0 \to \pi^+ + \pi^-$ is not possible, and why?

Solution

Decay (a) violates the conservation of baryon number (and angular momentum, since the spin of Λ^0 is $\frac{1}{2}$ and that of the pions is 0). Decay (b) is an observed weak interaction.

Problem

13. What happens to the sign of the wave functions proportional to the terms (a) xy^2t and (b) xy^2t^2 under the operation PT?

Solution

Under space-inversion, both terms (a) and (b) change sign (they are parity reversing): $P[xy^2t] \to (-x)(-y)^2t = -xy^2t$, and $P[xy^2t^2] \to -xy^2t^2$. Under time-inversion, the first (a) changes sign while the second (b) does not (the first is T-reversing, the second is T-conserving): $T[xy^2t] \to xy^2(-t) = -xy^2t$, and $T[xy^2t^2] \to xy^2t^2$. Under the combined transformation, the first term (a) is PT-conserving while the second (b) is PT-reversing: $PT[xy^2t] \to P[-xy^2t] \to xy^2t$, and $PT[xy^2t^2] \to P[xy^2t^2] \to -xy^2t^2$. (Note: $PT = TP$, i.e., P and T-transformations commute.)

Section 45-3   Quarks and the Standard Model

Problem
17. The J/ψ particle is an uncharmed meson that nevertheless includes charmed quarks. What is its quark composition?

Solution
The J/ψ must have quark content $c\bar{c}$ in order to have zero net charm.

Section 45-4   Unification

Problem
21. Repeat the preceding problem for the 10^{15} GeV energy of grand unification.

Solution
The temperature corresponding to the energy 10^{15} GeV = 10^{24} eV = $k_B T$, where the strong and electro-weak forces unify, is about $T = 10^{24}$ eV/$(8.617 \times 10^{-5}$ eV·K$^{-1} \approx 10^{28}$ K.

Section 45-5   Evolution of the Universe

Problem
25. Express the Hubble constant in SI units.

Solution
From Example 45-3, one sees that H_0 in SI units is the reciprocal of the Hubble time in seconds, i.e., $H_0 =$ (20 Gy $\times$ 3.156$\times 10^7$ s/y)$^{-1}$ = 1.58$\times 10^{-18}$ s^{-1}, where the value of H_0 used is the lower end of the currently accepted range. (Direct conversion of units, using 1 ly = $(3 \times 10^5$ km/s)(3.156$\times 10^7$ s) gives the same result: $H_0 =$ (15 km/s)/(9.47$\times 10^{28}$ km).) Another common distance unit used by astronomers is the parsec (1 pc = 3.26 ly, see Appendix D). $H_0 = 15$ km/s/Mly ≈ 50 km/s/Mpc.

Problem
29. A widely used value for H_0 is 17 km/s/Mly. What age does this value imply for the universe, under the simple assumptions of Example 45-3?

Solution
Alteration of the value of H_0 in Example 45-3 to 17 km/s/Mly changes the Hubble time to (17 km/s/Mly)$^{-1} \times$ $(3 \times 10^5$ km/s/(ly/y)) = 18 Gy.

Supplementary Problems

Problem
33. Muonium is a hydrogen-like atom consisting of a proton and a muon. What would be (a) the size and (b) the ground-state energy of muonium? The muon's mass is 207 times that of the electron.

Solution
We can use the results for the Bohr atom, with $m_\mu = 207 m_e$ replacing m_e (see Equations 39X-13 and 12a). For the ground state of muonium ($n = 1$), (a) $r_1 = a_0/207 = 0.0529$ nm $\div$ $207 = 256$ fm, and (b) $E_1 = 207 (-ke^2/2a_0) = 207 \times$ $(-13.6$ eV) = -2.81 keV.

Problem
37. Use Wien's law (Equation 39-2) to find the wavelength at which the 2.7-K cosmic microwave background has its peak intensity.

Solution
If the more accurate value for the temperature of the cosmic background radiation from Fig. 45-24 is used in Wien's displacement law (Equations 39-2), the wavelength at peak intensity found is $\lambda_{max} = 2.898$ mm·K/2.726 K = 1.063 mm.